AF353029

Micromanufacturing of Metallic Materials

Micromanufacturing of Metallic Materials

Editors

Jingwei Zhao
Zhengyi Jiang
Leszek Adam Dobrzański
Chong Soo Lee
Fuxiao Yu

MDPI • Basel • Beijing • Wuhan • Barcelona • Belgrade • Manchester • Tokyo • Cluj • Tianjin

Editors
Jingwei Zhao
Taiyuan University of
Technology
China

Zhengyi Jiang
University of Wollongong
Australia

Leszek Adam Dobrzański
Medical and Dental Engineering
Center for Research, Design and
Production ASKLEPIOS
Poland

Chong Soo Lee
Pohang University of Science
and Technology
South Korea

Fuxiao Yu
Northeastern University
China

Editorial Office
MDPI
St. Alban-Anlage 66
4052 Basel, Switzerland

This is a reprint of articles from the Special Issue published online in the open access journal *Materials* (ISSN 1996-1944) (available at: https://www.mdpi.com/journal/materials/special_issues/micromanu_metal).

For citation purposes, cite each article independently as indicated on the article page online and as indicated below:

LastName, A.A.; LastName, B.B.; LastName, C.C. Article Title. *Journal Name* **Year**, *Article Number*, Page Range.

ISBN 978-3-03943-509-8 (Hbk)
ISBN 978-3-03943-510-4 (PDF)

Contents

About the Editors

Jingwei Zhao is a Professor at Taiyuan University of Technology, China. He received his Ph.D. degree in Materials Processing Engineering in 2009. His main research interests include the microforming and micromanufacturing of metallic materials, metal rolling technology, the numerical modeling of manufacturing processes, and tribology and mechanics in materials manufacturing. He has accumulated extensive research experience in the field of materials processing and manufacturing engineering by working at the universities in Australia and South Korea, and has been very actively involved in the manufacturing of many kinds of metal alloys. He has published four scholarly books, two book chapters and over 150 research articles, which have had a significant impact on the field of materials processing and manufacturing, and have been acknowledged by a number of international academic communities.

Zhengyi Jiang is a Distinguished Professor at the University of Wollongong, Australia. He received his Ph.D. degree in Materials Processing Engineering in 1996. He is an expert in the area of materials processing and manufacturing, and has made a leading study on rolling mechanics and the micromanufacturing of metallic materials in Australia. He has successfully established the first research area based on micromanufacturing, and has developed a world-class research group and research laboratory in micromanufacturing. He has solid research experience and expertise on the theory, simulation and practice of microforming and micromanufacturing metals, as well as tribology in manufacturing processes. He has published three monographs, 27 edited books, four book chapters and over 500 peer-reviewed articles, with many in top journals in materials manufacturing, which have had a significant impact on materials manufacturing technology.

Leszek Adam Dobrzański Prof. DSc, PhD, MSc, Eng, Hon. Prof., M Dr HC is a Full Professor of Materials Engineering, Nanotechnology, Biomaterials, Medical, Dental, and Manufacturing Engineering, Management and Organization, and is the Director of the Science Centre ASKLEPIOS and the Supervisory Council Chairman of the Medical and Dental Engineering Centre for Research, Design and Production ASKLEPIOS Ltd in Gliwice, Poland. He is a Vice President and a Fellow of the Academy of Engineers in Poland, a foreign Fellow of the Ukrainian Academy of Engineering Sciences, and a Foreign Fellow of the Slovak Academy of Engineering Sciences as well as being the President of the World Academy of Materials and Manufacturing Engineering and a President of the International Association of Computational Materials Science and Surface Engineering. He is the Editor-in-Chief of the *Journal of Achievements in Materials and Manufacturing Engineering*, the *Archives of Materials Science and Engineering* and the *Open Access Library*. He is a creator and leader of the Scientific School, which resulted in him personally promoting a group of 62 completed PhD theses, as well as promoting ca. 1000 MSc and BSc theses. His scientific output includes about 2500 scientific publications with others currently in printing, including about 50 books and monographs and 50 chapters in books and monographs, 240 papers in English in journals covered in the Web of Science Core Collection, and 60 patents. He has been cited ca. 15000 times in worldwide scientific journals; h-index—51(GS), 31 (SC), 24 (WS).

Chong Soo Lee is a Professor at Pohang University of Science and Technology (POSTECH), South Korea. He received his Ph.D. degree from the Polytechnic Institute of New York University in

1985, and became a professor of POSTECH. His research interests are focused on the development of high-strength steels and Ti-alloys, plasticity, fatigue, hydrogen-delayed fracture resistance and the development of Ti-based bio-materials. He has published more than 300 SCI journals, and has served as an editorial board member of the *International Journal of Fatigue, Metals and Materials International, La Metallurgia Italiana and Current Nanomaterials*. Owing to his active academic contribution, he has been elected as a fellow of the European Academy of Science, the National Academy of Engineering of Korea (NAEK) and the Korean Academy of Science and Technology (KAST). He is currently a Professor in the Graduate Institute of Ferrous Technology, POSTECH.

Fuxiao Yu is a Professor at Northeastern University, China. He received his Ph.D. degree in Metallurgy in 1998 from the Indian Institute of Science, India. He is an expert in the area of the materials processing of aluminum and magnesium alloys through rapid solidification and subsequent hot deformation, as well as heat treatment. He has successfully established the research area of wrought Al-Si alloys on the theoretical basis of second phase particle-induced dynamic recrystallization, and industrialized wrought Al-Si alloys as structural materials. Recently, his research activity has also focussed on the micromanufacturing of aluminum alloys. He has published over 50 peer-reviewed articles.

Editorial

Recent Development in Micromanufacturing of Metallic Materials

Jingwei Zhao [1], Zhengyi Jiang [2,*], Leszek A. Dobrzański [3,*], Chong Soo Lee [4] and Fuxiao Yu [5]

[1] College of Mechanical and Vehicle Engineering, Taiyuan University of Technology, Taiyuan 030024, China; jzhao@tyut.edu.cn

[2] School of Mechanical, Materials, Mechatronic and Biomedical Engineering, University of Wollongong, Wollongong, NSW 2522, Australia

[3] Medical and Dental Engineering Center for Research, Design and Production ASKLEPIOS, ul. Krolowej Bony 13D, 44-100 Gliwice, Poland

[4] Graduate Institute of Ferrous Technology, Pohang University of Science and Technology (POSTECH), Pohang 37673, Korea; cslee@postech.ac.kr

[5] School of Materials Science and Engineering, Northeastern University, Shenyang 110819, China; fxyu@mail.neu.edu.cn

* Correspondence: jiang@uow.edu.au (Z.J.); leszek.dobrzanski@centrumasklepios.pl (L.A.D.)

Received: 20 August 2020; Accepted: 9 September 2020; Published: 11 September 2020

Abstract: Product miniaturization is a trend for facilitating product usage, enabling product functions to be implemented in microscale geometries, and aimed at reducing product weight, volume, cost and pollution. Driven by ongoing miniaturization in diverse areas including medical devices, precision equipment, communication devices, micro-electromechanical systems (MEMS) and microsystems technology (MST), the demands for micro metallic products have increased tremendously. Such a trend requires development of advanced micromanufacturing technology of metallic materials for producing high-quality micro metallic products that possess excellent dimensional tolerances, required mechanical properties and improved surface quality. Micromanufacturing differs from conventional manufacturing technology in terms of materials, processes, tools, and machines and equipment, due to the miniaturization nature of the whole micromanufacturing system, which challenges the rapid development of micromanufacturing technology. Against such a background, the Special Issue "Micromanufacturing of Metallic Materials" was proposed to present the recent developments of micromanufacturing technologies of metallic materials. The papers collected in the Special Issue include research articles, literature review and technical notes, which have been highlighted in this editorial.

Keywords: micromanufacturing; metallic materials; miniaturization; micro products

---

Quo Vadis, Munde? (in Latin—Where are you going world?) (It is a paraphrase of the words of the Apostle Peter "Quo Vadis, Domine", according to a legend fleeing Rome in fear of persecution and death, directed to Jesus Christ going in the opposite direction without fear. "Quo Vadis" is the title of the famous novel of the Polish writer Henryk Sienkiewicz, who received the Nobel Prize in Literature in 1905.). It seems to be one of the most critical questions that millions and maybe even billions of people are asking themselves. Just a few months ago, despite the areas of poverty and local wars, the answer was obvious. The idea was to make living conditions as pleasant as possible and to continuously improve them, and to do so to affect more and more of the world's citizens. The UN has defined these perspectives by formulating seventeen Sustainable Development Goals [1]. However, it turned out that the tiny creature of nature, the severe acute respiratory syndrome coronavirus 2 (SARS-CoV-2) with a diameter 100 trillion times smaller than the diameter of the Earth can completely block the course of world events and lead to its lockdown. This virus just caused the coronavirus disease (COVID-19) and the associated global pandemic. It, of course, will leave its mark on the coming years, maybe even decades.

Some development trends, however, seem to be still valid after a short stop. The Japanese government addressed the problem most widely when formulating the Society 5.0 program [2–7]. It concerns the development of civilization ranging from hunting and agriculture in primitive times to the developed IT society today. Previously created in Germany [8–10] and adopted in the European Union [11] and in many other countries, the program covers the next stages of the technological revolution. The current stage of Industry 4.0 marks the beginning of an era of digital industrial technology in which, in addition to people supporting automated production, systems, sensors, machines, metadata and IT, including the internet of people, things and services, which connects various participants in the value chain in the production process, through a network internally and inter-organizationally [12–15]. In the technological aspect, a key role is played by cyber-physical systems (CPS), similar to other cyber-social, cyber-business, and cyber-biological systems covering numerous algorithms, communication infrastructure, and high-computing cyberspace for developing and processing metadata. In industrial organizations, it results in improved security, operational efficiency, and data collection efficiency, as well as significant savings. CPS systems interact with each other and with neighboring smart components, communicating between machines and smart products. Smart production takes place in smart factories constituting production systems, where people, machines and sensors interact in real-time, and production decisions are made based on experiments and simulations of the real conditions of manufacturing products made in virtual reality. A virtual copy of the physical world in the form of a digital twin enables computer simulations and prediction, which is aided by metadata sets and self-education systems and the use of artificial intelligence methods. The nine technologies determining progress in Industry 4.0 include big data sets, autonomous robots, simulations, integration of horizontal and vertical systems, the internet of things, cybersecurity, cloud computing, additive manufacturing and augmented reality [16,17]. The model presented in this way, however, seems to be only a fragment of reality and is, therefore, a simplification. The full augmented holistic model of Industry 4.0 is pictured by a regular octahedron containing people as the causative agent, a technological platform with four complementary components, and smart products as the goal of the whole activity [18–20]. At the technological platform, the development of engineering materials and product manufacturing processes are taken into account, including additive technologies, development of technological machines, and CPS systems, which are in fact only one of the six components of the full augmented holistic model of Industry 4.0.

This approach to the problem clearly indicates the contemporary importance of both material and technological design [12–15,18–21]. However, it is not only additive technologies that determine technological success. Among modern products enabling universally expected improvement in the quality of life, health, and work efficiency are ever newer models of mobile phones and smartphones, CD and MP3 players, iPods, wide flat-screen displays and computers, including those installed in many market products, including for automotive, aviation and space applications, as well as in systems and manufacturing machines. The devices include pressure, thermal, temperature, gas, mass flow, speed, and sound sensors, injection nozzles, electrical, pneumatic and thermal micro-actuators, chemical micro-reactors, micro-motors, micro-gears, micro-valves, micro-fans, micro-tools, micro-molding, and micro-replication molds. These devices are examples of the now more and more miniaturized products, systems and devices, which also include, for example, micro-systems, including micro- and nano-electromechanical (MEMS and NEMS), micro-mechanical, micro-medical and micro-optical electronic-mechanical (MOES) systems, microreactors, fuel cells, and numerous micro-devices and sensors widely used among other applications, in automotive, aviation industries, telecommunications, and IT facilities. It is worth noting that the modern manufacturing of many of the above-mentioned micro- and even nano-elements requires compliance with the rules of the Industry 4.0 stage of the technological revolution, and on the other hand, their production is a prerequisite for the implementation of this modern Industry 4.0 technology, where different sensors and devices automatically identify raw materials, semi-finished products, and ready-made components and products. Micro-components are often used in various clinical areas of medicine and dentistry, including on cardiovascular sensors, microprocessor ceramic packaging, implantable devices, medical instruments, and implants, as well as coatings on micro polymers or metal implant components [22,23]. Currently, there is a significant variety of micro-products.

Nanomaterials, coatings and thin films coating the working surfaces of many products and components are becoming increasingly important, including for their use in micro- and nano-devices, sensors, in communication, medicine, and dentistry, but also in household appliances, sound and television sets, watches and many other products used daily by millions of people. Properly designed and manufactured material is often a determinant of technological progress and a prerequisite for the implementation of the design intention for micro-components and micro-devices. They play important roles in material processing and plastic deformation at the micro- and nano-scale, where their characteristics and properties include the structure of materials and nanomaterials, their phase composition, chemical purity, grain size, grain boundary structure, phase transformation micromechanisms, precipitation processes, dispersion and size of precipitates and intermetallic phases, plastic deformation micromechanisms, and surface layer deposition processes. The mechanical and physicochemical properties of these materials, as well as their mechanisms to resist material damage are also important for prevention of corrosion, abrasion, fatigue, the influence of elevated temperature, and even creep. These factors are important for determining the maintenance of micro-elements and micro-devices and require precise design before starting production and require consideration during their exploitation.

Micromanufacturing, together with nanotechnology, has now become a common technological practice [24–28]. Therefore, it is necessary to continually increase the knowledge of scientists, engineers, and managers involved in this production. Micromanufacturing includes production methods, technologies and systems that meet the harsh rigors of Industry 4.0, as well as development of suitable materials, machinery, tools, and organizational strategies and production management methods. The specificity of this production, mainly due to the miniaturized scale of products, requires separate design and manufacturing experience, the use of specific materials and adequate manufacturing equipment, which are often also miniaturized, and requires special care for the mechanization, automation, and computerization of the technologies used, and therefore this field is often combined and referred to as microsystems technology (MST). Among the materials used can be specified silicon, considered to be "technologically mature" and currently more and more often used rather than other engineering materials; materials used are very usually metal, but also composites, as well as polymers and ceramics. Generally, two main groups can be considered among MST products. MEMS can be regarded as a classic application; its uses include photolithography, plating, laser ablation, chemical etching, hybridized lithography and electroplating and molding known as LIGA (a German acronym for **L**ithographie, **G**alvanoformung, **A**bformung). The other group consists of other manufactured micro-products, not included in MEMS, using micro-mechanical cutting, laser cutting/drilling/patterning, micro-injection, micro-extrusion, micro-embossing, micro-molding, micro-stamping, as well as electrical discharge machining (EDM), also known as wire erosion, wire burning, die sinking, spark eroding or spark machining. This group of new processes for the production of micro products/features or so-called micro-machining also uses conventional technologies with a correspondingly reduced scale, which includes micro-machining (mechanical, thermal, electro-chemical, electrical discharge), micro-molding/replication, micro-formation methods (quick methods, electro molding, injection molding) and micro-joining.

Another goal of micromanufacturing is the production of products/devices using miniaturized machines/systems, usually instead of universal large-sized machines/systems. The production of micro-products and associated micro-production strategies differ from those used on a conventional scale. The individual components/products are manufactured at one or several manufacturing stations, employing subtractive machining or plastic deformation, as well as additive methods, to finally assemble or connect them by welding methods and/or joining materials. In the case of micro-products, deposition, layering, and/or pattering methods are used, with technological operations most often performed using one machine or technology platform in combination with integrated packaging and/or assembly operations. Conventional methods as well as hybrid production combining all or some of the subtractive, additive, forming, and/or joining technologies can also be used. Due to the use of various energy sources, the technologies used to produce micro-components/products can be mechanical, electrical, electro-chemical, chemical, laser, or electronic. As a rule, dedicated devices are used, enabling the implementation of a predetermined technological process, although more and more interest of producers of micro-products, and more importantly, manufacturers of technological machines, employ multi-process devices whose working platform allows a hybrid combination of various manufacturing processes, e.g., micro-milling, turning, grinding, buffing, polishing, micro-electro discharge machining, micro-electrochemical machining and/or laser machining, without having to clamp the workpiece/product each time. Favorable solutions in this area bode very well for the development of integrated micro-machining technologies.

The attractiveness of this avant-garde technology for micromanufacturing of metallic materials for the production of micro-components/products has prompted and encouraged us to prepare a Special Issue on "Micromanufacturing of Metallic Materials". Miniaturization of products is a trend that facilitates the use of the product, enabling the implementation of product functions in microscale geometry, aimed at reducing mass, volume, costs, and environmental pollution through the manufacture of this micro-product. From around 30 papers submitted, after careful selection as a result of the opinion-giving process, 14 papers were selected and published in this book. With the exception of one literature review and one other publication, the others are research papers. These papers have been prioritized.

Carpenter and Tabei [29] conducted a literature review on residual stress development, prevention and compensation in metal additive manufacturing, in which a holistic calculation framework was proposed to control the level of residual stresses by optimizing the conditions of the additive manufacturing processes. Also, various additive manufacturing technologies and residual stress sources, residual stress measurement techniques, and their dependence on additive manufacturing (AM) process conditions were briefly described, and current modeling methods were proposed to prevent permanent deformation of manufactured products using additive manufacturing methods.

Máthis et al. [30] have prepared a paper on the micro-tensile behavior of Mg-Al-Zn alloy processed by equal channel angular pressing (ECAP). In this research, AZ31 magnesium alloy was extruded four times in an equal rectangular channel. Deformation characteristics (yield stress, ultimate tensile stress, and uniform elongation) exhibit significant anisotropy as a consequence of different orientations between the stress direction and texture, and thus different deformation mechanisms were confirmed on the samples obtained using the electron backscatter diffraction (EBSD) technique.

The paper on bio-inspired functional surface fabricated by electrically assisted micro-embossing of AZ31 magnesium alloy was developed by Wang et al. [31]. This research focuses on the use of different current densities to perform embossed micro-channels in textured bulk metallic glass dies. These dies are prepared by thermoplastic forming based on the compression of photolithographic silicon molds. The results show that large areas of bio-inspired textures could be fabricated on magnesium alloy. Filling depth and depth–width ratio nonlinearly increases when higher current densities are used, and the temperature is kept below the temperature of the glass transition to avoid melting and to avoid an early breakage of the die. Electrically assisted micro-forming has demonstrated the ability to reduce size effects, improving formability and decreasing flow stress, making it a promising hybrid process to control the filling quality of micro-scale features.

Thangaraj et al. [32] conducted research to enhance the surface quality of a micro titanium alloy specimen in a wire electrical discharge machining (WEDM) process by adopting Taguchi–Grey relation analysis (TGRA)-based optimization. The surface measures of machined titanium alloys as dental materials can be enhanced by adopting a decision-making algorithm in the machining process. In the present study, Taguchi–Grey analysis-based criteria decision making was applied to the input process factors in the wire electric discharge machining (EDM) process. It was proved that the proposed method can enhance the efficacy of the process.

The team of Xue et al. [33] prepared a paper on an electrically-assisted rolling process of corrugated surface microstructure with T2 copper foil. Electrically-assisted forming is a low-cost and high-efficiency method to enhance the formability of materials. The electric current reduces the flow stress and the fracture strain. The study proves that the current can improve the forming quality of the corrugated foils and is a promising surface texture forming process.

The paper on adaptive spiral tool path generation for diamond turning of large aperture freeform optics was developed by Wang et al. [34]. This paper reports a novel adaptive tool path generation for slow tool servo slow tool servo (STS) iamond turning. Comparison of the surface generation of typical freeform surfaces with adaptive tool path generation and the commercial software DiffSys is conducted both theoretically and experimentally. The adaptive tool path generation can effectively reduce the volume of control points, decrease the vibration of side-feeding motion, and improve machining efficiency while surface quality is well maintained for large aperture freeform optics.

Liu et al. [35] conducted an experimental investigation on form error for slow tool servo diamond turning of micro lens arrays on a roller mold. In this study, a novel forming approach based on a slow tool servo is presented to fabricate microlens arrays on an aluminum alloy (6061) roller mold. Based on the different distribution patterns of the discrete points of the microlens, the equal-arc method and the equal-angle method are also proposed to generate the tool path. According to the kinematic analysis of the cutting axis, the chatter mark results from the overlarge instantaneous acceleration oscillations of the cutting axis during the slow tool servo diamond turning process of the microlens arrays. The results are acquired with fine surface quality. Slow tool servo assisted ultra-precision diamond turning is a promising machining process with high accuracy and low cost to generate the large-area microlens arrays on a roller mold.

The paper on microsheet metal deformation behaviors in ultrasonic-vibration-assisted uniaxial tension with aluminum alloy 5052 was prepared by Wang et al. [36]. In this investigation, ultrasonic-vibration-assisted uniaxial tensile experiments were carried out utilizing GB 5052 thin sheets of different thicknesses and grain sizes, respectively. The uniform deformation ability of thin sheets could be improved by increasing the hardening exponent with ultrasonic-vibration. The authors found that ultrasonic-vibration is helpful in improving the forming limit in micro sheet forming, e.g., microbulging and deep drawing processes.

Supercooled Zr35Ti30Cu8.25Be26.75 metallic glass is the topic of research in the paper entitled "Ultrasonic Vibration Facilitates the Micro-Formability of a Zr-Based Metallic Glass", which was written by Han et al. [37]. Ultrasonic vibration was introduced as an effective method to improve the micro-formability of metallic glasses, owing to its capabilities of improving the material flow and reducing interfacial friction. The more intriguing finding is that the micro formability of the Zr-based metallic glasses can be further improved by tuning the amplitude of the ultrasonic vibration. The results were demonstrated by the finite element method.

Ren et al. [38] conducted a numerical research on slip system evolution in ultra-thin stainless steel foil, in which a three-dimensional uniaxial tension model was established, based on the crystal plasticity finite element method and Voronoi polyhedron theory. The number and characteristics of active slip systems and the deformation degree of the grain are different due to the different initial grain orientations. The slip systems preferentially initiate at grain boundaries and cause slip system activity at the interior and free surface of the grain. The brass, S, and copper oriented 304 stainless steel foil exhibit a high strain hardening index, which is beneficial to strengthening. However, the cube and Goss oriented 304 stainless steel foil has a low deformation resistance and is prone to plastic deformation.

Suzuki et al. [39] elucidated the shearing mechanism of finish-type FB and extrusion-type FB for a thin foil of JIS SUS304 by numerical and EBSD analyses. In this research, a numerical analysis using finite element (FE) analysis was performed to clarify the shearing mechanism in the process of extrusion-type fine blanking (FB) for a thin foil of JIS SUS304. The principal stress near the shearing surface has mostly compressive components in extrusion-type FB due to its negative clearance, and the critical fracture value was also less than that in the finish-type FB, in which the principal stress near the shearing surface has mostly tensile components. Using SEM observations with electron backscatter diffraction (EBSD) analysis of the shearing surface, it was confirmed that the reductions in deformation-induced crystal orientation rotation and martensite transformation in extrusion-type FB are present in comparison with those in finish-type FB. Suzuki et al. [40] also prepared an article titled "Optimum Clearance in the Microblanking of Thin Foil of Austenitic Stainless Steel JIS SUS304 Studied from Shear Cut Surface and Punch Load". An extrusion-type fine blanking with a negative clearance was proposed by the authors instead of standard fine blanking, for creating a full-sheared surface in the micro blanking process. It was clarified that the clearance at which the cut surface does not fracture and minimization of the punch load is achieved is gained by the use of clearance—4 μm.

The paper on the fabrication of a micro-punch array by plasma printing for micro-embossing into copper substrates was prepared by Shiratori et al. [41]. Copper substrates were wrought to have micro-grooves for packaging by micro-stamping with the use of an AISI316 stainless steel micro-punch array. A negative pattern was printed directly onto the AISI316 die substrate, which was plasma nitrided. The printed surfaces were selectively sand-blasted to fabricate the micro-textured punch array for micro-embossing. Since the nitrogen supersaturated heads had sufficient hardness against the blasting media, the printed parts of AISI316 die were removed.

Machno et al. [42] contributed a paper on the subject "Impact of the Deionized Water on Making High Aspect Ratio Holes in the Inconel 718 Alloy with the Use of Electrical Discharge Drilling". This paper includes an analysis of the influence of the machining parameters (pulse time, current amplitude and discharge voltage) on the process performance (drilling speed, linear tool wear, taper angle, the hole's aspect ratio, and side gap thickness), during electrical discharge drilling (EDD) with the use of deionized water in the Inconel 718 alloy. An analysis of the results indicates increasing of the hole's aspect ratio by about 15% (above 30), decreasing the side gap thickness by about 40%, and enhanced surface integrity.

In summary, it is difficult to expect that in a scholarly book with a relatively limited volume, all issues raised by a modern approach to micromanufacturing could be included. This is simply impossible because of the very wide scope of this super modern technological approach. Therefore, it was only possible to give reasonably representative examples, and that is how it happened. We offer readers very interesting, in our opinion, research material. If at least one of the presented papers turns out to be interesting for the reader who reached for this book, we will have the basis to believe that as editors, we have achieved our goals. It remains for us to wish you fruitful reading.

Funding: This research was funded by the National Natural Science Foundation of China (grant number 51975398), and a research project supported by the Shanxi Scholarship Council of China (grant number 2020-037).

Conflicts of Interest: The authors declare no conflict of interest.

References

1. Park, Y.H. Build Capacity for International Health Agenda on the "Transforming Our World: The 2030 Agenda for Sustainable Development". *Heal. Policy Manag.* **2015**, *25*, 149–151. [CrossRef]
2. Japan Business Federation. *Society 5.0 in Co-Creating the Future (Excerpt)*; Keidanren: Tokyo, Japan, 2018.
3. Japan Business Federation. *Toward Realization of the New Economy and Society (Outline)*; Keidanren: Tokyo, Japan, 2016.
4. Harayama, Y. Society 5.0: Aiming for a New Human-Centered Society. *Hitachi Rev.* **2017**, *66*, 8–13.
5. Government of Japan Cabinet Office. *Society 5.0*; Government of Japan Cabinet Office: Tokyo, Japan, 2019.
6. Fukuyama, M. Society 5.0: Aiming for a New Human-Centered Society. *Japan Spotlight* **2018**, *27*, 47–50.
7. Fukuda, K. Science, technology and innovation ecosystem transformation toward society 5.0. *Int. J. Prod. Econ.* **2020**, *220*, 107460. [CrossRef]

8. Kagermann, H. Chancen von Industrie 4.0 Nutzen. In *Industrie 4.0 in Produktion, Automatisierung und Logistik*; Springer Fachmedien Wiesbaden: Wiesbaden, Germany, 2014; pp. 603–614.

9. Hermann, M.; Pentek, T.; Otto, B. *Design Principles for Industrie 4.0 Scenarios, A Literature Review*; Working Paper; Technische Universität: Dortmund, Germany, 2015.

10. Rüßmann, M.; Lorenz, M.; Gerbert, P.; Waldner, M.; Justus, J.; Engel, P.; Harnisch, M. *Industry 4.0: The Future of Productivity and Growth in Manufacturing Industries*; Boston Consulting Group: Boston, MA, USA, 2015.

11. European Commission. *Commission Sets Out Path to Digitise European Industry*; European Commission: Brussels, Belgium, 2016.

12. Dobrzański, L.A.; Dobrzańska-Danikiewicz, A.D. Why Are Carbon-Based Materials Important in Civilization Progress and Especially in the Industry 4.0 Stage of the Industrial Revolution. *Mater. Perform. Charact.* **2019**, *8*, 337–370. [CrossRef]

13. Dobrzański, L.A. Effect of heat and surface treatment on the structure and properties of the Mg-Al-Zn-Mn casting alloys. In *Magnesium and Its Alloys: Technology and Applications*; Dobrzański, L.A., Totten, G.E., Bamberger, M., Eds.; CRC Press: Boca Raton, FL, USA, 2019; pp. 91–202.

14. Dobrzański, L.A.; Dobrzańska-Danikiewicz, A.D. Applications of Laser Processing of Materials in Surface Engineering in the Industry 4.0 Stage of the Industrial Revolution. *Mater. Perform. Charact.* **2019**, *8*, 1091–1129. [CrossRef]

15. Beier, G.; Ullrich, A.; Niehoff, S.; Reißig, M.; Habich, M. Industry 4.0: How it is defined from a sociotechnical perspective and how much sustainability it includes—A literature review. *J. Clean. Prod.* **2020**, *259*, 120856. [CrossRef]

16. Boston Consulting Group. *Embracing Industry 4.0 and Rediscovering Growth*; Boston Consulting Group: Boston, MA, USA, 2019.

17. United Nations Industrial Development Organization (UNIDO). *Int. Organ.* **1967**, *21*, 511–520. [CrossRef]

18. Dobrzański, L.A.; Dobrzański, L.B. Approach to the Design and Manufacturing of Prosthetic Dental Restorations According to the Rules of Industry 4.0. *Mater. Perform. Charact.* **2020**, *9*, 394–476. [CrossRef]

19. Dobrzański, L.A.; Dobrzański, L.B. Dentistry 4.0 Concept in the Design and Manufacturing of Prosthetic Dental Restorations. *Processes* **2020**, *8*, 525. [CrossRef]

20. Dobrzański, L. Role of materials design in maintenance engineering in the context of industry 4.0 idea. *J. Achiev. Mater. Manuf. Eng.* **2019**, *1*, 12–49. [CrossRef]

21. Josed, R.; Seeram, R. Materials 4.0: Materials big data enabled materials discovery. *Appl. Mater. Today* **2018**, *10*, 127–132. [CrossRef]

22. Arsiwala, A.; Desai, P.P.; Patravale, V. Recent advances in micro/nanoscale biomedical implants. *J. Control. Release* **2014**, *189*, 25–45. [CrossRef] [PubMed]

23. Caldorera-Moore, M.; Peppas, N.A. Micro- and nanotechnologies for intelligent and responsive biomaterial-based medical systems. *Adv. Drug Deliv. Rev.* **2009**, *61*, 1391–1401. [CrossRef] [PubMed]

24. Jiang, Z.; Zhao, J.; Xie, H. *Microforming Technology: Theory, Simulation, and Practice*; Elsevier: Amsterdam, The Netherlands, 2017; pp. 1–24.

25. Zhao, J.; Huo, M.; Ma, X.; Jia, F.; Jiang, Z. Study on edge cracking of copper foils in micro rolling. *Mater. Sci. Eng. A* **2019**, *747*, 53–62. [CrossRef]

26. Chaubey, S.K.; Jain, N.K. State-of-art review of past research on manufacturing of meso and micro cylindrical gears. *Precis. Eng.* **2018**, *51*, 702–728. [CrossRef]

27. Oliaei, S.; Karpat, Y.; Davim, J.P.; Perveen, A. Micro tool design and fabrication: A review. *J. Manuf. Process.* **2018**, *36*, 496–519. [CrossRef]

28. Hasan, M.; Zhao, J.; Jiang, Z. A review of modern advancements in micro drilling techniques. *J. Manuf. Process.* **2017**, *29*, 343–375. [CrossRef]

29. Carpenter, K.; Tabei, A. On Residual Stress Development, Prevention, and Compensation in Metal Additive Manufacturing. *Materials* **2020**, *13*, 255. [CrossRef]

30. Máthis, K.; Kövér, M.; Stráská, J.; Trojanová, Z.; Dzugan, J.; Halmešová, K. Micro-Tensile Behavior of Mg-Al-Zn Alloy Processed by Equal Channel Angular Pressing (ECAP). *Materials* **2018**, *11*, 1644. [CrossRef]

31. Wang, X.; Xu, J.; Wang, C.; Egea, A.J.S.; Li, J.; Liu, C.; Wang, Z.; Zhang, T.; Guo, B.; Cao, J. Bio-Inspired Functional Surface Fabricated by Electrically Assisted Micro-Embossing of AZ31 Magnesium Alloy. *Materials* **2020**, *13*, 412. [CrossRef]

32. Muthuramalingam, T.; Annamalai, R.; Moiduddin, K.; Alkindi, M.; Ramalingam, S.; Alghamdi, O. Enhancing the Surface Quality of Micro Titanium Alloy Specimen in WEDM Process by Adopting TGRA-Based Optimization. *Materials* **2020**, *13*, 1440. [CrossRef]

33. Xue, S.; Wang, C.; Chen, P.; Xu, Z.; Cheng, L.; Guo, B.; Shan, D. Investigation of Electrically-Assisted Rolling Process of Corrugated Surface Microstructure with T2 Copper Foil. *Materials* **2019**, *12*, 4144. [CrossRef] [PubMed]

34. Wang, D.; Sui, Y.-X.; Yang, H.; Li, D. Adaptive Spiral Tool Path Generation for Diamond Turning of Large Aperture Freeform Optics. *Materials* **2019**, *12*, 810. [CrossRef] [PubMed]

35. Liu, Y.; Qiao, Z.; Qu, D.; Wu, Y.; Xue, J.; Li, D.; Wang, B. Experimental Investigation on Form Error for Slow Tool Servo Diamond Turning of Micro Lens Arrays on the Roller Mold. *Materials* **2018**, *11*, 1816. [CrossRef]

36. Wang, C.; Zhang, W.; Cheng, L.; Zhu, C.; Wang, X.; Han, H.; He, H.; Hua, R. Investigation on Microsheet Metal Deformation Behaviors in Ultrasonic-Vibration-Assisted Uniaxial Tension with Aluminum Alloy 5052. *Materials* **2020**, *13*, 637. [CrossRef]

37. Han, G.; Peng, Z.; Xu, L.; Li, N. Ultrasonic Vibration Facilitates the Micro-Formability of a Zr-Based Metallic Glass. *Materials* **2018**, *11*, 2568. [CrossRef]

38. Zhang, Z.; Fan, W.; Hou, J.; Wang, T. A Numerical Study of Slip System Evolution in Ultra-Thin Stainless Steel Foil. *Materials* **2019**, *12*, 1819. [CrossRef]

39. Suzuki, Y.; Shiratori, T.; Yang, M.; Murakawa, M. Elucidation of Shearing Mechanism of Finish-type FB and Extrusion-type FB for Thin Foil of JIS SUS304 by Numerical and EBSD Analyses. *Materials* **2019**, *12*, 2143. [CrossRef]

40. Suzuki, Y.; Yang, M.; Murakawa, M. Optimum Clearance in the Microblanking of Thin Foil of Austenitic Stainless Steel JIS SUS304 Studied from Shear Cut Surface and Punch Load. *Materials* **2020**, *13*, 678. [CrossRef]

41. Shiratori, T.; Aizawa, T.; Saito, Y.; Dohda, K. Fabrication of Micro-Punch Array by Plasma Printing for Micro-Embossing into Copper Substrates. *Materials* **2019**, *12*, 2640. [CrossRef] [PubMed]

42. Machno, M.; Bogucki, R.; Szkoda, M.; Bizoń, W. Impact of the Deionized Water on Making High Aspect Ratio Holes in the Inconel 718 Alloy with the Use of Electrical Discharge Drilling. *Materials* **2020**, *13*, 1476. [CrossRef] [PubMed]

Review

On Residual Stress Development, Prevention, and Compensation in Metal Additive Manufacturing

Kevin Carpenter and Ali Tabei *

School of Mechanical, Industrial and Manufacturing Engineering, Oregon State University, Corvallis, OR 97331, USA; carpekev@oregonstate.edu
* Correspondence: ali.tabei@oregonstate.edu

Received: 11 November 2019; Accepted: 15 December 2019; Published: 7 January 2020

Abstract: One of the most appealing qualities of additive manufacturing (AM) is the ability to produce complex geometries faster than most traditional methods. The trade-off for this advantage is that AM parts are extremely vulnerable to residual stresses (RSs), which may lead to geometrical distortions and quality inspection failures. Additionally, tensile RSs negatively impact the fatigue life and other mechanical performance characteristics of the parts in service. Therefore, in order for AM to cross the borders of prototyping toward a viable manufacturing process, the major challenge of RS development must be addressed. Different AM technologies contain many unique features and parameters, which influence the temperature gradients in the part and lead to development of RSs. The stresses formed in AM parts are typically observed to be compressive in the center of the part and tensile on the top layers. To mitigate these stresses, process parameters must be optimized, which requires exhaustive and costly experimentations. Alternative to experiments, holistic computational frameworks which can capture much of the physics while balancing computational costs are introduced for rapid and inexpensive investigation into development and prevention of RSs in AM. In this review, the focus is on metal additive manufacturing, referred to simply as "AM", and, after a brief introduction to various AM technologies and thermoelastic mechanics, prior works on sources of RSs in AM are discussed. Furthermore, the state-of-the-art knowledge on RS measurement techniques, the influence of AM process parameters, current modeling approaches, and distortion prevention approaches are reported.

Keywords: additive manufacturing; residual stress; thermal stress; distortion; prevention; modeling; computation

---

1. Metal Additive Manufacturing and Residual Stresses

1.1. Introduction

Additive manufacturing (AM) is quickly becoming a leading method for manufacturing components across many industries, including automotive, medical, and aerospace [1]. Compared to traditional manufacturing, also called subtractive manufacturing, where components are fabricated by removing material from a larger stock, AM involves layer-wise addition of a material to form a three-dimensional (3D) component by fusion of the layers. Immediately, one can imagine the benefits of AM for very complex geometries or materials, like titanium alloys, which are strong and very difficult to modify with subtractive methods. While AM technologies exist for a variety of material systems, the focus of this paper is on metal AM, referred to simply as "AM".

The costs to additively manufacture or 3D print a component may exceed that for traditional methods. For very simple geometries, it might be faster and easier to fabricate the part on a mill, lathe, or computer numerical control (CNC) machines. To 3D print a part, a computer-aided design (CAD) model must firstly be developed, and then sent to the AM machine, which slices the model into many very thin layers; then, each layer is deposited (either through powder or wire feedstock) onto the previous one and fused together, usually with the addition of heat. Comparisons were made for the cost of AM and traditional methods for several areas [1–4]; a general consensus is that, currently, traditional machining may remain more cost-effective than traditional machining in certain situations; however, as the technology is developing, AM is getting more efficient [2].

There are a few major obstacles preventing AM from fully surpassing traditional processes; AM is prone to microstructural defects and porosities that affect mechanical behavior of components, as well as residual stress (RS) formation, which can lead to geometric inaccuracies (part distortion) and deteriorate performance. Regarding geometrical accuracy, one must note that part distortions occur as a result of RS formation; the stresses that are generated tend to pull or push (depending on the direction of the RS) the material and deflect the part [5–7], as commonly seen in welding, hot rolling, and bending. In AM, deviations from the CAD geometry of as much as 2.1 mm were observed in a twin cantilever beam specimen of 11 cm by 1 cm [8,9]—an unacceptable amount of distortion for high-precision aircraft components. Regarding performance, RSs can adversely affect the structural reliability of the part in cyclic loading (fatigue) [10–13]. One study investigated RSs in electron beam AM (EBAM) titanium alloy, in as-built, stress-relieved, and hot isostatic pressed (HIPed) conditions, and the results indicated that hot isostatic pressing (HIPing) could cause microstructural changes to relieve RS and improve fatigue life of components, but neutron diffraction measurements suggested that most of the stresses were relieved during the EBAM build process at 600 °C. The improvements in fatigue life (over 100% between as-built conditions and HIPed, with respective fatigue strengths of 200 and 600 MPa at 10^7 cycles) were attributed to other microstructure and porosity defects [14].

In AM, the multiple layers of material are fused together by the addition of heat. As the end of this section explains in more detail, this localized source of heat creates massive temperature gradients in the material, both in-plane in the newly added layer and through the thickness of pre-existing layers. The large thermal gradients are the primary source of RS formation in the part. Not only does each AM technology have unique features, e.g., material feedstock, heat source, and atmospheric conditions, as discussed in this section, but operator-input process parameters, such as scanning strategy (heat source path pattern), scanning speed, laser power, and build orientation can also lead to varying characteristics of RSs (size, direction, distribution).

RSs are defined as the stresses that exist within a body without any externally applied loads (i.e. the body is in equilibrium with its surroundings) [5,15]. They are also referred to as "internal" or "locked-in" stresses, and they can either strengthen a material, like toughened glass [16], or weaken a part. In AM, RSs might often be referred to as thermal stresses, since their origin is the steep thermal gradients in the manufacturing process. According to Shorr [17], thermal stresses are generated when a non-uniform temperature field causes localized thermal expansion that is interfered by non-expanding (or less-expanding) surrounding material, bodies, or parts. To fully grasp how RSs are formed, one must understand the underlying physics and governing equations for mechanical and thermal loads. Section 2 provides brief explanations of these concepts; for greater detail, the reader is referred to common engineering textbooks [15,17,18]. RSs are classified based on the size of their effects: macro- vs. micro-stresses [15]; or type I, type II, and type III stresses [5,12,13]. Type I stresses, or "macro" stresses, act over large lengths, with respect to the dimensions of the part. Type II stresses act over distances at the grain-size level and are often associated with phase transformations, while type III stresses are at the atomic scale (e.g., dislocation stress fields and crystal lattice defects) [12,13].

It was established that AM steel often exhibits a large portion of retained austenitic microstructure due to relatively rapid cooling [19]. However, the thermal stresses are believed to act as the driving mechanism for the phase formations of austenite to ferrite or martensite, in processes such as laser beam melting (LBM) [20].

Figure 1 shows a transmission electron microscopy (TEM) image of selective laser melted precipitation-hardened (PH) stainless steel, and it indicates that both martensitic and austenitic phases are present [21]. It was further demonstrated that a post-process heat treatment of LBM stainless-steel components partially transformed the austenite to martensite [20].

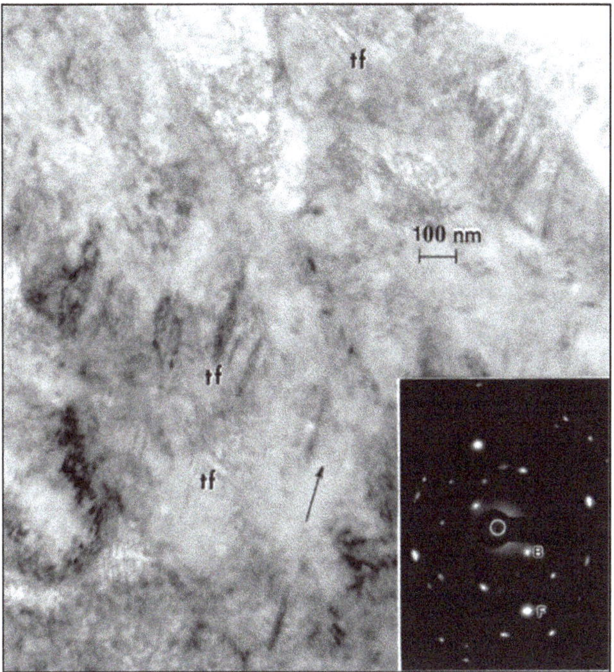

Figure 1. TEM image revealing dislocations, stacking fault traces, and deformation twin faults (tf) in austenite grains mixed with martensite in the build direction of selective laser melting (SLM) of a PH stainless steel. The black arrow indicates the ($\bar{1}$12) FCC direction, and the selected-area electron diffraction pattern image (bottom right) shows mixed diffraction spots (martensite: B = bcc-α, austenite: F = fcc-γ) [21].

It was further shown by Uhlmann et al. [22] that a post-process stress-relieving heat treatment can change the microstructure in Ti–6Al–4V. It was observed that an inhomogeneous microstructure developed in the as-built SLM specimens. Yet, after HIPing or a stress-relief heat treatment, a more homogeneous microstructure developed [22], as shown in the scanning electron microscope (SEM) images in Figures 2 and 3.

Because of their impact on part performance, it is important to be able to experimentally measure or computationally model and predict RSs. Since RSs vary with respect to the size of the area over which they act, the selected measurement technique must have sufficient spatial resolution to capture the effects of the stress. Although stress is obtained indirectly from strain measurements, the methods are commonly referred to as residual stress measurements. A key concept for the determination of RSs is that the strains measured are elastic strains (i.e., stress and strain are related through Hooke's law, discussed later). These strains can be measured either destructively (where the part experiences significant, irreversible alterations) or non-destructively (where the part retains most of its original integrity) (refer to Figure 4). An important effect that is the basis of measuring RSs destructively is that residual stresses, while forming, cause the part to distort, and, by removing the material which contains the RSs, the part relaxes to dimensions that would exist without RSs [12]. Section 3 provides summaries of experimental measurement of residual stresses.

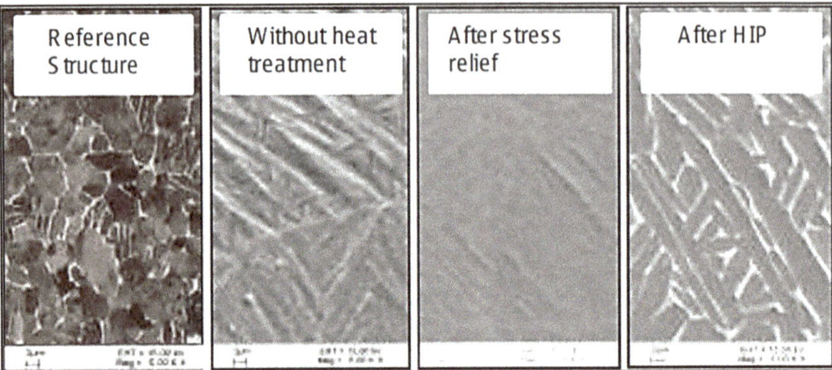

Figure 2. SEM images of Ti–6Al–4V microstructure for the reference structure (cast), and the SLM process without heat treatment, after a stress-relief heat treatment, and after hot isostatic pressing (HIPing) [22].

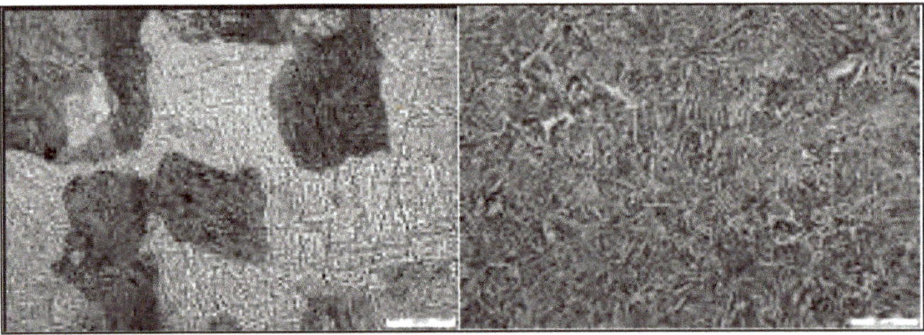

Figure 3. SEM images of Ti–6Al–4V microstructure of SLM specimens without a stress-relief heat treatment (**left**) and after HIPing (**right**) [22].

Because experiments can be costly, particularly for AM, it is often desirable to simulate a process numerically. A physically realistic and validated simulation can allow for rapid, inexpensive investigations into the individual effect of the technology features and process parameters on RSs in AM. Section 4 serves to report on modeling techniques in AM, with a focus on prediction of RSs. Finally, Section 5 discusses current reports on strategies to prevent RS formation and distortion mitigation in AM. One of the primary concerns is that many researchers are seeking to achieve the geometric accuracy by utilizing the effects of RS (shrinkage/warpage), but the RSs still play a detrimental role in mechanical properties and still exist in the part. As one can imagine, the challenges introduced above would prevent many parts requiring high-precision or structural integrity, such as aircraft components and medical devices, from being additively manufactured. These drawbacks were an area of investigation for many years, and this review paper serves to collect and report, specifically, on the most recent advancements toward RS prediction and prevention.

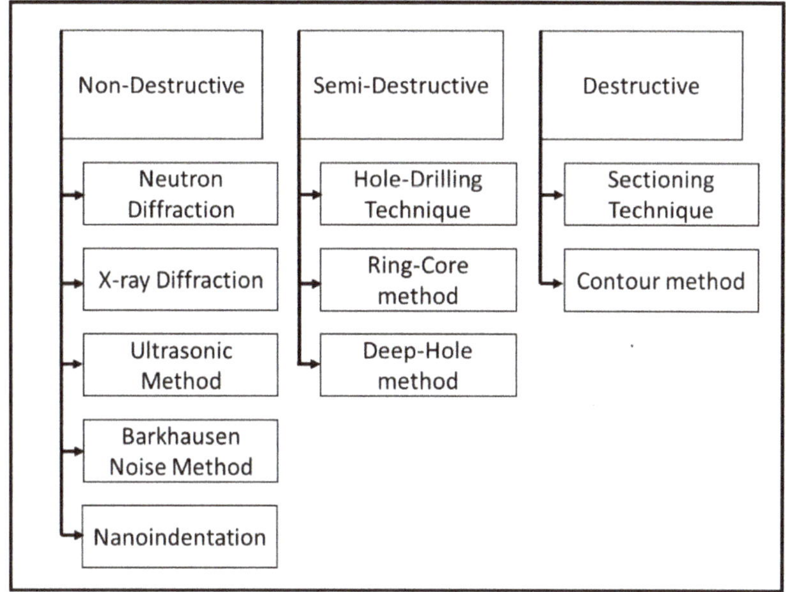

Figure 4. Categories of residual stress (RS) measurement techniques (adopted from Reference [13]).

1.2. AM Technologies

AM technologies are typically categorized by their feedstock and heat source. Figure 5 illustrates the various AM technologies, according to Nickels [23]. As seen in Figure 5, there are two primary categories: powder bed fusion (PBF) and directed energy deposition (DED). PBF encompasses those AM technologies that utilize a bed of powder particles; an arm called a "rake" or "roller" slides a thin layer of powder across the baseplate, and the layer geometry is scanned with a laser or electron beam, melting the powder particles into a solid layer, as shown in Figure 6. DED includes AM technologies that use either a powder or a wire feedstock; the powder is blown from a nozzle into the path of the heat source (a laser, typically), as seen in Figure 7, and the wire-fed AM technology is mostly related to welding processes. The wire material is fed and melted layer by layer, until the part is completely fabricated, as shown in Figure 8 [24]. The sections below discuss technologies that share common heat sources, as opposed to feedstock.

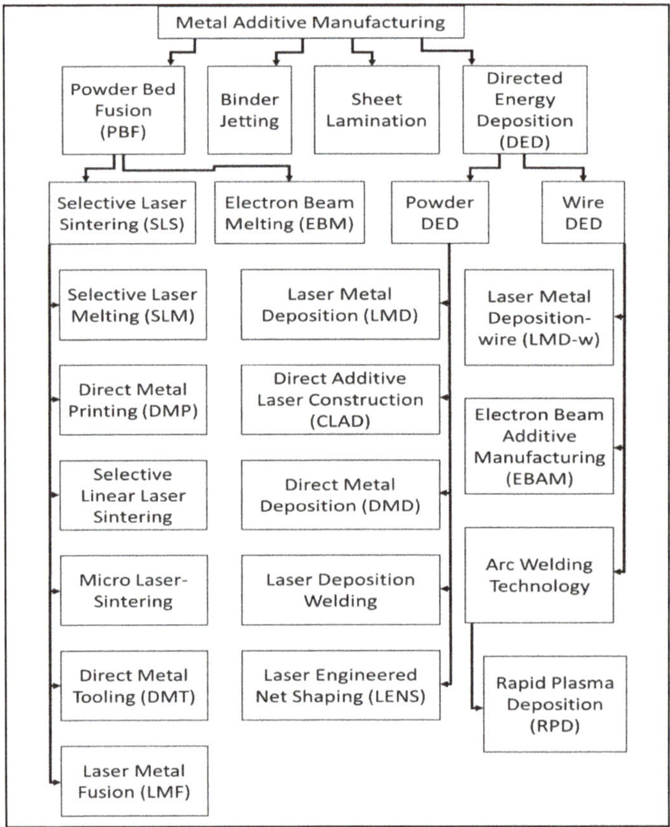

Figure 5. Categorization of the different additive manufacturing (AM) technologies (adopted from Reference [23]).

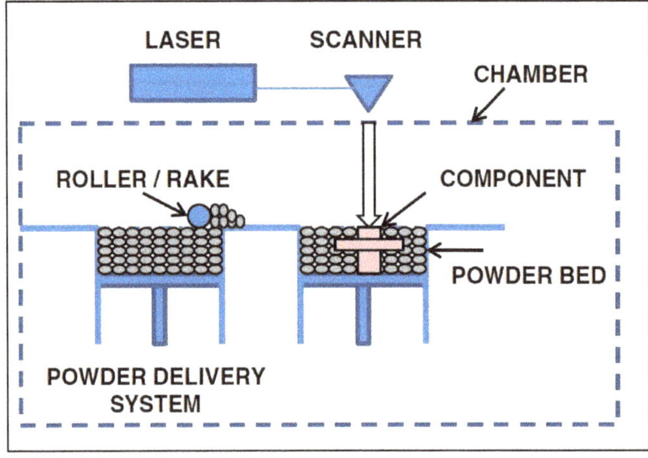

Figure 6. Schematic of typical powder bed fusion (PBF) technology [24].

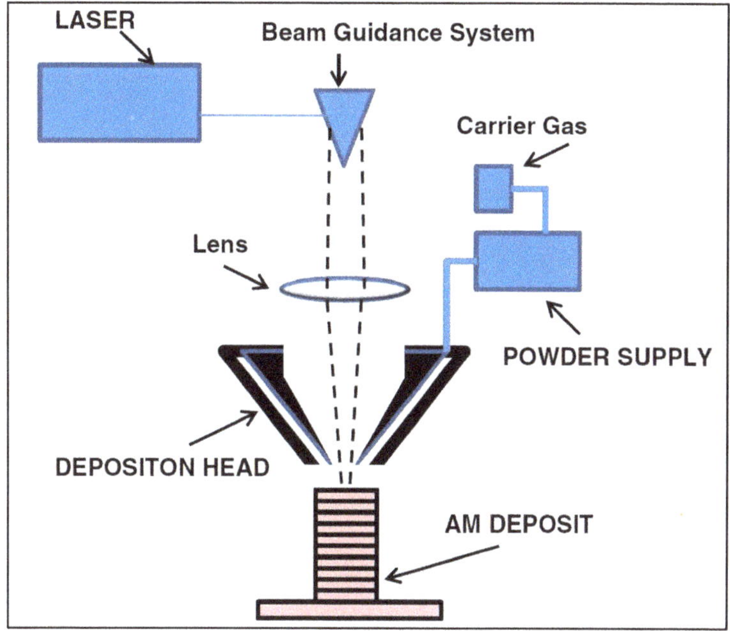

Figure 7. Schematic of typical directed energy deposition (DED) powder-based technology [24].

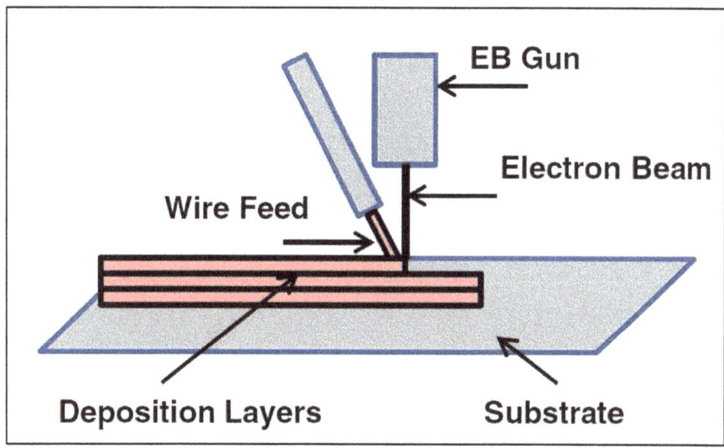

Figure 8. Schematic of typical DED wire-fed technology [24].

1.2.1. Laser Melting

Application of a laser beam to melt the powder or wire feedstock is prevalent in AM; examples include selective laser sintering (SLS), selective laser melting (SLM), direct metal laser sintering (DMLS), laser-engineered net shaping (LENS), direct metal deposition (DMD), laser powder deposition (LPD), and selective laser cladding (SLC). The most common laser-based AM processes include SLS, SLM, DMD, and LPD [25].

In SLS, the bed of fine powders is heated to just below the material's melting temperature, and a laser beam traces out the layer geometry, with sufficient power to sinter the powders to fuse together.

SLM is very similar to SLS, except, instead of sintering the powder, it is fully melted. The difference between SLS and SLM is only a minor technicality, but the consequence is that the SLM laser source is usually of higher power [25].

In LENS, DMD, and LPD, instead of a *bed* of powder, a stream of powder material is delivered through a nozzle directed into the focused laser beam at the region of interest. LENS melts the metal powder that is delivered, usually by a pressurized inert gas, circumferentially around the laser head. DMD can either sinter or melt the powder, which is delivered through a number of nozzles in a similar fashion as LENS. The laser automatically positions itself to aim at defined points from a 3D model of the part. LPD also uses a stream of powder directed onto the part in the laser beam [25].

1.2.2. Extrusion

Extrusion processes do not involve materials in powder form; instead, the material is in the form of a wire. Typically, the nozzle through which the wire is fed is heated, and the material is softened or melted. From the nozzle, the material is deposited onto the build plate, layers are added, and they solidify upon cooling, as shown in Figure 9.

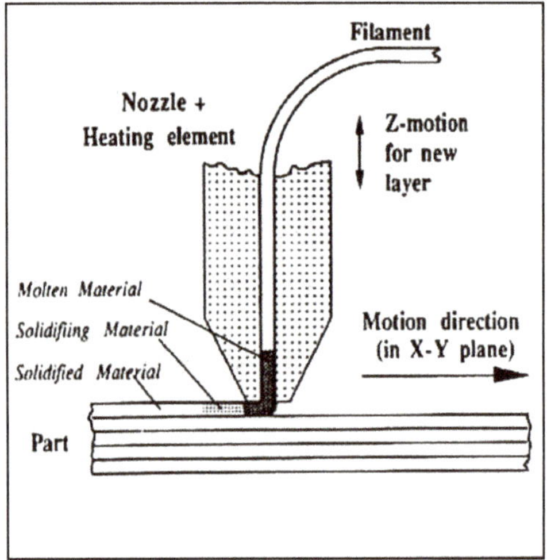

Figure 9. Illustration for a typical fused deposition modeling (FDM) process [26].

FDM (fused deposition modeling) uses a moveable deposition head and deposits the wire material according to the computer-sliced 3D model. The heated extrusion nozzle generally heats the material 1 °C above its melting temperature, so that it solidifies immediately after deposition and fuses with prior layers. Typically, FDM has two deposition heads—one for the build material and the other for support structures [25].

1.2.3. Material Jetting

Material jetting involves the controlled spraying of molten material or adhesive (called a binder), such that the particles bind to each other into a solid part. The binder holds the powder particles together, and no phase change occurs.

1.2.4. Electron Beam

Electron beam additive manufacturing (EBAM) is identical to the laser melting methods discussed above, except that the energy source is an electron beam and not a laser beam [27]. A difference from laser melting is that EBAM must be carried out in a vacuum chamber to avoid oxidation, thus restricting the size of the part to the vacuum chamber dimensions. EBM processes may also involve heating the powder bed during the build to reduce temperature gradients [28].

1.3. Sources of Residual Stresses in AM

As discussed previously, the various metal AM processes involve localized heating and cooling of top surfaces and the re-melting of preceding layers. As a consequence of these non-homogeneous thermal loads, RSs are generated, which may result in distortion of the part [5–7,28] and also deteriorate performance. As discussed by Mercelis and Kruth [29], Kruth et al. [30], Withers and Bhadeshia [5], and Withers [10], the main source of RSs in processes with the melting, solidification, and re-melting thermal cycle can be described by the temperature gradient mechanism (TGM) model. In the TGM model, the heat source is often a high-intensity point source, and the material temperature at the location of the heat source quickly elevates with respect to the surrounding material. The hot material expands, but is restricted by the less expanding, cooler material around it. This restriction creates a compressive stress in the heat source region. As the hot material cools, it contracts, but, again, the contraction is restricted by the less expanding surrounding material, which results in permanent tensile residual stresses in the part, and it often leads to a deflection or warped end product, as shown in Figure 10.

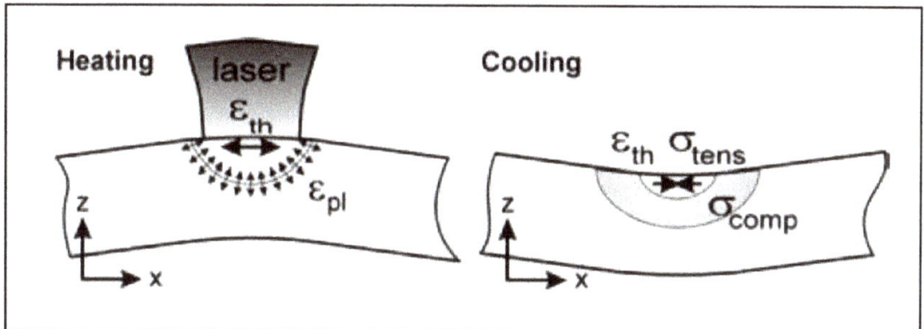

Figure 10. Illustration of the temperature gradient mechanism and distortion, where ε_{th} is the thermal strain, ε_{pl} is the plastic strain, and σ_{tens} and σ_{comp} are the tensile and compressive stresses, respectively [29].

The re-melting and re-solidifying of the pre-existing layers of material also contributes to the RSs in the part. When a new layer is deposited and melted, the recently solidified layers are likely to re-melt (depending on the process technology and parameters) or at least reach high temperatures again [31]. The previous layers cool and shrink underneath the new top layer. The shrinkage of pre-existing layers pulls and stretches the top layer, resulting in permanent tensile RSs. Another source of RSs in AM processes is the inhomogeneous lattice spacing. Because of the non-equilibrium process of AM, the microstructure is non-homogeneous. The inconsistent microstructure makes the lattice spacing spatially dependent, which makes the RS directions and magnitudes dependent on location [28].

Formation of RSs depends greatly on the AM process parameters. According to an investigation by Wu et al. [32] with stainless-steel 316L triangular prisms and L-shaped bars produced by SLM, various process parameters were studied and a reduction of RSs was achieved by decreasing the scan island size (discussed below) from about 650 MPa to about 400 MPa, as shown in Figure 11, and by increasing the laser power and speed; overall, compressive RSs typically existed in the center of the bar, and tensile stresses existed near the surfaces [32].

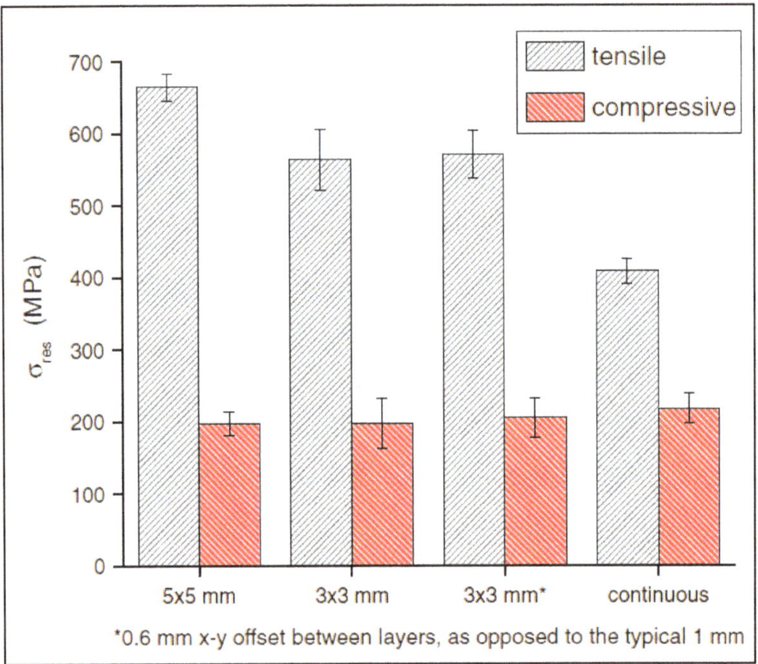

Figure 11. Measured residual stresses utilizing different scan island sizes [32].

Kruth et al. [30] also investigated scanning strategy on RS formation and deformations during SLM of iron-based powders. Directional scanning, where the laser scanned the entire surface back and forth in one direction, resulted in large deflections in the perpendicular direction. Sector-wise scanning divided the surface into grid-like sections, called islands, and scanned each island either successively (adjacent grid sectors) or far apart to prevent heat influence from previous scans, called least heat influence (LHI), as shown in Figure 12. It was determined that the size of the islands did not greatly affect the distortion, but the successive scan strategy resulted in lower RS formation than both the directional scanning and the least heat influence sector scan strategy, likely due to the lower thermal gradients between the current island and the surrounding islands that were melted previously [30].

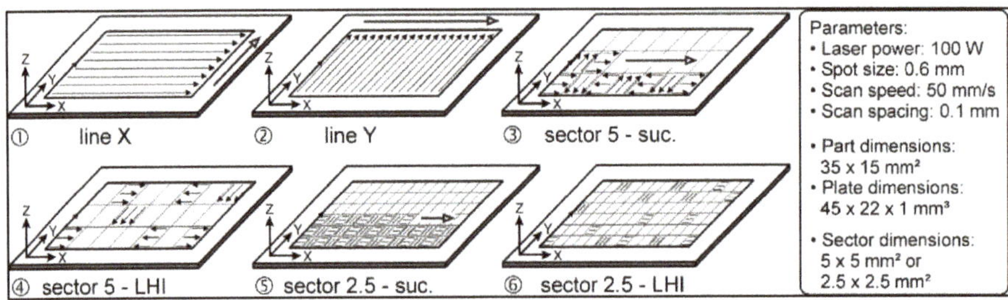

Figure 12. Scanning strategies described in Kruth et al. [30].

Lu et al. [33] investigated the island size on SLM of Inconel 718. It was concluded that the island size influenced the RS formation, contrary to the findings of Kruth et al. [30] for a different material. Island sizes of 2×2, 3×3, 5×5, and 7×7 mm were investigated. The 2×2-mm islands resulted in the lowest RS (around 100 MPa), but this was accompanied by the formation of cracks, which would have relieved much of the internal stresses. With consideration given to density, mechanical properties, and slightly lower RSs, the island size of 5 × 5 mm was determined to be optimal for Inconel 718 produced by SLM, yielding about 150 MPa [33].

Similar to Wu et al. [32], Mercelis et al. [29] showed theoretically and experimentally that, in SLM processes of stainless-steel 316L powder, tensile RSs formed at the top of the sample and the bottom baseplate interface, and compressive stresses formed in the center of the part. It was also determined that directional scanning resulted in high RSs in the transverse direction (about 110 MPa at the surface), and a small island, successive section scanning strategy resulted in low RS formation (about 65 MPa at the surface) [29], in agreement with the findings of Kruth et al. [30].

Liu et al. [34] also investigated the effects of energy input and scanning track length on RS formation in stainless-steel 316L bars produced by SLM. It was determined that compressive stresses existed in the center, while tensile stresses existed in the top layers. The energy input was controlled by the scanning speed; slower scanning speeds related to higher energy inputs. It was shown that the distribution of RSs was not affected (compressive in the center and tensile on the top), but the magnitude of the RSs increased with the energy input (slower scanning speeds generated larger RSs). For the greatest heat input, RSs of about 210 MPa were observed, while, for the lowest heat input, they were as low as 30 MPa. Finally, the length of the scanning track was investigated, leading to the conclusion that a longer track length led to larger RSs—nearly 200 MPa for the long track length, and as low as 45 MPa for the sector scanning. The relationship between track length and RS magnitude lies in the concept that the track shrinks in the scan direction upon cooling, and longer track lengths have less shrinkage compensation, leaving RSs [34].

According to van Belle et al. [35], powder thickness and cooling time between layers affects RS formation in a maraging steel produced by SLM. Consistent with previously discussed studies, tensile RSs were observed on the top layers. It was found that a thin, 20-μm layer of powder with a long cooling time (34 s) resulted in larger RS magnitudes (by three times) than samples produced with 40-μm-thick powder layers and 8-s cooling times between layers. In a study by Gusarov et al. [31], it was determined that RSs in SLM parts of alumina are dependent on the shape of the re-melted domain, but not the size. It was also modeled and experimentally shown that the maximum tensile RSs of about 75 MPa existed in the laser scanning (longitudinal) direction, and the transverse tensile stresses had about one-half the magnitude as the longitudinal stresses [35].

Wang and Chou [36] investigated the RSs formed in Ti–6Al–4V produced by EBAM and Inconel 718 produced by SLM. The material systems were of different geometries (and, thus, scanning strategies); therefore, a side-by-side comparison could not be made. The magnitudes of the RSs in the Ti alloy were reported as lower than the Inconel, which the authors attributed to the fact that EBAM takes place in a vacuum at high temperatures, resulting in a slower cool-down rate than SLM and, thus, stress-relieved components; furthermore, the EBAM process pre-heated the baseplate prior to melting the powder and, thus, had less steep thermal gradients. Another difference was that the titanium exhibited compressive RSs in the build and transverse directions, while the Inconel had compressive stresses in the transverse direction, but tensile stresses in the build direction. Stresses in the longitudinal direction were not reported. The reason that the Inconel had tensile stresses in the build direction was the unique scanning strategy; using 100 × 100-μm scan islands, the RSs were minimized, and the solidified islands pulled surrounding islands and previous layers in tension [36].

In a study by Cottam and Wang [37], H13 tool steel was fabricated using DMD and investigated for RS formation and microstructure characterization. It was determined that the RS distribution was inconsistent with other reports, attributed to the low-temperature phase transformation of the H13 steel. The RS in the top and bottom of the sample was reported to be compressive at 250 MPa, with a narrow band of tensile stress just above the central region in the build direction of about 150 MPa. Drawing parallels to welding processes, low-temperature martensitic phase transformations have compressive stresses, as shown in Reference [37].

The RS formation in FDM of acrylonitrile butadiene styrene (ABS) plastic was investigated by Saphronov et al. [6]. It was determined that, contrary to the literature, compressive stresses existed at the top and bottom of the specimen, and tensile stresses existed in the center. The disparity with the literature was attributed to the fact that the build plate used in this study was flexible, rather than rigid, which sets up an interesting question of the effects of build plate material on RSs in metal components [6].

2. Residual Stresses: Mechanics Background

The fundamental governing equation for elasticity (i.e., reversible deformation) is Hooke's Law.

$$\sigma_{ij} = C_{ijkl}\varepsilon_{kl}, \tag{1}$$

where σ is the applied stress, C is the material's stiffness matrix, ε is strain, and i, j, and k denote 1, 2, and 3, independently. In the 3D Cartesian coordinate system, 1 corresponds to the x-axis, 2 is the y-axis, and 3 is the z-axis.

In addition to mechanical loads, thermal loads can be related to strains in a body [17]. In this case, a change in temperature can cause a material to expand or contract, governed by the following equation:

$$\varepsilon_{th} = \alpha\Delta T, \tag{2}$$

where α is the material coefficient of thermal expansion, ΔT is the change in temperature, and ε_{th} is the thermal strain. The principle of strain superposition dictates that the mechanical strains (ε_σ) and thermal strains (ε_{th}) are summed to a total strain value.

$$\varepsilon = \varepsilon_\sigma + \varepsilon_{th}. \tag{3}$$

The total strain, given in the above equation, can used to obtain the stress in a part through the following constitutive equation [38]:

$$\sigma_{ij} = \frac{E}{(1+\nu)(1-2\nu)}\left[\nu\delta_{ij}\varepsilon_{kk} + (1-2\nu)\varepsilon_{jj} - (1+\nu)\alpha\Delta T\delta_{ij}\right], \tag{4}$$

where E is the modulus of elasticity, ν is Poisson's ratio, and δ_{ij} is the Kronecker delta, taking values of 0 for $i \neq j$ and 1 for $i = j$.

3. Measuring Residual Stresses

3.1. Destructive Methods

Destructive measurement techniques are often referred to as stress-relaxation methods or mechanical methods. Common measurement techniques of this kind include hole-drilling, ring-core, deep hole, sectioning, and contour methods. By removing material which contains RSs and measuring the degree of the material relaxation (the deformation), the RS values can be determined.

3.1.1. Hole Drilling

In a thorough report on measurement techniques for residual stresses [13], the hole-drilling method is described as the removal of material (a drilled hole) of relatively small size—typically on the order of 1.8 mm in diameter and 2 mm in depth—in the region where RSs are to be measured. Prior to drilling, strain gauges are arranged on the surface around the hole location (according to test standards), and, once the material is removed, the part relaxes, and the corresponding relaxation strains are measured; these strains are used to calculate the associated stresses [39].

This method is not very complex and offers fast results, making it a very common technique in practice. Note that the size of the hole is often not large enough to significantly impact the integrity of the part, and it can be repaired easily, if necessary. Issues exist, however, with this method, including the concerns of stresses induced by the machining process, a non-cylindrical hole, and non-circular (elliptical) shape. Despite these sources of error in the stress measurement, the hole-drilling method remains an established method for determining RSs.

3.1.2. Ring Core

The ring-core method involves the measurement of strains of a surface induced by removing the material around the outside of it. If one considers the hole-drilling method, the ring-core method can be thought of as its inverse; a ring of material is removed to a certain depth, and the inner material is allowed to relax. Strain gauges on this inner material capture the relaxation strains. The strains are used to obtain the stresses associated with the relaxation. This method is superior to the hole-drilling method, because it offers much larger surface strains, but often causes significant damage to the part, which makes it far less desirable for use in practice.

It is important to note that RSs may not be uniform through the thickness of the part; thus, many researchers employ an incremental hole-drilling or ring-core method. These methods remove material at incremental depths, so as to record stress values at various depths and build a stress profile through the thickness [13].

3.1.3. Deep-Hole Drilling

Deep-hole drilling involves a combination of hole-drilling and ring-core methods. Firstly, a hole is drilled through the thickness of the part, and the diameter of that hole is accurately measured. The ring-core method is then introduced to remove an amount of material around that hole. The material between the ring and the hole relaxes as the RSs are removed, and the diameter of the hole is measured again. The change in diameter is used to calculate the stresses that were removed. Again, the incremental depth of the ring core in this method is used to build a stress profile through the thickness of the part. While this technique is largely destructive, it is found to be useful in instances where the part is thick and very large macro-stresses are expected to exist [13].

3.1.4. Sectioning

Sectioning is a highly destructive method for measuring RSs. Sections of the specimen are removed, and the deformation is measured. It is important that the sectioning process be conducted without inducing plastic deformation or heat, which would interact with the RSs. As before, strain gauges are used on the specimen to measure the relaxation strains, which are used, in turn, to obtain the stresses [40].

The deformation that occurs as a result of the relaxation may be axial deformation or curvature; the axial deformation is a result of membrane RSs (surface stresses), and curvature is a result of bending RSs (through-thickness stresses). A common assumption is that the bending stresses vary linearly through the thickness [13].

3.1.5. Contour

Finally, the contour method is a newer method for measuring RSs. In this method, the specimen is cut, and then the surface contour is measured, followed by data reduction and analysis. Arguably, the cutting of the specimen is the most important, since the quality of the cut (flatness, constant width, no discontinuities) influences the contour measurement and, thus, the data reduction and analysis steps. Commonly, a wire electric discharge machine (EDM), which uses sparks to remove material, is used in the material cutting process for uniform width and flat cutting.

The contour measurement is carried out on the contoured surfaces created from the cut (contoured due to the relaxation from the removal of RSs). A coordinate measuring machine (CMM) is often employed in this step for high precision and accuracy. Data reduction is conducted by averaging each pair of points (the mirrored point on both cutting faces). Finally, the smoothed data (from the data reduction step) is used as an input (displacement boundary conditions) into a finite element model to calculate the original stress [13].

3.1.6. Other Methods

Other destructive methods exist for measuring RSs: excision, splitting, and curvature (layer removal). Commonly used with thin plates, excision involves the removal of some material around a strain gauge to back out the stress in the part. Splitting is often used in thin-walled tubes, and it involves the sawing of a deep cut into the specimen, and the opening or closing of the cut by surrounding material (during the relaxation) can be related to RSs. The curvature method is often used in thin plates, where, by removing (or adding) a thin layer of material, the plate deflects upward or downward; the extent of the deflection is related to the RS within the part [13].

3.2. Non-Destructive Methods

The destructive methods discussed so far are advantageous because they provide drastic stress relaxation for measurements, but they are not ideal when dealing with parts for use. By definition, the induced damage degrades the integrity of the part and therefore, the part cannot be used for service after the measurement. There can be cases that the geometry allows for devising coupons to be separated from the functioning part, specifically for destructive RS measurement. On the other hand, non-destructive methods, where the part is not significantly altered, are much more favorable with expensive components. The non-destructive techniques include X-ray diffraction, neutron diffraction, and magnetic, ultrasonic, and thermoelastic methods.

3.2.1. X-ray Diffraction

X-ray diffraction (XRD) for measuring RSs makes use of the fact that, under stress, interplanar spacing in a crystal lattice can change [12,13]. In XRD measurements of RS, an X-ray beam is focused onto the sample, and the reflected beam is captured by a detector, which measures the intensity. Typically, the detector is moving with respect to the sample and X-ray source, capturing different reflective angles, and the XRD pattern of the sample is displayed as intensity (of the reflected beam) vs. twice the angle of reflection. By comparing the stressed XRD pattern to an unstressed pattern—typically a powder sample—and employing the Bragg's law (see Equation (5)), the change in interplanar spacing can be related to elastic strains and, thus, to RSs, explained by the following equations [41]:

$$n\lambda = 2d\sin\theta, \tag{5}$$

and

$$d_{\varphi\psi} = \left(\frac{1+\nu}{E}\right)\sigma_\varphi \sin^2\psi - \left(\frac{\nu}{E}\right)(\sigma_{11} + \sigma_{22}), \tag{6}$$

where d represents the interplanar spacing, which corresponds to certain (hkl) planes or, equivalently, a specific 2θ Bragg angle. λ is the wavelength of X-ray beam, and n is the order of diffraction. $d_{\varphi\psi}$ denotes the stressed interplanar spacing for the same (hkl) planes, while the incident beam (or equivalently the sample) is tilted by ψ and rotated by φ. The elastic strain associated with the change in the interplanar spacing is directly related to the RS, σ_φ, which is normal to the (hkl) planes. Since other components of stress (e.g., σ_{11} or σ_{22}) are not of interest for RS measurement, the slope of a plot of measured $d_{\varphi\psi}$ vs. $\sin^2\psi$, yields $\left(\frac{1+\nu}{E}\right)\sigma_\varphi$, from which the RS, i.e., σ_φ, can be obtained straightforwardly if elastic properties of the material (ν and E) are known [41]. Measurement of RS by XRD is quite common for AM parts, as demonstrated by Mishurova et al. [42] and Yan et al. [43], amongst other researchers, and as shown in Figure 13.

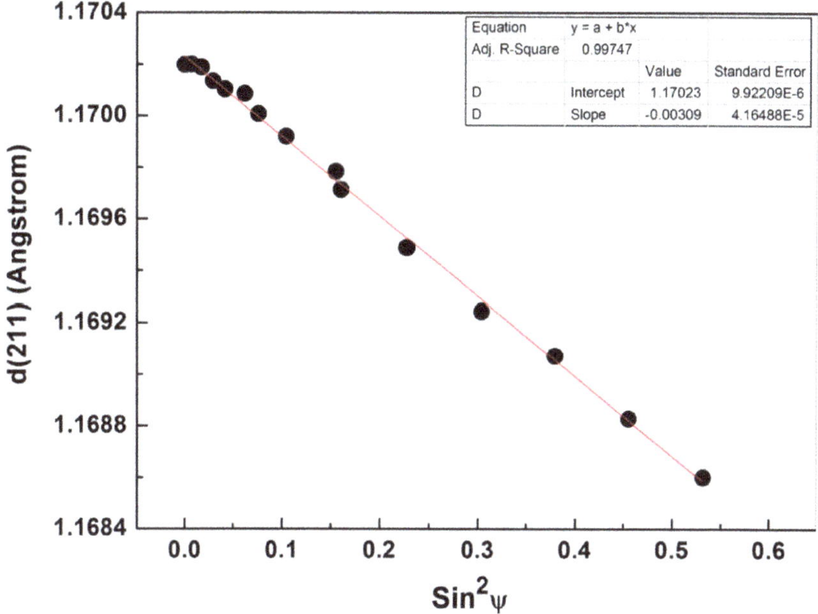

Figure 13. Plot of the interplanar spacing *d* for (211) planes vs. *sin*$^2\psi$ to obtain RS in H13 steel made by SLM [43].

A requirement of this method is that the material is crystalline; it must have many randomly oriented grains sufficient to produce an XRD pattern in any orientation of the surface. This method is often restricted to small geometries which can fit in the XRD machine, while also not interfering with the reflected beam to the detector. Furthermore, XRD is limited to measurements near the surface of the material (a few microns), and it cannot provide information regarding through-thickness stresses. This method could, however, be combined with some type of layer removal technique to generate the stress profile through the thickness, but it would then be considered destructive [13].

3.2.2. Neutron Diffraction

Very similar to the XRD method is neutron diffraction. Changes in the lattice spacing due to elastic strains can be related to RSs within the material [13]. Neutron diffraction can be employed in different ways; a constant wavelength neutron source is used, and Bragg's law determines the interplanar spacing from the diffraction patterns (similar to XRD); another option is to use a pulsed beam in conjunction with a time-of-flight method. The time-of-flight method holds the angle of incidence and reflection constant, but it varies the wavelength of the neutron wave; it shows precision measurements of $\Delta d/d$ on the order of 10^{-5} for interplanar distances [44]. With the wide range of neutron energies (from the varied wavelengths), the neutrons with the highest energy arrive at the target specimen first. The strain is determined by the time between the source and detection as follows: $\varepsilon = \Delta t/t$ [44]. An advantage of neutron diffraction over XRD is the greater penetration depth, as deep as 100 mm in aluminum and 25 mm in steel [13]. Major drawbacks are the size and cost of the equipment.

3.2.3. Barkhauser Noise Method

Methods exist to determine RSs within ferromagnetic materials; one such method is the Barkhauser noise method. The theory behind this method is that the small magnetic regions within the material (which are magnetized along the crystallographic magnetization axes) have boundaries called domain walls. These domains are oriented with respect to one another such that the total magnetization of the material is zero (unless it is a permanent magnet). When an external magnetic field is applied to the ferromagnetic material, the total magnetization of the workpiece changes due to the rotation of the domains: the domains align parallel to the magnetic field direction. The movement of the domain walls is impeded by grain boundaries, dislocations, second-phase materials, and other impurities in the material. The restrictive forces on the domain walls from the defects and impurities can be overcome by applying a larger magnetic force. As the individual domains are suddenly rotated, the total magnetization also increases in jumps. The sudden increases in magnetization can induce electrical pulses in a coil. The Barkhauser noise is the combination of all of the electrical pulses from all of the domain movements. With an appropriate set-up, an alternating magnetic field can be generated, and stress can be determined from the magnetic force [12,13]. As mentioned, this method is only applicable in ferromagnetic materials, and it is really only useful for surface stresses (up to 0.2 mm on parts that were surface-hardened). While not as deep as neutron diffraction, magnetic methods also have greater penetration than XRD [12,13].

3.2.4. Ultrasonic Methods

According to the acoustic elasticity effect, the velocity of an elastic wave propagating through a solid material has a dependency on mechanical stresses on the material. This effect allows for the RS measurement technique known as the ultrasonic method [12,13]. One common approach is the pulse-echo technique, where a transmitting transducer sends a wave through the material, and it detects the wave after it propagates through the material. The average of the RSs present in the material is determined from the time between pulse and detection. The biggest advantage of this technique over many of the others is its universality; this method is applicable to any solid medium. Also, the depth is on the order of millimeters (much deeper penetration than XRD), and the equipment is easy to set up, portable, safe (no radiation hazards), and inexpensive [12,13].

3.2.5. Thermoelastic Methods

Understanding that deformation in a material can generate changes in temperature allows one to use temperature maps of a material to determine the stresses present. With the appropriate infrared camera set up, the temperature profile of a build can be obtained, and the internal stresses can be obtained. Because the temperature–stress effect is quite small, the resolution of infrared cameras must be fine enough for accurate determinations of stress, making this technique limited [12].

3.2.6. Nanoindentation Techniques

Localized mechanical properties such as hardness and elastic modulus are influenced by RSs. Nanoindentation (NI) is a process via which a very small indent is made in the material while recording the applied force and penetration depth, in order to determine material properties. Many researchers used NI to measure the affected localized properties, and they calibrated the measurements to obtain localized surface RSs for polycrystalline materials and metallic glasses [45–50]. Suresh and Giannakopoulos [47] developed a standard method for estimating the surface RSs, and they showed mathematically and experimentally that tensile RSs allow for a larger contact area between material and indenter (and, thus, greater penetration depth) than compressive stresses for a given load. As described by Suresh and Giannakopoulos [47], an important consideration in using NI to measure RSs is that only surface/localized RSs can be measured, proportional to the depths and diameters of the indenter probes. Nevertheless, the NI methods remain effective non-destructive measurement techniques applicable to the surfaces of large components and thin films. Several research groups applied NI methods to AM parts for the determination of bulk material properties [51–53].

3.3. Common Residual Stress Measurement Approaches for Additively Manufactured Parts

Of the RS measurement techniques presented above, the non-destructive techniques are favorable because they do not cause major alterations to the build which degrade the part's integrity; however, the destructive methods are the most standardized and allow for through-thickness stress measurements. While either approach can be applied to additively manufactured parts, residual stresses in these components are often measured non-destructively.

Stainless steel 316L fabricated by laser PBF was investigated by Wu et al. [32], and RSs were measured with neutron diffraction, coupled with the sectioning method. AM process parameters were investigated for their effect on RS formation. It was determined that compressive stresses existed within the center of the parts, and tensile stresses were observed at the surfaces [32].

The additive manufacturing of Ti–6Al–4V using a modified gas tungsten arc welding and an automated wire addition in a layer-by-layer process was investigated by Hoye et al. [54]. Residual strains were measured using the non-destructive neutron diffraction technique, and it was determined that, in the longitudinal direction (parallel to the weld direction), RSs were the most prominent, with values of 565 ± 35 MPa 1 mm below the surface of the baseplate. RSs of almost 70% of the yield strength of the titanium alloy were reported, and they were located near the surface at the centerline of the weld. Also, at the interface region between the deposited material and the baseplate, great variations in the principle stresses were observed [54]. An important consideration for this study was the fact that the build was machined after the additive manufacturing process to achieve the desired part geometry, which may have contributed to the magnitude of the RSs measured. The stress profile is illustrated in Figure 14 [54].

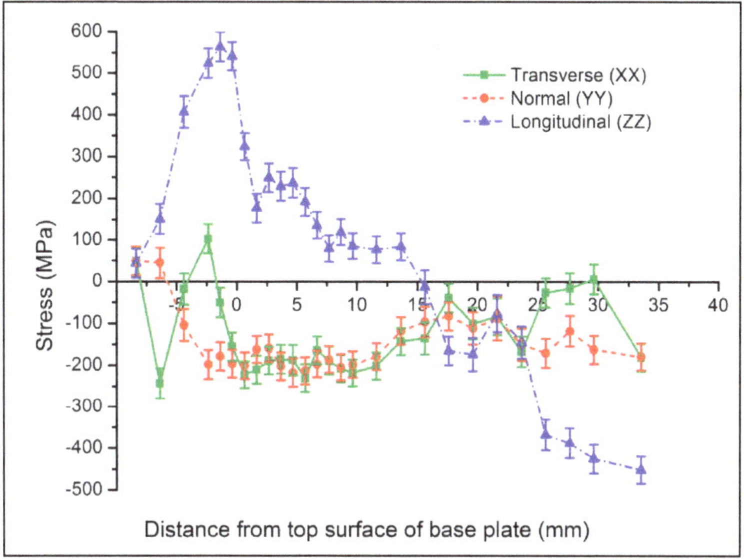

Figure 14. Residual stress measurements in wire-fed additive manufacturing of Ti–6Al–4V at the weld centerline distributed in the vertical z-direction [54].

Ding et al. [55] manufactured a thin wall of mild steel using the wire and arc additive manufacturing (WAAM) process with a modified gas metal arc welding heat source. RSs in the sample were measured via neutron diffraction and the time-of-flight approach. An important finding was that the stresses in the longitudinal direction (parallel to the weld direction) were dominant over normal and transverse-directed stresses, with values as high as 450 MPa in the longitudinal direction. It was also observed that distortions and stress redistribution occurred after the sample was unclamped from the baseplate [55].

The WAAM process was also investigated by Colegrove et al. [56], who also used the neutron diffraction method to characterize the residual stress formed. Samples of mild steel were fabricated similarly to Ding et al. [55], as shown in Figure 15. After adding each layer, rollers were applied to relieve RSs and deformation; a profiled roller matched the surface shape of the deposited layer, and a slotted roller prevented lateral distortion. Similar results to Ding et al. [55] for the RS distribution were reported, reaching nearly 600 MPa [56] for the unrolled or as-built case, as shown in Figure 16.

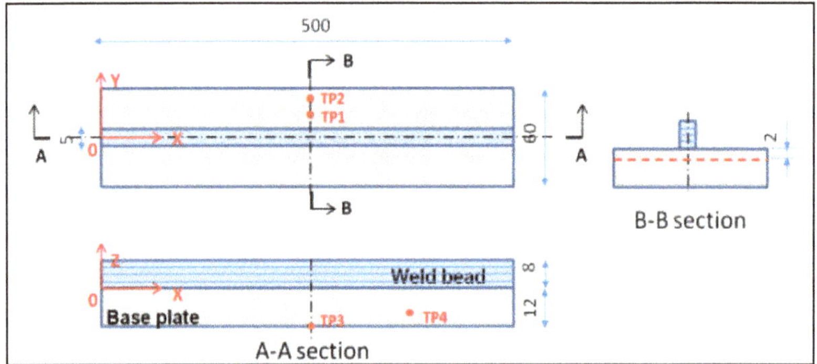

Figure 15. Geometry used by Ding et al. and Colegrove et al., with locations of thermocouple probes identified [55,56].

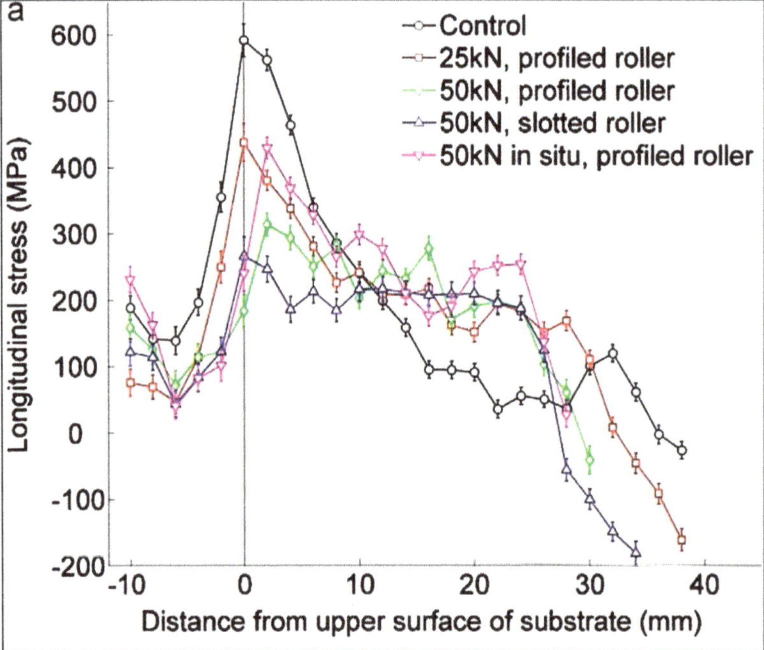

Figure 16. Longitudinal residual stress distribution through the depth of the sample prepared by Ding et al. and Colegrove et al. [55,56].

Work was conducted by An et al. [57] using neutron diffraction methods for determining RS in Inconel 625 thin-walled, curved samples fabricated by laser PBF. It was determined that, at the top of the sample and near the baseplate, tensile hoop stresses of about 270 MPa existed. Also, tensile axial stresses of as much as 500 MPa existed around the edges, and compressive axial stresses of around 200 MPa existed in the middle of the sample. It was shown that the applicability of neutron diffraction was possible for Inconel 625 produced by laser PBF [57].

A titanium alloy, Ti–6Al–4V, was additively manufactured via EBAM by Cao et al. [38], and RSs were also measured with neutron diffraction; the geometry and measurement orientation are shown in Figure 17. As before, it was determined that the most significant RS was in the longitudinal direction, parallel to the beam path, reaching about 320 MPa. In the same work, the hole-drilling method was used and shown to be an effective method in additively manufactured parts [38].

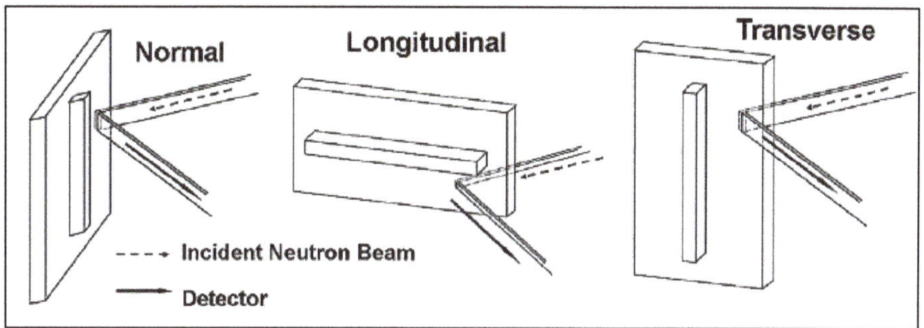

Figure 17. Geometry used by Cao et al., showing the neutron diffraction measurement orientations [38].

Brice et al. [58] also used neutron diffraction to investigate 2219-T8 aluminum samples fabricated with EBAM. It was apparent that an RS concentration was located at the interface between the deposited material and the baseplate, reaching values of nearly 30 MPa along the entire interface. It was reported that the RSs were all compressive in the principle directions. It was also noted that, after the deposited material was un-clamped from the baseplate, the structure rebalanced, and the RS profile changed to accommodate the new equilibrium boundary conditions [58].

Simson et al. [59] explored RSs in SLM samples of 316L stainless steel. Using XRD, RSs were measured at the surface, and layers were removed via elecropolishing for additional XRD measurements at different depths. RSs measured on the top surface of the sample exhibited the largest magnitude in the scanning direction, and the largest RSs measured on the side of the sample were in the build-direction reaching about 220 MPa. The difference in the RS orientation for the different layers measured was attributed to the thermal gradient mechanism affecting the top surface residual stresses, and to the cool-down mechanism, affecting the build-direction stresses [59].

Ahmad et al. [60] investigated RSs in two materials manufactured by SLM: Ti–6Al–4V and Inconel 718. Through the contour method and numerical simulation, RS profiles were developed. The authors reported that significant distortion was observed in the titanium samples, but little noticeable distortion was seen in the Inconel. The RS distribution was similar between the materials, with the highest tensile RSs at the corners and surface of the specimens (titanium: 920 MPa, Inconel: 837 MPa) and compressive RSs at the center region (titanium: 335 MPa, Inconel: 459 MPa) [60].

An SLM titanium alloy (Ti–6Al–4V) was also investigated by Knowles et al. [61] with the destructive hole-drilling method. To capture the RSs through the thickness of the part, the hole was drilled incrementally, and the procedure followed the standard ASTM E837-08. The RSs measured in the samples were reported to be greatest at the surface, in some cases exceeding the material's yield strength. The maximum measured value was 1508 MPa, and the minimum measured value was 135 MPa. The author concluded with the suggestion that work be performed to mitigate or relieve these stresses during manufacture [61].

In general, it is observed that, in AM components, tensile stresses are present on part surfaces, while compressive stresses exist in the center of the parts [29,30,32,34,60]. RSs of the greatest magnitudes are found in the direction of the scan line, compared to the transverse and build directions [38,54–56,59]. It is also frequently observed that the scan strategy affects RS formation, with small successive islands having lower magnitudes of RS than directional scanning [30,32,33]. Furthermore, the scanning speed was shown to affect the magnitude of the RSs; slower scanning speeds (and therefore greater energy input) resulted in larger RS than faster scanning speeds [34]. It was also observed that a large inter-layer dwell time results in larger RSs than with a short time between layers [35]. Finally, it was shown that pre-heating the baseplate of the build can reduce the thermal gradients and, thus, the RSs formed [36]. The EBAM process was suggested to generate RSs lower in magnitude, when compared to SLM, but a concrete comparison was not made; different geometries and scanning strategies were investigated between the AM technologies [36].

4. Computer Modeling of Residual Stresses

4.1. General Approaches for Modeling Residual Stresses

Essentially, to model RSs formed by AM processes, a thermal analysis is performed, followed by a mechanical analysis. The thermal analysis introduces the temperature distribution in the build, and the mechanical analysis determines distortions due to the thermal loads. These two frameworks are either coupled or uncoupled. If they are coupled, the mechanical analysis is performed at every time step of the thermal analysis, in order to account for the heat generated by deformations. The state of the system at each time step is calculated either implicitly or explicitly. In explicit methods, the current time step is used to calculate later time steps, and implicit methods use the current and the later state to determine the later state. Implicit methods can handle any time step size, while numerical instability may occur for explicit methods with large time steps. For highly non-linear processes, such as those in AM (temperature-dependent materials, plastic deformations, etc.), implicit methods may not converge, while explicit methods do, but they may require small time steps, increasing the computation cost. The non-linearity often forces an explicit approach, and convergence requires small time steps, resulting in large computational burdens for these types of simulations [62].

Zohdi [63] presented a computational methodology for the evolution of RSs from the deposition of hot particles, as seen in AM. The approach involved an implicit-staggered, coupled thermo-mechanical analysis, with an adaptive time step implemented and a finite difference time domain. The adaptive time step was to enhance computation time; small time steps were used when the system was changing rapidly, and larger time steps were used when the process was slower [63].

AM computer models are either at the micro or macro level, each of which provides different information. Megahed et al. [64] reported that micro-scale models of AM processes involve the interaction between the heat source and feedstock (powder particles, for example). The micro-scale models provide information about the melt pool size, temperature distribution, and material consolidation quality, whereas macro-scale models use the heat-affected zone dimensions and thermal cycle to calculate the RSs in the part. Typically, simulations performed for RS calculations are on the macro scale, and they assume some heat source distribution [65]. For the most physically accurate model of the AM process, one would model the heat source interaction with the feedstock at the micro level, and then model the fluid flow of the molten material, before finishing the simulation with a macro model for RS and part distortion prediction using the generated temperature distribution, as suggested by Ganeriwala [66] and Ganeriwala and Zohdi [67].

Regardless of the spatial consideration of the model, the heat transfer in AM processes must be accounted for—the heat source for melting (laser or electron beam, typically), and the heat sinks for cooling mechanisms (heat dissipation through convection and radiation from free surfaces, as well as conduction through the material), as seen in Figure 18.

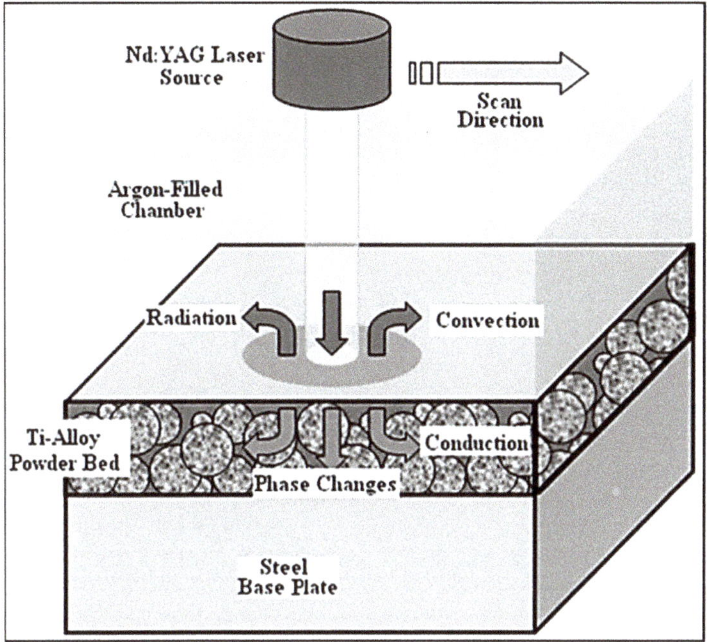

Figure 18. Laser melting of Ti alloy powder on a steel baseplate with heat transfer mechanisms shown [68].

No matter what modeling approach or time integration is used, conservation laws have to be obeyed and can be found in many heat transfer textbooks [69]. Generally, the energy balance for a closed system resembles the following Equation:

$$Q_{flux} = Q_{cond.} + Q_{conv.} + Q_{rad.,} \tag{7}$$

where Q_{flux} is the total heat flux, and $Q_{cond.}$, $Q_{conv.}$, and $Q_{rad.}$ are the conduction, convection, and radiation heat transfer mechanisms, respectively [68]. The heat conduction Equation can be expressed as follows:

$$\frac{\partial}{\partial x}\left(k\frac{\partial T}{\partial x}\right) + \frac{\partial}{\partial y}\left(k\frac{\partial T}{\partial y}\right) + \frac{\partial}{\partial z}\left(k\frac{\partial T}{\partial z}\right) + \dot{q} = \rho C_P \frac{\partial T}{\partial t}, \tag{8}$$

according to Fourier's law, where k is the thermal conductivity, T is the temperature, $\dot{q}$ is the rate of heat transfer into the system, ρ is the density, C_p is the constant pressure heat capacity of the material, x, y, and z are the spatial coordinates, and t is the time. Enthalpy, H, is often taken into consideration to capture the phase-change information.

$$dH = C_P dT. \tag{9}$$

Finally, in the case of heat input by the beam (modeled as a heat flux) and heat losses from convection and radiation, according to Labudovic et al. [70], the boundary conditions are treated as follows:

$$k\frac{\partial T}{\partial n} - \dot{q}_s + h(T - T_0) + \sigma\epsilon\left(T^4 - T_0^4\right) = 0, \tag{10}$$

where T_0 is the initial ambient temperature, n is the surface normal vector, $\dot{q}_s$ is the rate of heat input, h is the convection heat transfer coefficient, σ represents the Stefan–Boltzmann constant, and ε is the emissivity. The stress calculations are more or less the same throughout the literature, but the assumptions and simplifications made regarding heat flux, temperature distribution, and boundary conditions affect the RS predictions. Heigel et al. [71] stated that accurate RS calculations from computational models require detailed knowledge of the surface heat transfer; thus, any over-simplifications in the thermal analysis can result in inaccurate stress calculations.

4.2. Specific Approaches of Modeling Residual Stresses in AM

The work of Ghosh and Choi [72] reported a methodology for the finite element analysis (FEA) prediction of microstructure formation, as well as RSs, in laser-aided DMD of H13 tool steel on a mild steel substrate. Using the same heat conduction equation and boundary conditions described by Equations (8)–(10), a transient thermal analysis was performed, assuming a Gaussian heat flux distribution, given by the following equation [73]:

$$Q = \frac{AP}{\pi r^2 d_0} e^{\left[-2\frac{((x-v_x t)^2 + (y-v_y t)^2)}{r^2}\right]} \frac{1}{5}\left[-3\left(\frac{z}{d_0}\right)^2 - 2\frac{z}{d_0} + 5\right], \tag{11}$$

where A is the laser absorption coefficient of the powder, P is the laser power, r is the laser spot radius, d_0 is the penetration depth, x, y, and z are the coordinates of the laser spot center, v_x and v_y are the laser spot speeds in x- and y-directions, respectively, and t is time. The temperature results were provided to a user-defined subroutine to determine the fraction of each phase of the material; strains from transformation plasticity and volume change were also computed. The uncoupled mechanical analysis then computed elastic, plastic, and thermal strains at each time step. It was reported that the time steps were much smaller during the rapid initial heating than for the slow cooling process in order to converge. It was also stated that the time steps for the thermal and mechanical analyses were independent of each other. The stresses generated were obtained from the calculated strains, assuming a linearly elastic and linearly plastic relationship. Experiments were carried out, and RSs were measured with XRD. Discrepancies between experiment and simulation were attributed to X-ray beam size relative to the interface size between laser passes, as well as the formation of "peaks" and a "valley" due to the overlap between laser passes; the XRD likely missed data points in the "valley" [72].

An investigation by Zaeh et al. [74] explored the transient physical effects in SLM processes for 1.2709 tool steel. Two types of coupled thermo-mechanical simulations were carried out: a layer-based detailed model and a part-based global model. The layer-based model utilized sufficiently accurate heat source models that mimicked the thermal interaction between the laser beam and the powder bed. This model allowed for process parameter optimization (scan strategy and speed, as well as baseplate temperature) regarding RS and deformation on single layers. The part-based global model allowed for the entire part quality. The heat source was applied uniformly to entire layers, without concern for the scan strategy. The part-based global model calculated maximum tensile residual stresses of about 1000 MPa in the horizontal plane and 416 MPa in the build direction, and maximum compressive stresses of 1100 MPa in the horizontal plane and 804 MPa in the build direction at the outer supports; the layer-based model predicted tensile stresses of 86 MPa in the longitudinal direction and compressive stresses of about 628 MPa. Experimental validation of the simulations using neutron diffraction revealed tensile stresses of 305 MPa in the longitudinal direction at the top surface and a compressive stress of 184 MPa at the bottom of the part. The predicted stresses in the longitudinal direction were about 187 MPa at the top surface of the part. The simulations predicted deformations and RSs that were within measured values [74].

Zaeh and Branner [75] extended the work of Zaeh et al. [74] by investigating RS and deformation in SLM of 1.2709 tool steel through simulations. A coupled thermo-mechanical system was employed, and calculated RSs were verified with neutron diffraction. Simplifications were made to prevent long computation times; for instance, the thermal load was applied to an entire layer for 20 ms, instead of modeling the individual scanning vector. Radiative and convective boundary conditions were applied to the part during the cool-down phase, as well as conduction to the baseplate. Tensile stresses were calculated within the horizontal plane to have a maximum of 1000 MPa at the edge of the structure and 416 MPa in the build direction. For a specimen fabricated with the parameters used in the model, the experimentally measured RSs were lower than the simulation predictions. Discrepancies were attributed to the simplifications of the model and the differences in the support structure geometry between model and experiment [75].

Krol et al. [76] compared simulation results to experimental measurements of RSs formed in AM processes; the geometry can be seen in Figure 19. The heat flux was applied to the entire layer, without modeling scan strategies. An experimental investigation of process parameters on residual stress formation revealed, with neutron diffraction, that the finite element model must be of sufficient detail to capture RS progression as a function of process parameters [76].

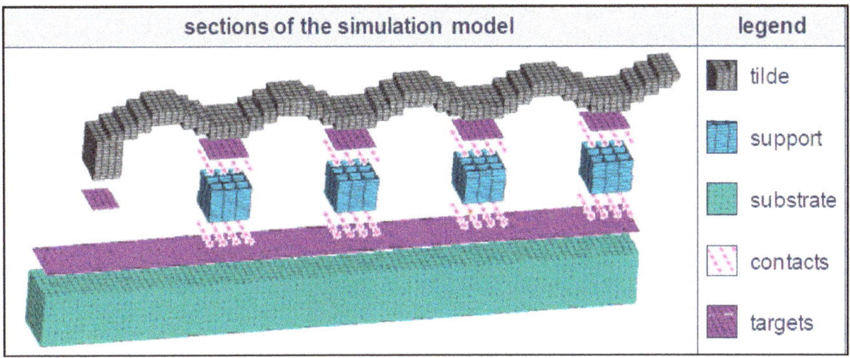

Figure 19. Geometry used by Krol et al. for finite element simulations [76].

The SLM of titanium and nickel powder was simulated by Gu and He [77] for the purpose of RS predictions. The calculated stress values were compared to qualitative experimental results. The simulation presented was a coupled thermo-mechanical analysis with a Gaussian heat source distribution, and the results indicated that the maximum RSs occurred at the end of the scanning track, with values of 86.3 MPa in the scanning direction, 95.7 MPa in the transverse direction, and 23.2 MPa in the build direction. The qualitative experimental results were based on visual crack formation in the manufactured part, indicating locations of large RSs, which agreed with simulated stress distributions [77].

EBAM was simulated by Cao et al. [38] for the purpose of predicting RS and part distortion. Experimental measurements using neutron diffraction and the hole-drilling method were used to validate the simulations for Ti–6Al–4V specimens. After simulating different types of heat sources (Gaussian heat distribution, simple point heat source, double ellipsoid, and uniform heat source), a uniform heat distribution most closely fit the shape of the molten pool in the experiments. The temperature distribution was obtained through the heat conduction equation and heat transfer equation discussed previously. To account for the effect of fluid flow of the molten material on heat transfer (the Marangoni flow), the thermal conductivity was artificially increased by a factor of three, based on other reported works [78]. The RSs were calculated in a coupled mechanical analysis. RSs were measured at five points in the longitudinal direction and five points in the transverse direction, showing maximum values of about 320 MPa; comparing the measured values to the simulation, good agreement was found, but it was noted that five data points were insufficient to fully validate the model. It was also noted that the model predicted zero distortion near the plate extremities, but measurements with a coordinate measurement machine (CMM) indicated small distortions [38]. A final comment indicated that the effects of microstructure evolution were not simulated, which would have impacted the predicted RSs, as explained by Myhr et al. [79].

Denlinger et al. [80] modeled an electron beam-deposited Ti–6Al–4V AM part, and they explored RSs generated. The uncoupled thermo-mechanical analysis assumed a Goldak double ellipsoid heat source distribution [81], given as follows:

$$Q = \frac{6\sqrt{3}P\eta f}{abc\pi\sqrt{\pi}}e^{-[\frac{3x^2}{a^2} + \frac{3y^2}{b^2} + \frac{3(z+v_w t)^2}{c^2}]},$$ (12)

where P is the power of the electron beam, η is the absorption frequency, f is the process scaling factor, x, y, and z are the coordinates, and a is the transverse dimension, b is the melt pool depth, and c is the longitudinal dimension of the ellipsoid, v_w is the scanning speed, and t is the time. Distortion measurements were compared against the simulated distortion, and the magnitudes of the simulated distortions were larger than those measured, but agreed reasonably well with a maximum error of 29%. The magnitude of the displacement predicted by the simulation are shown in Figure 20 [80].

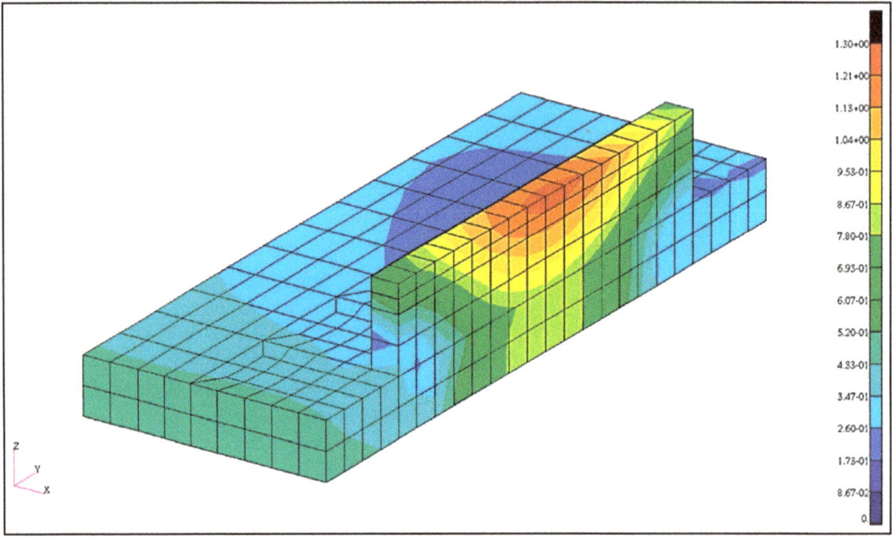

Figure 20. Displacement magnitudes (in mm) at the end of the simulation: maximum displacement of 1.3 mm, minimum displacement of 0 mm [80].

A finite element method was developed by Ding et al. [55] for the thermo-mechanical response during a WAAM process of mild steel, employing a steady-state thermal model, which was compared to a conventional transient model and experimental measurements of residual stresses. In both models, the thermal and mechanical systems were coupled, and the Goldak double ellipsoid heat source [81] was applied. The models did underestimate the value of RSs in a one-layer wall, which was attributed to the microstructure evolution not being captured by the model. Overall, the steady state and thermal models agreed well with experimental measurements of residual stress, with the steady-state model showing an 80% advantage in terms of computation time [55].

Chae [82] investigated RS evolution in DMD of low alloy steel by numerical simulation of laser–powder interaction and molten fluid flow, and then the temperature distribution was used to calculate the thermal strains, volumetric strains (from martensitic phase transformations), and plastic strains due to the phase transformations. The strains were used to calculate RSs from three-dimensional Hooke's law. XRD was used to measure residual strains for a comparison to the simulation predictions, and it was determined that the stresses at the top surface were over-predicted by 8.6%, and those at the melt pool interface were over-predicted by 35.7%. The discrepancies were attributed to the extrapolated values of AISI 4340 steel material properties at high temperatures, due to limitations of access in the database [82].

Parry et al. [83] explored the effect of scan strategy on RS formation through simulations of SLM of Ti–6Al–4V alloy. The coupled thermo-mechanical analysis utilized the Goldak double ellipsoid heat source model, and time-independent plasticity was modeled with a von Mises yield criterion to capture the cyclic non-linear work hardening effect (the Baushinger effect). It was not explicitly stated whether or not microstructure evolution was modeled [83].

Li et al. [9] developed a multiscale finite element approach for prediction of part distortion and RSs generated during SLM of AlSi10Mg powder. The proposed simulation involved the micro-scale modeling of a single track to obtain a temperature history, and then a meso-scale model, which extended the micro-scale model to a deposited powder layer, as shown in Figure 21; finally, a macro-scale model was used to apply the thermal load calculated from the meso-scale model to an entire part. Applying the Gaussian heat source distribution to the surface of a powder layer, the micro-scale model was solved. The temperature history was used as an input to the meso-scale model, where an equivalent body heat flux was applied to the entire layer as follows:

$$q = \frac{AP}{d_s d_m H},$$ (13)

where d_s is the laser spot diameter, d_m is the melt pool depth, and H is the scan spacing (hatch spacing). Finally, the macro-scale model was a coupled thermo-mechanical analysis, and the part was divided into 12 layers, with each layer given the equivalent body heat flux [9].

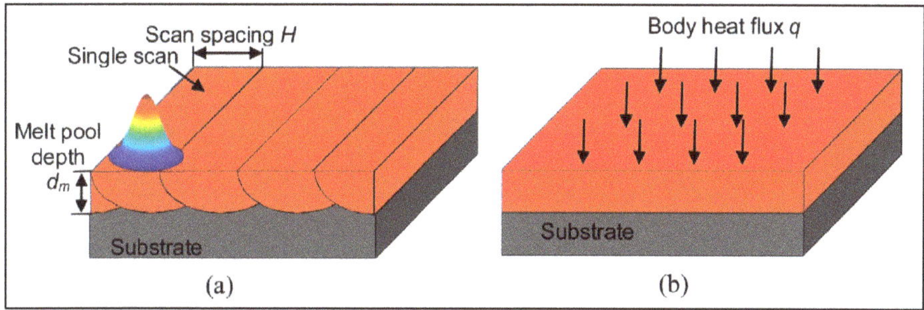

Figure 21. Micro-scale simulation of scan strategy (**a**), to obtain an equivalent layer heat flux (**b**) [9].

Li et al. [84] simulated SLM of iron-based powders to predict RSs and part distortion. Due to a lack of availability of material properties, temperature-independent properties were used in the model. The thermal loads were similar to those described in Li et al. [9]. Comparisons to experimental data were not discussed for model validation, but it was suggested that temperature-dependent material properties would be investigated in future works [84].

Denlinger et al. [85] modeled an EBAM process of Ti–6Al–4V. RSs and distortions were calculated and compared to measured stresses from the hole-drilling method. The thermo-mechanical analysis was uncoupled, and the heat source was modeled as the Goldak double ellipsoid [81]. A unique methodology in this work was the implementation of stress relaxation, where the stress and strains were reset to 0 when the temperature exceeded a defined stress relaxation temperature. Different values of relaxation temperature were investigated, and it was reported that the absence of relaxation effects resulted in over-prediction of distortion by more than 500%, and a simulated relaxation temperature of 690 °C matched measured values most closely (within 25%) [85].

Heigel et al. [71] modeled directed energy deposition (DED) of Ti–6Al–4V using the Goldak double ellipsoid heat source model [81]. The stress relaxation technique described by Denglinger et al. [85] was employed. Free and forced convection as boundary conditions were investigated in the simulations, and RS predictions were compared to experimental results. The forced convection model matched most closely to experimental results [71].

Wang et al. [86] used neutron diffraction to measure RSs in Inconel 625 manufactured by DED, and they compared measurements to a finite element model of the same process. The thermo-mechanical analysis was coupled, employing the Goldak double ellipsoid heat source model [81], with radiation and convection boundary conditions. The heat treatment applied to the sample may have relieved RSs, leading to an error in the computation of the residual strains and stresses [86].

EBAM of Inconel 718 was modeled by Prabhakar et al. [87]. In order to save computation time, the uncoupled thermo-mechanical analysis assumed the heat transfer of the process to be uniform across the entire layer, due to the rapid process, and radiation effects were also ignored. Distortions caused by RSs were qualitatively compared to experiment, and they were observed to be in agreement [87].

Denlinger et al. [88] developed a finite element model to predict RS formation in Inconel 718 produced by laser PBF processes. The heat source was modeled with the Goldak double ellipsoid model [81], but the thermal boundary conditions were treated in a way that accounted for heat source–particle interactions; the thermal conductivity of the powder, k_p, was calculated as follows [88]:

$$k_p = k_f \left[\left(1 - \sqrt{1 - \varnothing}\right)\left(1 + \varnothing \frac{k_r}{k_f}\right) + \sqrt{1 - \varnothing}\left(\frac{2}{1 - \frac{k_f}{k_s}}\left(\frac{2}{1 - \frac{k_f}{k_s}}\ln\frac{k_s}{k_f} - 1\right) + \frac{k_r}{k_f}\right)\right], \tag{14}$$

where k_f is the thermal conductivity of the argon gas surrounding the particles, ϕ is the fractional porosity of the powder bed, k_s is the conductivity of the solid, and k_r is the heat transfer of the radiation between the individual particles [88], calculated as follows:

$$k_r = \frac{4}{3}\sigma T^3 D_p , \tag{15}$$

where D_p is the average diameter of the particles, σ is the Stefan–Boltzmann constant, and T is the temperature. The emissivity of the powder is also calculated as follows [88]:

$$\varepsilon_p = A_H \varepsilon_H + (1 - A_H)\varepsilon_s , \tag{16}$$

where A_H is the porous area fraction of the powder surface, ε_H is the emissivity of the powder surface vacancies, and ε_s is the emissivity of the solid, defined as follows [88]:

$$A_H = \frac{0.908\,\varnothing^2}{1.908\,\varnothing^2 - 2\varnothing + 1} , \tag{17}$$

$$\varepsilon_H = \frac{\varepsilon_s\left[2 + 3.082\left(\frac{1-\varnothing}{\varnothing}\right)^2\right]}{\varepsilon_s\left[1 + 3.082\left(\frac{1-\varnothing}{\varnothing}\right)^2\right] + 1}. \tag{18}$$

A mesh coarsening strategy was also implemented for computation time considerations, which allowed elements below the deposited layer to merge as the heat sources moved in the build direction. The thermal and mechanical analyses were uncoupled, and the distortion predictions caused by RSs agreed strongly with measurements (maximum error of 5%) [88].

Zhao et al. [89] modeled RS formation in direct metal laser sintering of Ti–6Al–4V with a coupled thermo-mechanical analysis. The heat source was modeled as two distributions for comparison. Firstly, a uniform heating pattern was used, given as follows [89]:

$$Q = \frac{P}{\pi r^2}. \tag{19}$$

Next, a semi-spherical power distribution model was used, given as follows [89]:

$$Q = \frac{3P}{2\pi r^2} \sqrt{\left| 1 - \frac{(x - x_0)^2}{r^2} - \frac{(y - y_0)^2}{r^2} \right|}, \qquad \text{for } |x - x_0| < r \text{ and } \left| y - y_0 \right| < r, \tag{20}$$

where x_0 and y_0 are coordinates of the laser spot center. The properties of the powder were estimated from the density, ρ_0, heat capacity, c_{p0}, and thermal conductivity, k_0, of the solid material as follows [89]:

$$\frac{\rho}{\rho_0} = 1 - \varnothing, \tag{21}$$

$$\frac{c_p}{c_{p0}} = 1 - \varnothing, \tag{22}$$

$$\frac{k}{k_0} = \frac{1 - \varnothing}{1 + 11\,\varnothing^2}. \tag{23}$$

The emissivity of the powder was also assumed, as discussed above [88]. Calculated residual stresses in Reference [89] were not compared to experimental results.

5. Residual Stress-Induced Distortion Prevention and Compensation

5.1. Approaches to Prevent Deflection

As discussed previously, adjusting AM process parameters can reduce RSs, but it may not completely eliminate them from forming; therefore, strategies exist to take advantage of the distortions induced by RSs to achieve accurate part dimensions. Mukherjee et al. [90] showed, analytically, that the process variables from AM influence thermal strains (and, thus, thermal stresses), to guide researchers investigating the mitigation of part distortion. It was shown that low heat input can reduce thermal strains, the combined effect of a decrease in laser power and layer height can reduce the distortion, and the combination of a decrease in laser power and an increase in scanning speed can reduce the thermal strains.

Denlinger and Michaleris [91] investigated three distortion mitigation techniques in EBAM with wire feedstock of Ti–6Al–4V. Three distortion mitigation techniques were investigated. Firstly, the part was heated after the deposition to relax the thermal stresses; the other two methods involved the deposition of additional material across the neutral axis of the build, extending beyond the part geometry, to be machined after completion, either after the completion of each layer or after the completion of all layers. These methods were examined with finite element analysis, and the most successful was implemented into experiment. It was determined that depositing additional material after the completion of each layer was the most successful, with 91% of the bending distortion eliminated (see Figures 22 and 23) [91].

Colegrove et al. [92] investigated bulk deformation processes applied during the AM process for property, RS, and distortion control. Rolling was applied to the WAAM of Ti–6Al–4V to apply plastic deformation to relax the RSs formed. Depending on the orientation of the roller with respect to the part (shown in Figure 24), the distortion and RS could be nearly eliminated from about 550 MPa to less than 200 MPa [92].

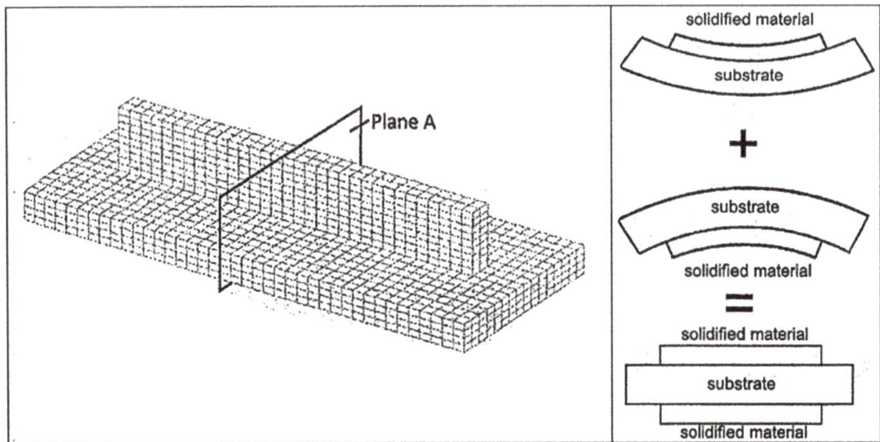

Figure 22. Desired geometry with cross-sectional plane for residual stress measurements (**left**), and proposed concept (**right**) by Denlinger and Michaleris—upward and downward distortion mitigated by a balance of sacrificial material [91].

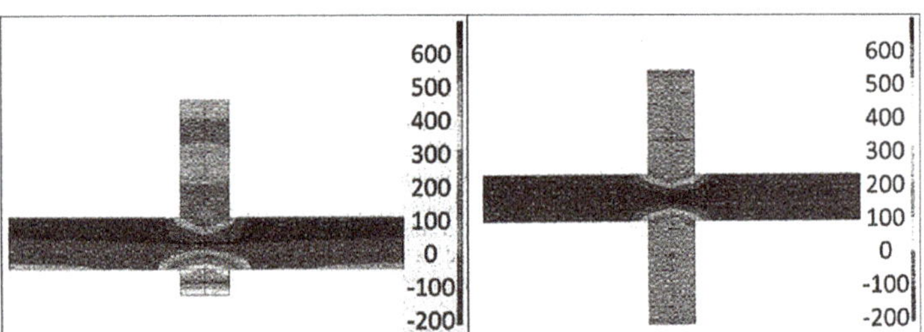

Figure 23. Residual stress distribution with only four sacrificial layers added (**left**) and with 12 balanced layers (**right**); the balanced sacrificial material yields a uniform distribution, reducing the distortion of the part [91].

Aggarangsi and Beuth [93] investigated the effects of localized pre-heating in AM for reducing the RSs generated. The researchers mentioned that uniform pre-heating was used to reduce RSs in small parts, but it was impractical for very large parts; thus, they proposed a localized pre-heating method. The two proposed methods involved a second heat source (lower power) in front of or behind the melt pool, or the pre-heating of the top surface to a predetermined temperature. Uncoupled thermal–structural simulations were used to investigate the effects on RS and part distortion in AISI 304 stainless steel. It was determined that the additional heat source following or leading the primary beam did not significantly reduce the thermal gradient and RSs. However, pre-heating the top-most layer to 673 K did reduce the maximum stress by 18% [93].

In a study by Stucker et al. [94], RapidSteel 2.0 and LaserForm ST-100 (specialty steels) parts were fabricated via SLM, and the process parameters during the post-AM furnace/infiltration stage were investigated for their effect on final part geometry. It was determined that the LaserForm ST-100 was sensitive to the temperature ramp rate in the furnace, and it had more uniform shrinkage between the axes and better feature definition (sharper corners); however, it had much larger absolute shrinkage than the RapidSteel 2.0, which was more sensitive to the amount of infiltrant used. It was suggested that a furnace cycle, which heated the part very rapidly, held it for the least amount of time necessary for infiltration, and then cooled the part very slowly, would minimize shrinkage [94].

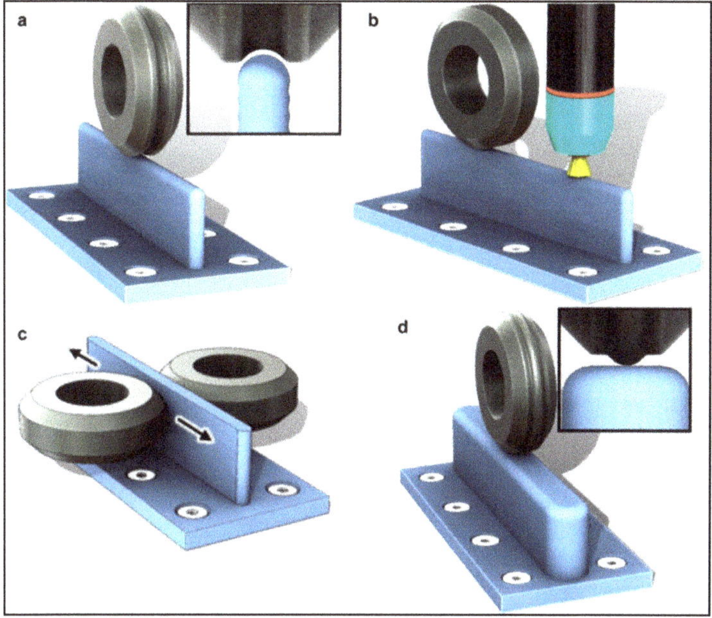

Figure 24. Investigations of bulk deformation processes (rolling) applied to AM components during the build process as a slotted roller in the build direction (**a**), flat rolling ahead of the heat source in the build direction (**b**) or in the transverse direction (**c**), and an inverted profile roller for thick regions (**d**) [92].

Krol et al. [76] compared simulation results with neutron diffraction-measured RSs in AlSi12. Process parameters were investigated for their influence on RS formation, including pre-heating the specimen, increasing the scanning velocity, and changing the support pattern from a block to an optimized pattern (with respect to the RSs simulated). The stress measurements indicated that the optimized support pattern consistently resulted in a lower RS state (about 70 GPa vs. about 10 GPa in some instances). Interestingly, the pre-heated (200 °C) specimens had higher tensile stresses than the room-temperature specimens, and the increase in scanning velocity reduced the RSs formed [76].

The effect of inter-layer dwell time on RS formation in Inconel 718 and Ti–6Al–4V by a laser-based directed energy deposition was investigated by Denlinger et al. [95]. It was determined that the dwell time between layers had a significant effect on the formation of RSs. The dwell time serves to allow additional cooling during the AM process. It was found that, for Inconel 718, an increase in dwell time from 0 to 40 s showed very small changes in distortion and magnitude of RSs; however, for the titanium alloy, the longer dwell times increased distortion by as much as 54%, and RSs by as much as 122%. Distortion measurements were made using a coordinate measurement machine (CMM), and RS was measured with the hole-drilling method [95].

Vastola et al. [73] explored a finite element simulation of Ti–6Al–4V alloy manufactured by electron beam melting to investigate the effects of beam size, beam power, scanning speed, and powder bed temperature on RS development. A small beam size was found to generate larger RSs within a smaller heat-affected zone (HAZ), and larger beam sizes had a larger HAZ with a more uniform stress distribution. The beam power was increased by 20%, and the HAZ was observed to increase by 15% and, thus, increase the distribution of generated stresses. The scanning speed was seen to affect the depth of the HAZ; lower scan speeds had a deeper HAZ and deeper thermal gradients, thereby leading to deeper RS generation. Finally, an increase in powder bed temperature was seen to reduce the magnitudes of the RSs more than the other parameters [73].

In studies by Li et al. [96] and Li et al. [97], a multiscale finite element model was developed to accurately predict part distortion caused by RS formation in SLM processes. The simulation was compared against experimental data for an iron-based powder. The simulation started with a micro-scale model of the heat source–powder particle interaction, to obtain the temperature distribution of the molten material. An equivalent heat flux was developed from the temperature data in the micromodel and applied to a meso-scale model to obtain a RS field. Finally, the stress field was fed into a macro-scale model to obtain part distortion with different scan strategies. The simulation results were compared with experimental data and found to agree. Furthermore, the different scan strategies revealed that a successive island strategy resulted in lower RS formation than the least heat influence island strategy (see Figure 25) [96,97].

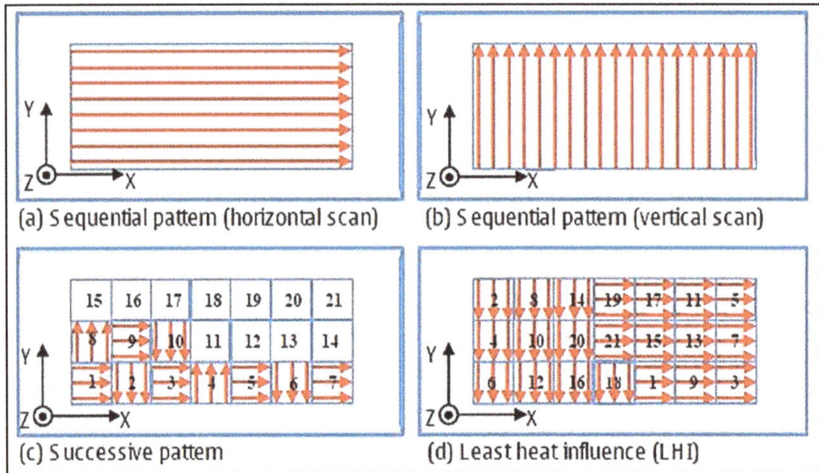

Figure 25. Different scan patterns investigated for residual stress formation in SLM [96].

5.2. Corrective Design to Mitigate Residual Stresses in AM

Afazov et al. [98] developed a method to compensate for the RSs and the subsequent distortions in AM parts. The proposed method used a mathematical model to pre-distort the CAD geometry based on a 3D scan of the distorted, as-built part. The workflow, shown in Figure 26, firstly prints the part from the CAD file, scans the part, and compares the 3D scan data with the CAD geometry; then, it inverts the distortion (while interpolating areas where measurement data are not available) and re-creates a corrected surface mesh for printing. Using this distortion-compensation technique on an impeller geometry (110 mm diameter, 40 mm height, with turbine blade height of 80 mm), Inconel 718 parts fabricated in laser powder bed fusion were successful with tolerances of ±65 μm [98].

Afazov et al. [99] also implemented a distortion correction, to take advantage of the generated RSs, in order to obtain the correct part geometry. SLM was modeled with finite element analysis using Ti–6Al–4V powder. The distortion was predicted, and the displacements of the mesh were inverted. The coordinates of the mesh were updated based on the inverted distortion prediction. It was shown that the non-corrected geometry had distortion of ±200 m, and the corrected geometry from the FEA resulted in distortions of ±45 m [99].

Similar to the work of Afazov et al. [98] and Afazov et al. [99], Xu et al. [100] developed a computational framework to compensate for the distortion in AM processes, to manufacture dimensionally/geometrically accurate parts. The developed process is shown in Figure 27. The part is additively manufactured from a CAD model, and the as-built (distorted) geometry is measured with a CMM or 3D scanner. The distorted geometry is compared to the original CAD model, and a new, "corrected" CAD model is generated by inverting the displacements between the geometry and the

CAD model. The new CAD model, then, has dimensions that, when printed, would distort into the correct final geometry. The authors reported an improvement in shape deformation of 55% when the compensated framework was implemented [100].

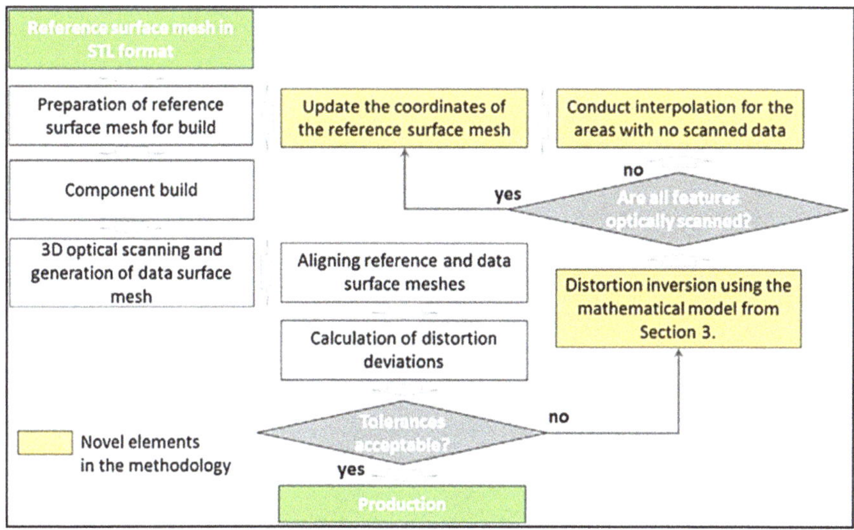

Figure 26. Workflow of proposed distortion prevention method in AM [98].

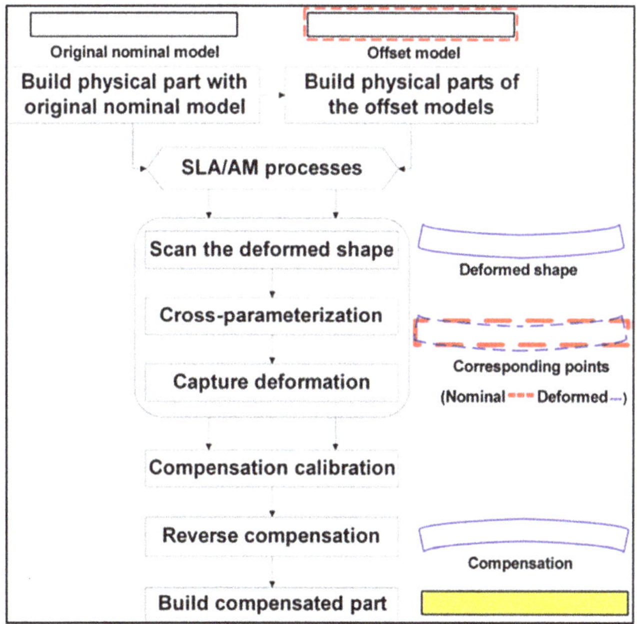

Figure 27. Workflow of the proposed distortion prevention technique [100].

In work by Yaghi et al. [101], a method for mitigating distortion in AM parts was presented. Two stainless-steel 316 impellers of 48.6 mm in height, with the largest diameter being 109 mm, were manufactured via a laser powder bed fusion AM process and then machined to the final geometry, as shown in Figure 28. Measurements of distortion and RSs were taken on the parts using optical measurements and the hole-drilling method combined with the contour method. In Figure 28, the surface RSs changed from ~550 to ~250 MPa from (a) to (b). Point (c) had residual stresses of ~150 MPa about 0.5 mm below the surface. A finite element model was developed to predict distortion, which was validated with the measured data. The predicted distortions were applied to the finite element model in the negative direction, and the surface mesh was re-mapped, resulting in a pre-distorted CAD model for printing. Using the distortion compensation method, the part distortion was reduced from 200 μm to less than 100 μm [101].

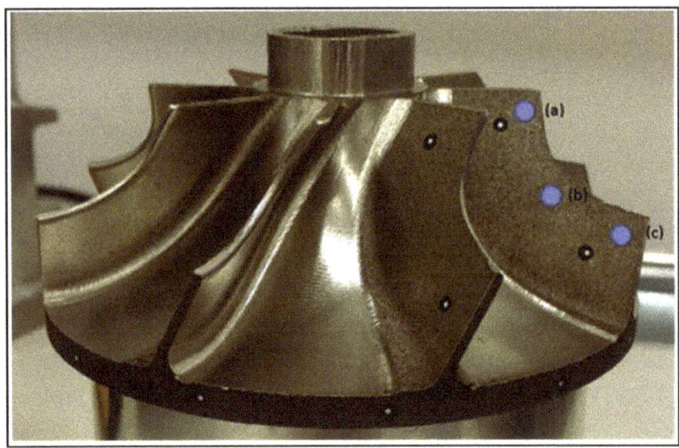

Figure 28. Printed impeller discussed in Yaghi et al. [101]. The measured surface RSs for (**a**,**b**) were around 550 and 250 MPa, respectively; (**c**) had an RS of 150 MPa about 0.5 mm below the surface.

It is important to note that the distortion prevention techniques in which the CAD geometry is altered, such that the distortion during manufacture results in an accurate geometry, as seen in References [98–101], only enables dimensional accuracy; the RSs still remain in the parts. While the final geometry of the part is correct, the fatigue life and structural integrity of the part might still be affected [10]. According to Li et al. [28], the most common approach currently employed to reduce the residual stresses in AM processes is to pre-heat the feed stock or substrate to decrease the thermal gradients during manufacture. Furthermore, a post-manufacturing heat treatment of AM parts was shown to reduce the dislocation density and, thus, type III RSs.

6. Conclusions and Future Works

AM is a growing technology, but there is a huge demand for RS and part distortion prevention in additively manufactured components, as one of the most appealing qualities of AM is the ability to produce complex near-net shape geometries usually faster than most other techniques. Any distortion in precision parts can be catastrophic in use, and RSs can negatively impact the fatigue life and other mechanical performance characteristics. It was demonstrated that the microstructure of various material systems can change during AM processes, and inhomogeneous microstructure evolution, as well as non-uniform phase transformations, can generate RSs within the material. In this regard, stress-relieving heat treatments can help create a more uniform and stress-free microstructure.

As discussed, there are a number of methods for measuring RS, each with advantages and disadvantages, different spatial resolutions, and capabilities regarding the size of the part. There are also a variety of AM techniques and variable parameters available, each of which was shown to induce RSs in the part, exceeding the yield strength of a material in some cases. It was shown that track length, island size, laser power, and scanning speed need to be optimized to reduce the generated RS formation. Much work was conducted to explore the effects of process parameters on RS formation, both experimentally and computationally.

In this paper, different AM technologies were described, and recent investigations on RS formation were reported. As governed by the temperature gradient mechanism, compressive stresses typically form in the center of the part, and tensile stresses form on the top layers. The largest magnitudes of RSs are observed in the direction of the scan line. Small, successive scan islands result in lower RSs than directional scanning. Faster scanning was observed to result in lower RSs than slower scanning, due to the energy input that goes with the scanning speed. The dwell time between layers affects the RSs formed, with large dwell times causing larger RSs. Finally, pre-heating the baseplate prior to build can decrease the thermal gradients and lower the magnitudes of RSs. The EBAM process was shown generate RSs lower in magnitude, compared to SLM, but SLM remains one of the most common technologies due to its affordable and simple (no vacuum chamber) set-up.

The benefits of computational models of AM processes include the rapid and inexpensive investigations of process parameters on RS, but the downside is the computational burden of the simulations, as well as the correct capture of the physics and the validity of the assumptions made. Experimental investigations are, of course, desirable, but they are expensive and time-consuming. The research community needs a physically accurate and computationally inexpensive AM process simulation. Another important and often neglected source of RS formation is the microstructure evolution; this impacts the fundamental material properties, and it plays a direct role in performance, but it is often missed in the simulations.

In addition to a computational framework to accurately capture the physics of AM processes, further experimental investigations into residual stress prevention techniques must be conducted, not just the distortion prevention techniques. While the geometry is preserved in the distortion prevention methods, the RSs (which are the source of the distortion) remain in the part, affecting the fatigue life and structural integrity of the components. Furthermore, in order to more accurately capture the development and evolution of RSs in AM by computer simulations, micro-RS contributions need to be superimposed to the current macro-scale frameworks. In this regard, strong coupling of phase field schemes to the current formalisms mentioned in detail is a must in any robust AM modeling framework, in order to account for dendritic growth and other solidification phenomena and the associated expansions and contractions of the semi-solid and re-solidified regions around the melt pool. Additionally, considering the effects of anisotropy of elastic properties in the semi-solid and re-solidified regions in the simulation of RS development is yet to be studied.

Funding: This research received no external funding.

Conflicts of Interest: The authors declare no conflict of interest.

References

1. Attaran, M. The rise of 3-D printing: The advantages of additive manufacturing over traditional manufacturing. *Bus. Horiz.* **2017**, *60*, 677–688. [CrossRef]
2. Allen, J. An Investigation into the Comparative Costs of Additive Manufacture vs. Machine from Solid for Aero Engine Parts. In *Cost Effective Manufacture via Net-Shape Processing*; Research and Technology Organisation (NATO): Neuilly-sur-Seine Cedex, France, 2006; pp. 1–10.
3. Thomas, D. Costs, Benefits, and Adoption of Additive Manufacturing: A Supply Chain Perspective. *Int. J. Adv. Manuf. Technol.* **2016**, *85*, 1857–1876. [CrossRef] [PubMed]
4. Thomas, D.S.; Gilbert, S.W. Costs and Cost Effectiveness of Additive Manufacturing. *NIST Spec. Publ.* **2014**, *12*, 1176.
5. Withers, P.J.; Bhadeshia, H.K.D.H. Residual stress. Part 2—Nature and origins. *Mater. Sci. Technol.* **2001**, *17*, 366–375. [CrossRef]
6. Saphronov, V.A.; Khmyrov, R.S.; Grigoriev, S.N.; Gusarov, A.V. Distortions and Residual Stresses at Layer-by-Layer Additive Manufacturing by Fusion. *J. Manuf. Sci. Eng.* **2016**, *139*, 031017. [CrossRef]

7. Roy, T.D.; Wei, H.L.; Zuback, J.S.; Mukherjee, T.; Elmer, J.W.; Milewski, J.O.; Beese, A.M.; Wilson-Heid, A.; De, A.; Zhang, W. Additive manufacturing of metallic components—Process, structure and properties. *Prog. Mater. Sci.* **2018**, *92*, 112–224.

8. Buchbinder, D.; Meiners, W.; Pirch, N.; Wissenbach, K.; Schrage, J. Investigation on reducing distortion by preheating during manufacture of aluminum components using selective laser melting. *J. Laser Appl.* **2014**, *26*, 012004. [CrossRef]

9. Li, C.; Liu, J.F.; Fang, X.Y.; Guo, Y.B. Efficient predictive model of part distortion and residual stress in selective laser melting. *Addit. Manuf.* **2017**, *17*, 157–168. [CrossRef]

10. Withers, P.J. Residual stress and its role in failure. *Rep. Prog. Phys.* **2007**, *70*, 2211–2264. [CrossRef]

11. Zhang, J.; Wang, X.; Paddea, S.; Zhang, X. Fatigue crack propagation behaviour in wire + arc additive manufactured Ti-6Al-4V: Effects of microstructure and residual stress. *Mater. Des.* **2016**, *90*, 551–561. [CrossRef]

12. Withers, P.J.; Bhadeshia, H.K.D.H. Residual stress. Part 1—Measurement techniques. *Mater. Sci. Technol.* **2001**, *17*, 355–365. [CrossRef]

13. Rossini, N.S.; Dassisti, M.; Benyounis, K.Y.; Olabi, A.G. Methods for Measuring Residual Stresses in Components. *Mater. Des.* **2012**, *35*, 572–588. [CrossRef]

14. Hrabe, N.; Gnäupel-herold, T.; Quinn, T. Fatigue properties of a titanium alloy (Ti–6Al–4V) fabricated via electron beam melting (EBM): Effects of internal defects and residual stress. *Int. J. Fatigue* **2017**, *94*, 202–210. [CrossRef]

15. Dieter, G. *Metallurgy and Metallurgical Engineering Series*; McGraw Hill Book Company: New York, NY, USA, 1961.

16. Todd, R.I.; Boccaccini, A.R.; Sinclair, R.; Yallee, R.B.; Young, R.J. Thermal residual stresses and their toughening effect in Al_2O_3 platelet reinforced glass. *Acta Mater.* **1999**, *47*, 3233–3240. [CrossRef]

17. Shorr, B.F. *Thermal Integrity in Mechanics and Engineering*; Springer: Berlin/Heidelberg, Germany, 2015.

18. Philpot, T.A. *Mechanics of Materials: An Integrated Learning System*; Wiley Global Education: Hoboken, NJ, USA, 2012.

19. Herzog, D.; Seyda, V.; Wycisk, E.; Emmelmann, C. Additive manufacturing of metals. *Acta Mater.* **2016**, *117*, 371–392. [CrossRef]

20. LeBrun, T.; Nakamoto, T.; Horikawa, K.; Kobayashi, H. Effect of retained austenite on subsequent thermal processing and resultant mechanical properties of selective laser melted 17-4 PH stainless steel. *Mater. Des.* **2015**, *81*, 44–53. [CrossRef]

21. Murr, L.E.; Martinez, E.; Hernandez, J.; Collins, S.; Amato, K.N.; Gaytan, S.M.; Shindo, P.W. Microstructures and properties of 17-4 PH stainless steel fabricated by selective laser melting. *J. Mater. Res. Technol.* **2012**, *1*, 167–177. [CrossRef]

22. Uhlmann, E.; Kersting, R.; Klein, T.B.; Cruz, M.F.; Borille, A.V. Additive Manufacturing of Titanium Alloy for Aircraft Components. *Procedia CIRP* **2015**, *35*, 55–60. [CrossRef]

23. Nickels, L. *Additive Manufacturing: A User's Guide*; Elsevier Ltd.: Amsterdam, The Netherlands, 2016.

24. Frazier, W.E. Metal additive manufacturing: A review. *J. Mater. Eng. Perform.* **2014**, *23*, 1917–1928. [CrossRef]

25. Bikas, H.; Stavropoulos, P.; Chryssolouris, G. Additive manufacturing methods and modelling approaches: A critical review. *Int. J. Adv. Manuf. Technol.* **2016**, *83*, 389–405. [CrossRef]

26. Kruth, J.P. Material Incress Manufacturing by Rapid Prototyping Techniques. *CIRP Ann.* **1991**, *40*, 603–614. [CrossRef]

27. Gong, X.; Anderson, T.; Chou, K. Review on powder-based electron beam additive manufacturing technology. In Proceedings of the ASME/ISCIE 2012 International Symposium on Flexible Automation, St. Louis, MO, USA, 18–20 June 2012; pp. 507–515.

28. Li, C.; Liu, Z.Y.; Fang, X.Y.; Guo, Y.B. Residual Stress in Metal Additive Manufacturing. *Procedia CIRP* **2018**, *71*, 348–353. [CrossRef]

29. Mercelis, P.; Kruth, J.P. Residual stresses in selective laser sintering and selective laser melting. *Rapid Prototyp. J.* **2006**, *12*, 254–265. [CrossRef]

30. Kruth, J.P.; Froyen, L.; van Vaerenbergh, J.; Mercelis, P.; Rombouts, M.; Lauwers, B. Selective laser melting of iron-based powder. *J. Mater. Process. Technol.* **2004**, *149*, 616–622. [CrossRef]

31. Gusarov, A.V.; Pavlov, M.; Smurov, I. Residual stresses at laser surface remelting and additive manufacturing. *Phys. Procedia* **2011**, *12*, 248–254. [CrossRef]

32. Wu, A.S.; Brown, D.W.; Kumar, M.; Gallegos, G.F.; King, W.E. An Experimental Investigation into Additive Manufacturing- Induced Residual Stresses in 316L Stainless Steel. *Metall. Mater. Trans. A* **2014**, *45*, 6260–6270. [CrossRef]
33. Lu, Y.; Wu, S.; Gan, Y.; Huang, T.; Yang, C.; Junjie, L.; Lin, J. Study on the microstructure, mechanical property and residual stress of SLM Inconel-718 alloy manufactured by differing island scanning strategy. *Opt. Laser Technol.* **2015**, *75*, 197–206. [CrossRef]
34. Liu, Y.; Yang, Y.; Wang, D. A study on the residual stress during selective laser melting (SLM) of metallic powder. *Int. J. Adv. Manuf. Technol.* **2016**, *87*, 647–656. [CrossRef]
35. Van Belle, L.; Vansteenkiste, G.; Boyer, J.C. Investigation of Residual Stresses Induced during the Selective Laser Melting Process. *Key Eng. Mater.* **2013**, *554–557*, 1828–1834. [CrossRef]
36. Xiaoquing, W.; Kevin, C. Residual Stress in Metal Parts Produced by Powder-Bed Additive Manufacturing Processes. *J. Chem. Inf. Model.* **2013**, *53*, 1689–1699.
37. Cottam, R.; Wang, J.; Luzin, V. Characterization of microstructure and residual stress in a 3D H13 tool steel component produced by additive manufacturing. *J. Mater. Res.* **2014**, *29*, 1978–1986. [CrossRef]
38. Cao, J.; Gharghouri, M.A.; Nash, P. Finite-element analysis and experimental validation of thermal residual stress and distortion in electron beam additive manufactured Ti-6Al-4V build plates. *J. Mater. Process. Technol.* **2016**, *237*, 409–419. [CrossRef]
39. Olabi, A.G.; Hashimi, M.S.J. Stress relief procedures for low carbon steel (1020) welded components. *J. Mater. Process. Technol.* **1996**, *56*, 552–562. [CrossRef]
40. Tebedge, N.; Alpsten, G.; Tall, L. Residual-stress Measurement by the Sectioning Method. *Exp. Mech.* **1973**, *13*, 88–96. [CrossRef]
41. Cullity, B.D.; Stock, S.R. *Elements of X-ray Diffraction*; Prentice Hall: Upper Saddle River, NJ, USA, 2001; Volume 3.
42. Mishurova, T.; Cabeza, S.; Artzt, K.; Haubrich, J.; Klaus, M.; Genzel, C.; Requena, G.; Bruno, G. An assessment of subsurface residual stress analysis in SLM Ti–6Al–4V. *Materials* **2017**, *10*, 4. [CrossRef]
43. Yan, J.J.; Zheng, D.L.; Li, H.X.; Jia, X.; Sun, J.F.; Li, Y.L.; Yan, M. Selective laser melting of H13: Microstructure and residual stress. *J. Mater. Sci.* **2017**, *52*, 12476–12485. [CrossRef]
44. Santisteban, J.R.; Edwards, L.; Steuwer, A.; Withers, P.J. Time-of-flight neutron transmission diffraction. *J. Appl. Crystallogr.* **2001**, *34*, 289–297. [CrossRef]
45. Tsui, T.Y.; Oliver, W.C.; Pharr, G.M. Influences of stress on the measurement of mechanical properties using nanoindentation: Part 1. Experimental studies in an aluminum alloy. *J. Mater. Res.* **1996**, *11*, 752–759. [CrossRef]
46. Bolshakov, A.; Oliver, W.C.; Pharr, G.M. Inlfuences of stress on the measurement of mechanical properties using nanoindentation: Part II. Finite Element Simulations. *J. Mater. Res.* **1996**, *11*, 760–768. [CrossRef]
47. Suresh, S.; Giannakopoulos, A.E. A new method for estimating residual stresses by instrumented sharp indentation. *Acta Mater.* **1998**, *46*, 5755–5767. [CrossRef]
48. Ghidelli, M.; Sebastiani, M.; Collet, C.; Guillemet, R. Determination of the elastic moduli and residual stresses of freestanding Au-TiW bilayer thin films by nanoindentation. *Mater. Des.* **2016**, *106*, 436–445. [CrossRef]
49. EHerbert, G.; Oliver, W.C.; de Boer, M.P.; Pharr, G.M. Measuring the elastic modulus and residual stress of freestanding thin films using nanoindentation techniques. *J. Mater. Res.* **2009**, *24*, 2974–2985. [CrossRef]
50. Wang, L.; Bei, H.; Gao, Y.F.; Lu, Z.P.; Nieh, T.G. Effect of residual stresses on the hardness of bulk metallic glasses. *Acta Mater.* **2011**, *59*, 2858–2864. [CrossRef]
51. Wang, X.; Gong, X.; Chou, K. Scanning Speed Effect on Mechanical Properties of Ti-6Al-4V Alloy Processed by Electron Beam Additive Manufacturing. *Procedia Manuf.* **2015**, *1*, 287–295. [CrossRef]
52. Li, W.; Liu, W.; Qi, F.; Chen, Y.; Xing, Z. Determination of micro-mechanical properties of additive manufactured alumina ceramics by nanoindentation and scratching. *Ceram. Int.* **2019**, *45*, 10612–10618. [CrossRef]
53. Gong, X.; Lydon, J.; Cooper, K.; Chou, K. Microstructural Analysis and Nanoindentation Characterization of Ti–6Al–4V Parts from Electron Beam Additive Manufacturing. In Proceedings of the ASME 2014 International Mechanical Engineering Congress and Exposition, Montreal, QC, USA, 14–20 November 2014; pp. 1–8.
54. Hoye, N.; Li, H.J.; Cuiuri, D.; Paradowska, A.M. Measurement of Residual Stresses in Titanium Aerospace Components Formed via Additive Manufacturing. *Mater. Sci. Forum* **2014**, *777*, 124–129. [CrossRef]

55. Ding, J.; Colegrove, P.; Mehnen, J.; Ganguly, S.; Almeida, P.S.; Wang, F.; Williams, S. Thermo-mechanical analysis of Wire and Arc Additive Layer Manufacturing process on large multi-layer parts. *Comput. Mater. Sci.* **2011**, *50*, 3315–3322. [CrossRef]

56. Colegrove, P.A.; Coules, H.E.; Fairman, J.; Martina, F.; Kashoob, T.; Mamash, H.; Cozzolino, L.D. Microstructure and residual stress improvement in wire and arc additively manufactured parts through high-pressure rolling. *J. Mater. Process. Technol.* **2013**, *213*, 1782–1791. [CrossRef]

57. An, K.; Yuan, L.; Dial, L.; Spinelli, I.; Stoica, A.D.; Gao, Y. Neutron residual stress measurement and numerical modeling in a curved thin-walled structure by laser powder bed fusion additive manufacturing. *Mater. Des.* **2017**, *135*, 122–132. [CrossRef]

58. Brice, C.A.; Hofmeister, W.H. Determination of bulk residual stresses in electron beam additive-manufactured aluminum. *Metall. Mater. Trans. A Phys. Metall. Mater. Sci.* **2013**, *44*, 5147–5153. [CrossRef]

59. Simson, T.; Emmel, A.; Dwars, A.; Böhm, J. Residual stress measurements on AISI 316L samples manufactured by selective laser melting. *Addit. Manuf.* **2017**, *17*, 183–189. [CrossRef]

60. Ahmad, B.; van der Veen, S.O.; Fitzpatrick, M.E.; Guo, H. Residual stress evaluation in selective-laser-melting additively manufactured titanium (Ti–6Al–4V) and inconel 718 using the contour method and numerical simulation. *Addit. Manuf.* **2018**, *22*, 571–582. [CrossRef]

61. Knowles, C.R.; Becker, T.H.; Tait, R.B. Residual Stress measurements and structural integrity implications for selective laser melted Ti-6Al-4V. *S. Afr. J. Ind. Eng.* **2012**, *23*, 119–129. [CrossRef]

62. Schoinochoritis, B.; Chantzis, D.; Salonitis, K. Simulation of metallic powder bed additive manufacturing processes with the finite element method: A critical review. *Proc. Inst. Mech. Eng. Part B J. Eng. Manuf.* **2017**, *231*, 96–117. [CrossRef]

63. Zohdi, T.I. Modeling and simulation of cooling-induced residual stresses in heated particulate mixture depositions in additive manufacturing. *Comput. Mech.* **2015**, *56*, 613–630. [CrossRef]

64. Megahed, M.; Mindt, H.; Dri, N.N.; Duan, H.; Desmaison, O. Metal additive-manufacturing process and residual stress modeling. *Integr. Mater. Manuf. Innov.* **2016**, *5*, 4. [CrossRef]

65. Luo, Z.; Zhao, Y. A survey of fi nite element analysis of temperature and thermal stress fi elds in powder bed fusion Additive Manufacturing. *Addit. Manuf.* **2018**, *21*, 318–332. [CrossRef]

66. Ganeriwala, R.K. *Multiphysics Modeling of Selective Laser Sintering/Melting*; Unversity of California: Berkeley, CA, USA, 2015.

67. Ganeriwala, R.; Zohdi, T.I. Multiphysics modeling and simulation of selective laser sintering manufacturing processes. *Procedia CIRP* **2014**, *14*, 299–304. [CrossRef]

68. Roberts, I.A.; Wang, C.J.; Esterlein, R.; Stanford, M.; Mynors, D.J. A three-dimensional finite element analysis of the temperature field during laser melting of metal powders in additive layer manufacturing. *Int. J. Mach. Tools Manuf.* **2009**, *49*, 916–923. [CrossRef]

69. Incropera, F.P.; Lavine, A.S.; Bergman, T.L.; DeWitt, D.P. *Fundamentals of Heat and Mass Transfer*; Wiley: Hoboken, NJ, USA, 2007; Volume 16.

70. Labudovic, M.; Hu, D.; Kovacevic, R. A Three Dimensional Model For Direct Laser Metal Powder Deposition and Rapid Prototyping. *J. Mater. Sci.* **2003**, *38*, 35–49. [CrossRef]

71. Heigel, J.C.; Michaleris, P.; Reutzel, E.W. Thermo-mechanical model development and validation of directed energy deposition additive manufacturing of Ti–6Al–4V. *Addit. Manuf.* **2015**, *5*, 9–19. [CrossRef]

72. Ghosh, S.; Choi, J. Modeling and Experimental Verification of Transient/Residual Stresses and Microstructure Formation in Multi-Layer Laser Aided DMD Process. *J. Heat Transf.* **2006**, *128*, 662. [CrossRef]

73. Vastola, G.; Zhang, G.; Pei, Q.X.; Zhang, Y. Controlling of residual stress in additive manufacturing of Ti6Al4V by finite element modeling. *Addit. Manuf.* **2016**, *12*, 231–239. [CrossRef]

74. Zaeh, M.F.; Branner, G.; Krol, T.A. A three dimensional FE-model for the investigation of transient physical effects in Selective Laser Melting. In *Innovative Developments in Design and Manufacturing*; CRC Press: Boca Raton, FL, USA, 2009; pp. 433–442.

75. Zaeh, M.F.; Branner, G. Investigations on residual stresses and deformations in selective laser melting. *Prod. Eng.* **2010**, *4*, 35–45. [CrossRef]

76. Krol, T.A.; Seidel, C.; Schilp, J.; Hofmann, M.; Gan, W.; Zaeh, M.F. Verification of structural simulation results of metal-based additive manufacturing by means of neutron diffraction. *Phys. Procedia* **2013**, *41*, 849–857. [CrossRef]

77. Gu, D.; He, B. Finite element simulation and experimental investigation of residual stresses in selective laser melted Ti–Ni shape memory alloy. *Comput. Mater. Sci.* **2016**, *117*, 221–232. [CrossRef]

78. Lampa, C.; Kaplan, A.F.H.; Powell, J.; Magnusson, C. An analytical thermodynamic model of laser welding. *J. Phys. D Appl. Phys.* **1997**, *30*, 1293–1299. [CrossRef]

79. Myhr, O.R.; Klokkehaug, S.; Grong, O.; Fjaer, H.G.; Kluken, A.O. Modeling of microstructure evolution, residual stresses and distortions in 6082-T6 aluminum weldments. *Weld. J. N. Y.* **1998**, *77*, 286.

80. Denlinger, E.R.; Irwin, J.; Michaleris, P. Thermomechanical Modeling of Additive Manufacturing Large Parts. *J. Manuf. Sci. Eng.* **2014**, *136*, 061007. [CrossRef]

81. Goldak, J.; Chakravarti, A.; Bibby, M. A new finite element model for welding heat sources. *Metall. Trans. B* **1984**, *15*, 299–305. [CrossRef]

82. Chae, H.M. *A Numerical and Experimental Study for Resdiual Stress Evolution in Low Alloy Steel during Laser Aided Additive Manufacturing Process*; University of Michigan: Ann Arbor, MI, USA, 2013.

83. Parry, L.; Ashcroft, I.A.; Wildman, R.D. Understanding the effect of laser scan strategy on residual stress in selective laser melting through thermo-mechanical simulation. *Addit. Manuf.* **2016**, *12*, 1–15. [CrossRef]

84. Li, C.; Liu, J.F.; Guo, Y.B. Prediction of Residual Stress and Part Distortion in Selective Laser Melting. *Procedia CIRP* **2016**, *45*, 171–174. [CrossRef]

85. Denlinger, E.R.; Heigel, J.C.; Michaleris, P. Residual stress and distortion modeling of electron beam direct manufacturing Ti-6Al-4V. *Proc. Inst. Mech. Eng. Part B J. Eng. Manuf.* **2015**, *229*, 1803–1813. [CrossRef]

86. Wang, Z.; Denlinger, E.; Michaleris, P.; Stoica, A.D.; Ma, D.; Beese, A.M. Residual stress mapping in Inconel 625 fabricated through additive manufacturing: Method for neutron diffraction measurements to validate thermomechanical model predictions. *Mater. Des.* **2017**, *113*, 169–177. [CrossRef]

87. Prabhakar, P.; Sames, W.J.; Dehoff, R.; Babu, S.S. Computational modeling of residual stress formation during the electron beam melting process for Inconel 718. *Addit. Manuf.* **2015**, *7*, 83–91. [CrossRef]

88. Denlinger, E.R.; Gouge, M.; Irwin, J.; Michaleris, P. Thermomechanical model development and in situ experimental validation of the Laser Powder-Bed Fusion process. *Addit. Manuf.* **2017**, *16*, 73–80. [CrossRef]

89. Zhao, X.; Iyer, A.; Promoppatum, P.; Yao, S. Numerical modeling of the thermal behavior and residual stress in the direct metal laser sintering process of titanium alloy products. *Addit. Manuf.* **2017**, *14*, 126–136. [CrossRef]

90. Mukherjee, T.; Manvatkar, V.; Debroy, T. Mitigation of thermal distortion during additive manufacturing. *Scr. Mater.* **2017**, *127*, 79–83. [CrossRef]

91. Denlinger, E.R.; Michaleris, P. Mitigation of distortion in large additive manufacturing parts. *Proc. Inst. Mech. Eng. Part B J. Eng. Manuf.* **2017**, *231*, 983–993. [CrossRef]

92. Colegrove, P.A.; Donoghue, J.; Martina, F.; Gu, J.; Prangnell, P.; Hönnige, J. Application of bulk deformation methods for microstructural and material property improvement and residual stress and distortion control in additively manufactured components. *Scr. Mater.* **2017**, *135*, 111–118. [CrossRef]

93. Aggarangsi, P.; Beuth, J.L. Localized preheating approaches for reducing residual stress in additive manufacturing. In Proceedings of the Solid Freeform Fabrication Symposium, Austin, TX, USA, 14–16 August 2006; pp. 709–720.

94. Stucker, B.; Malhotra, M.; Qu, X.; Hardro, P.; Mohanty, N. RapidSteel Part Accuracy. In Proceedings of the International Solid Freeform Fabrication Symposium, Austin, TX, USA, 7–9 August 2000; pp. 133–140.

95. Denlinger, E.R.; Heigel, J.C.; Michaleris, P.; Palmer, T.A. Effect of inter-layer dwell time on distortion and residual stress in additive manufacturing of titanium and nickel alloys. *J. Mater. Process. Technol.* **2015**, *215*, 123–131. [CrossRef]

96. Li, C.; Fu, C.H.; Guo, Y.B.; Fang, F.Z. Fast Prediction and Validation of Part Distortion in Selective Laser Melting. *Procedia Manuf.* **2015**, *1*, 355–365. [CrossRef]

97. Li, C.; Fu, C.H.H.; Guo, Y.B.B.; Fang, F.Z.Z. A multiscale modeling approach for fast prediction of part distortion in selective laser melting. *J. Mater. Process. Technol.* **2016**, *229*, 703–712. [CrossRef]

98. Afazov, S.; Okioga, A.; Holloway, A.; Denmark, W.; Triantaphyllou, A.; Smith, S.A.; Bradley-Smith, L. A methodology for precision additive manufacturing through compensation. *Precis. Eng.* **2017**, *50*, 269–274. [CrossRef]

99. Afazov, S.; Denmark, W.A.D.; Toralles, B.L.; Holloway, A.; Yaghi, A. Distortion prediction and compensation in selective laser melting. *Addit. Manuf.* **2017**, *17*, 15–22. [CrossRef]

Materials **2020**, *13*, 255

100. Xu, K.; Kwok, T.H.; Zhao, Z.; Chen, Y. A reverse compensation framework for shape deformation control in additive manufacturing. *J. Comput. Inf. Sci. Eng.* **2017**, *17*, 021012. [CrossRef]
101. Yaghi, A.; Ayvar-Soberanis, S.; Moturu, S.; Bilkhu, R.; Afazov, S. Design against distortion for additive manufacturing. *Addit. Manuf.* **2019**, *27*, 224–235. [CrossRef]

Article

Micro-Tensile Behavior of Mg-Al-Zn Alloy Processed by Equal Channel Angular Pressing (ECAP)

Kristián Máthis [1,2], Michal Köver [3], Jitka Stráská [1], Zuzanka Trojanová [1,*], Ján Džugan [3] and Kristýna Halmešová [3]

[1]   Department of Physics of Materials, Faculty of Mathematics and Physics, Charles University, Ke Karlovu 5, 121 16 Praha 2, Czech Republic; mathis@met.mff.cuni.cz (K.M.); jitka.straska@mff.cuni.cz (J.S.)
[2]   Nuclear Physics Institute of the CAS, 250 68 Řež, Czech Republic
[3]   COMTES FHT, Průmyslová 995, 334 41 Dobřany, Czech Republic; kover_m@centrum.sk (M.K.); jdzugan@comtesfht.cz (J.D.); kristyna.halmesova@comtesfht.cz (K.H.)
*   Correspondence: ztrojan@met.mff.cuni.cz; Tel.: +420-9-5155-1658

Received: 22 August 2018; Accepted: 5 September 2018; Published: 7 September 2018

Abstract: Commercially available AZ31 magnesium alloy was four times extruded in an equal rectangular channel using three different routes (A, B, and C). Micro tensile deformation tests were performed at room temperature with the aim to reveal any plastic anisotropy developed during the extrusion. Samples for micro tensile experiments were cut from extruded billets in different orientations with respect to the pressing direction. Information about the microstructure of samples was obtained using the electron back-scatter diffraction (EBSD) technique. Deformation characteristics (yield stress, ultimate tensile stress and uniform elongation) exhibited significant anisotropy as a consequence of different orientations between the stress direction and texture and thus different deformation mechanisms.

Keywords: magnesium alloy; equal channel angular pressing; processing route; miniaturized tensile tests; slip systems; twinning

---

1. Introduction

The drive for product miniaturization in various application fields, including biomedicine, the watchmaking industry and communication technologies has significantly increased the demand for metallic micro-parts. There are two approaches for their manufacturing: (i) Micro-machining (electron discharge-, laser- or focused ion beam micro-machining or etching techniques) or (ii) micro-forming when products have a high surface quality and a near-net shape can be obtained in few steps. Consequently, the advantage of micro-forming technology in comparison to micro-machining is the lower cost, given by a higher production rate [1]. In both cases, the so-called size-effect has to be taken into the account. At the microscale, materials cannot be regarded as homogeneous as the microstructural size can be similar to or larger than the parts' dimension. This means that only a few grains are present in the cross-section of a semi-product or micro-part. This makes the downscaling of the conventional forming techniques difficult as the role of the orientation and size of every single grain substantially influences the process flow. The micro-machined products can also behave unexpectedly; deformation anisotropy, scatter in flow stress, size-dependent tribology properties or non-linear increases to the specific cutting energy have been observed [2].

The above-listed difficulties can be overcome by using ultra-fine-grained (UFG) materials. The submicron or nanometer grain size of UFGs, which gains a large number of grains per cross-section of a component, ensures the reproducibility of the physical properties. Further, UFG materials usually exhibit higher strength (cf. the Hall-Petch relation) and superplastic behavior at elevated temperatures [3].

Magnesium alloys are very popular in the structural applications as they are among the lightest structural materials. The application of severe plastic deformation (SPD) methods results in the formation of an ultra-fine microstructure. In the last two decades, equal channel angular pressing (ECAP) has become the most popular and the most intensively studied SPD technique for magnesium alloys. As it was shown by Kim [4] and Estrin [5], ECAP and the micro-extrusion can be integrated into a single processing flow, which results in the production of high-strength micro-parts. Despite the popularity of the ECAP process, deformation behavior on the micro-scale is rarely studied. There are only a few papers (Al [6,7], Mg [8], Ti [9]) that deal with this topic.

In this work, the mechanical properties of ECAP processed AZ31 magnesium alloy was studied using micro tensile tests (M-TT) [10]. The deformation behavior as a function of the processing route and specimen orientation with respect to the pressing direction is discussed in detail. The EBSD data were used for the interpretation of the achieved results.

2. Experimental Section

Sand cast AZ31 magnesium alloy with a nominal composition of Mg—3 wt% Al—1 wt% Zn was investigated in this study. The material was subjected to standardization annealing for 18 h at 390 °C. The microstructure of the alloy before and after annealing is reported in Figure 1a,b. Small precipitates. visible in Figure 1a, are the $Mg_{17}Al_{12}$ electron compound. After standardization annealing, these particles were mostly dissolved as is obvious from Figure 1b. The grain size of the annealed alloy was 300 μm. Samples depicted hereafter "as cast" are after standardization annealing.

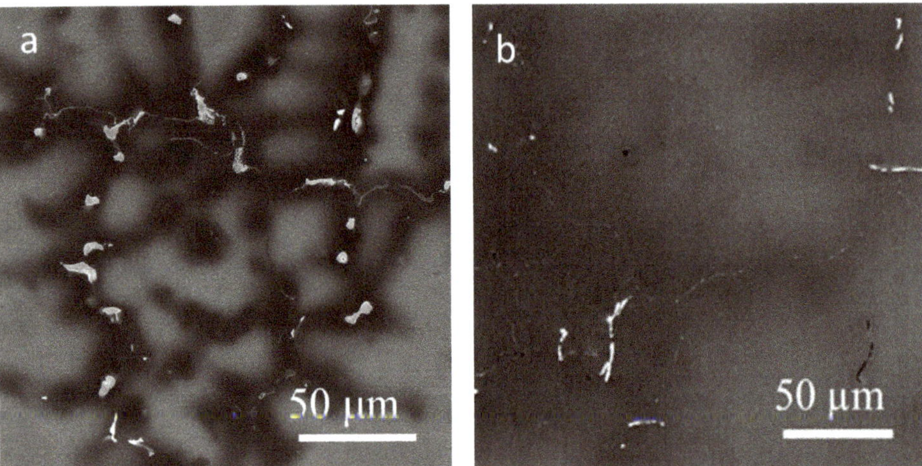

Figure 1. SEM micrographs of the cast alloy before annealing (**a**) and after annealing (**b**).

The ECAP was performed on $10 \times 10 \times 100$ mm^3 billets at a temperature of 250 °C and up to 4 passes, following routes A, B$_c$ and C [3]. All samples were pressed at a speed of 10 mm/min through a die consisting of rectangular channels (inner angle $\varphi = 90°$, outer curvature $\psi = 0°$) with the same cross-section of 10×10 mm^2. In order to study the orientation dependence of micro-tensile properties, samples with their longitudinal axis parallel to the ED-, TD- and ND-planes, this was parallel to the extrusion (pressing), transversal- and normal-directions, respectively, were machined from the billets (Figure 2a). The sample shape ($l = 3$ mm, $w = 1.5$ mm, $t = 0.5$ mm) is depicted in Figure 2b.

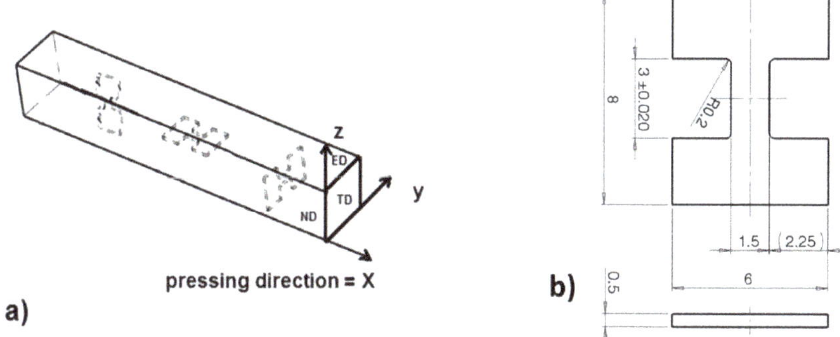

Figure 2. (**a**) The sample coordinate system and the orientation of the specimens. (**b**) Dimensions of the specimens used for micro-tensile tests.

The tensile tests were performed at room temperature with a strain rate of 10^{-3} s^{-1} in a servo-electric test machine M-TT (Material testing Technology. Wheeling, IL, USA) with a 5 kN load capacity. The strain was measured using the Digital Image Correlation (DIC) technique (Dantec Dynamics AVS, Skovlunde, Denmark), implemented in Aramis software. The sample coordinate system used for texture representation is depicted in Figure 2a. The samples were grinded by SiC papers using diamond suspensions down to 0.25 μm. A final surface treatment for EBSD was performed by ion polishing using a Gatan Precision Ion Polishing System (PIPS ion mill, (Gatan, Pleasanton, CA, USA)) at 4 kV and an incidence angle of 6°. The examination of microstructures and textures was carried out using a Quanta field emission gun (FEG) scanning electron microscope (Thermo Fisher Scientific, Waltham, MA, USA) operated at 10 kV equipped with the electron back-scattered diffraction (EBSD) camera. Areas of approximately 250×250 μm^2 with a step size of 0.2 μm were examined. The average grain size was estimated form the EBSD data.

3. Results and Discussion

3.1. Initial Microstructure and Texture

The inverse pole figure maps of the initial microstructures after four passes are shown in Figure 3. Areas smaller (80 μm × 80 μm) than the scanned one are presented in order to see the details. It is obvious that the ECAP process resulted in a substantial refinement of the microstructure—small grains were formed on all planes for every processing route. The average grain sizes and the fraction of high angle grain boundaries (HAGB, >15°) are listed in Table 1.

The average grain size on all planes was almost the same for route B$_c$. In contrast, the grain refinement was orientation dependent for the two other processing routes. For both routes A and C, it is obvious (Figure 3) that the microstructures have a significant character of bimodality—the area fractions of larger grains exceed that for route B$_c$. This effect can be characterized by a log-normal distribution of the grain sizes (only grains with misorientation angles larger than 15° were considered). The narrowest distribution in all examined directions was found for route B$_c$ (Figure 4), whereas the widest for route C. Further, the fraction of HAGBs was also the largest for route B$_c$. This result is in good agreement with many previous works, which have reported that route B$_c$ leads to the most rapid evolution of ultrafine microstructure [3,11].

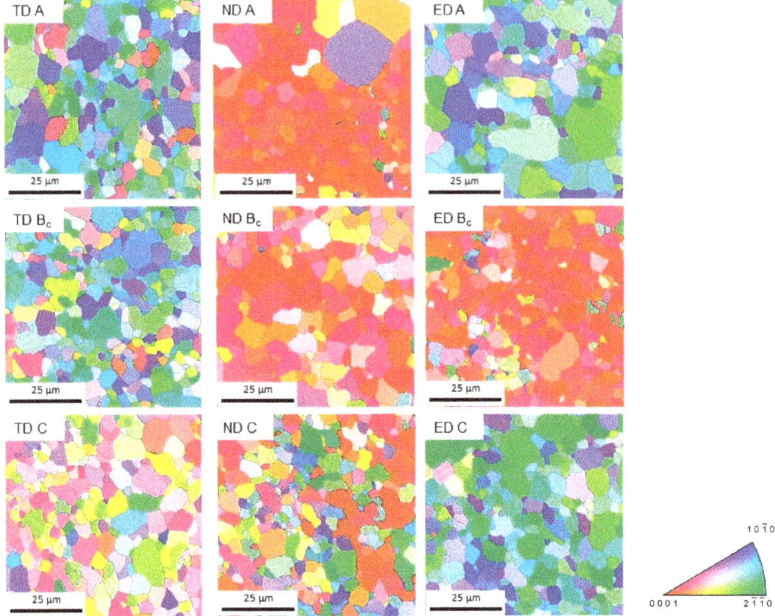

Figure 3. Inverse pole figure maps obtained for the particular planes and ECAP routes.

Table 1. Average grain sizes and fractions of high angle grain boundaries for particular ECAP routes and planes, as estimated from EBSD measurements.

Route/Plane	TD		ND		ED	
ECAP Route	Grain Size (μm)	Fraction of HAGBs	Grain Size (μm)	Fraction of HAGBs	Grain Size (μm)	Fraction of HAGBs
A	3.57	85.5%	4.20	89.7%	3.98	91.0%
B_C	3.43	92.1%	3.48	93.8%	3.37	90.2%
C	3.97	86.9%	4.33	89.5%	4.81	89.6%

During the ECAP process, simple shear was the dominant deformation mode, which is accompanied with large crystal rotations [12]. Their magnitude depended on the orientation of the active slip system with respect to the shear stress. The maximum value was reached when the Burgers vector of active dislocations was perpendicular to the shear direction. The grains tended to reach a stable orientation as the strain increased. This can be simply fulfilled for route C as between two consecutive passes only the shear strain sign was changed, which did not cause further rotations. However, this was not the case for route A nor for route B_c, where every orientation reached in the nth pass was unstable with respect to shear in the $(n + 1)$th pass [12]. This led to activation of various slip systems, particularly the non-basal ones, as was experimentally proved in our previous work [11]. The non-basal dislocations tended to form sessile configurations, which were the seeds for new grain nuclei. This effect was most pronounced at the grain boundaries, where the stress concentrations were the largest. Therefore, the larger grains were surrounded with a chain of smaller grains, according to the observations of Figueiredo [13]. The (0001) pole figures showing the texture of the TD plane (this is perpendicular to the extrusion direction), are shown in Figure 5. Following the notation of Beausir [14] and Krajňák [15], two texture components A (route A and B_c) and B (route C) were observed. As it is discussed in detail in these papers, both texture components form after the first ECAP pass. The type A texture component was a result of extension twinning activated during the compression of the billet in the feed-in (vertical) channel and the second-order pyramidal ($<c+a>$) slip. The type B texture appeared as a consequence of the increased activity of the basal slip. As the number of passes increased, the different routes developed different textures owing to the mutual orientation of the shear planes of the consecutive passes. For route A, when no sample rotation took place, the repeated pressing always caused extension twinning and a <*c+a*>-slip in the vertical channel. This led to the strengthening of the A-type texture component. For route B$_c$, the increased activity of a <*c+a*>-slip has been observed [11], which led to the preservation of the A-type texture component. However, some B-type components also remained. For route C, after the first pass, the grains rotated into an orientation favorable for the basal slip. As it is discussed above, in this case only the shear sign, not the direction changed. Consequently, the B-type texture component became dominant. The prevailing orientations of grains within the billet are exemplified by the hexagonal prisms in Figure 5.

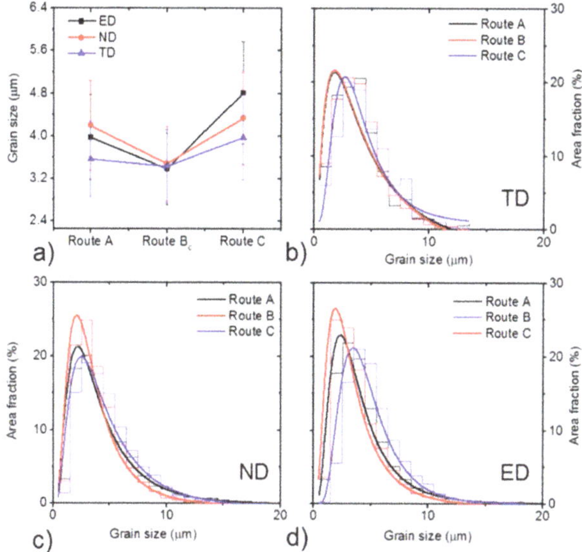

Figure 4. Average grain sizes of the particular samples (**a**). Grain size distributions for the particular ECAP routes in the transversal plane (**b**), normal plane (**c**) and extrusion plane (**d**).

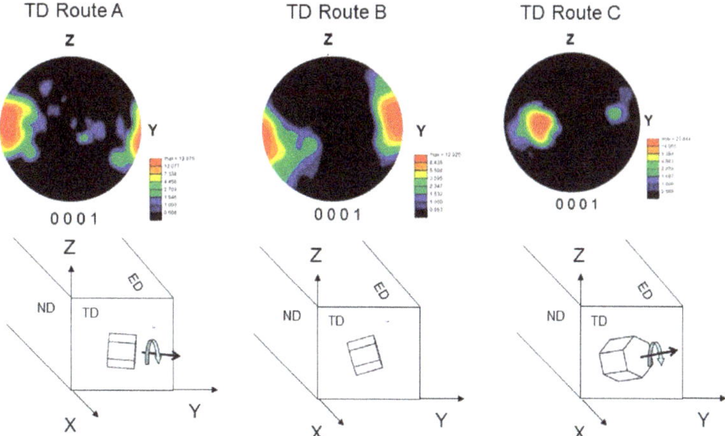

Figure 5. (0001) EBSD pole figures for the TD plane and the corresponding schemes of the prevailing grain orientations.

3.2. Mechanical Properties

The stress-strain curves for the particular routes and directions are plotted in Figure 6. The characteristic parameters yield stress (YS), ultimate tensile strength (UTS) and uniform elongation are listed in Table 2. All tests were repeated twice and the scatter between the values was below 2%. It is obvious that the ECAP process increased the yield stress and tensile strength in comparison to the initial state.

Table 2. Mechanical characteristics as a function of the processing route and direction of the tensile testing.

Route	Plane	Yield Stress (σ_{02})	Ultimate Tensile Strength (σ_{max})	Uniform Elongation (ε_u)
		[MPa]	[MPa]	[%]
Route A	TD	201	275	4.4
	ND	117	264	9.4
	ED	144	253	8.8
Route B_c	TD	160	261	5.9
	ND	131	242	8.7
Route C	TD	231	295	3.9
	ND	103	216	9.9
	ED	112	236	11.8
Initial	-	57	195	8.1

The main contribution to this increment was given by the refinement of the microstructure according to the Hall-Petch rule [16,17]. The results of the micro tensile tests on the ECAPed samples clearly showed that both the processing routes and the direction of the tensile tests had a substantial impact on the mechanical properties. Since the grain sizes were similar, the texture was the foremost parameter responsible for the variance of the stress values [18]. The texture determined the Schmid factor (SF) of the particular deformation mechanism. Based on the EBSD measurement, the SFs for uniaxial tension, parallel to the particular direction, were calculated by TSL OIM Analysis 7 software for $(0001) \langle 11\bar{2}0 \rangle$ basal, $\{10\bar{1}0\} \langle 11\bar{2}0 \rangle$ prismatic, $\{10\bar{1}2\} \langle 11\bar{2}3 \rangle$ 2nd order pyramidal and $\{10\bar{1}2\} \langle 10\bar{1}1 \rangle$ extension twinning systems (Table 3).

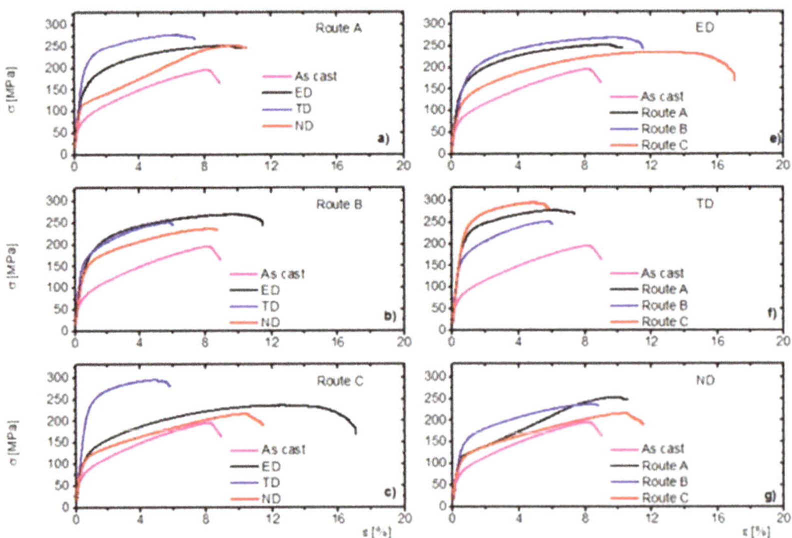

Figure 6. Stress-strain curves (**a–c**) for particular routes as a function of the tensile direction and (**e–g**) for the particular directions as a function of ECAP routes.

Table 3. Schmid factors for the particular deformation mechanisms in particular billet planes calculated for uniaxial tension.

Plane_Route	Basal Slip	Prismatic Slip	Pyramidal $<c+a>$ Slip	Extension Twinning
TD_A	0.19	0.45	0.44	0.03
TD_B	0.31	0.13	0.43	0.02
TD_C	0.17	0.45	0.43	0.15
ND_A	0.24	0.39	0.42	0.35
ND_B	0.23	0.40	0.41	0.24
ND_C	0.19	0.45	0.44	0.29
ED_A	0.32	0.17	0.43	0
ED_B	0.20	0.42	0.43	0.36
ED_C	0.37	0.22	0.38	0

It was obvious that the SF for the pyramidal $<c+a>$-slip was almost the same for all routes and directions. The activation of twinning was generally easier in ND, and for ED after route B processing. The SF for the prismatic $<a>$-slips also had high values, except in ED for routes A and C or in TD for route B, respectively. There was a large scatter in the SF values for the basal $<a>$-slip. The activation of this mechanism was facilitated rather in ED. In the following two sections we discuss separately the influence of the ECAP routes and tensile directions on the mechanical properties in terms of the determined SFs.

3.2.1. Influence of the Tensile Direction on Mechanical Properties for the Particular ECAP Routes

Generally, it can be concluded that the highest YS and the lowest ultimate elongation were observed in TD for all routes. In contrast, the lowest YS value was observed for ND. The uniform elongations had similar values for ED and ND. The ultimate tensile strength was also the lowest for ND, except route A, where its value was a bit higher than that for ED.

Route A (Figure 6a)—the SF_{basal} and $SF_{twinning}$ values were the lowest in the TD direction. The deformation was realized mainly by the prismatic $<a>$-slip and the pyramidal $<c+a>$-slip. The critical resolved shear stress for these mechanisms at room temperature was several orders higher compared to those for the basal slip and extension twinning [19]. In ND, the deformation curve had a characteristic S-shape, which was clear evidence of the activation of extension twinning [20]. One can argue that in fine-grained structures the probability of twinning is low [21]. However, we have to keep in mind that the structure was bimodal—larger grains ($d > 10$ µm) were also present. In these grains, the extension twins were easily nucleated [22]. As a consequence of the rapid twin growth, the ultimate elongation exceeded that for TD. The UTS value was similar to that for ED owing to the secondary hardening when the twin growth terminates [20]. In ED there was the highest SF_{basal} value but the $SF_{twinning}$ value was zero. Consequently, the activation of the $<c+a>$-slip was required for plastic deformation, according to the von Mises rule [23]. The YS was between the ED and ND owing to the interaction of basal and non-basal dislocations. The uniform elongation was quite large owing to the activity of the basal slip [15].

Route B (Figure 6b)—in the ND direction, the ratio of the SF values was similar to that of route A, thus YS and UTS were the lowest. The activity of twinning was lower than that of the previous case (cf. the shape of the curves), which was most probably given by a more uniform grain size. Despite the highest SF_{basal} and lowest $SF_{prismatic}$ values, the YS slightly exceeded of that for ED, where the ratio of these SFs was exactly the opposite. The reason can be given in the fact that the higher $SF_{twinning}$ in ED lowers the YS and increases the ultimate elongation.

Route C (Figure 6c)—in this case, this situation is almost analogical to route A. The highest YS and UTS and the lowest ultimate elongation were observed for TD owing to the low SF_{basal} and high $SF_{prismatic}$ and $SF_{pyramidal}$ values. The ND again had the lowest strength value and the curves indicated (but not as evidently as for route A) activation of twinning. In ED, the basal slip and the

<c+a>-dislocation activity was expected, whereas the probability of twinning was close to zero. Thus, the ultimate elongation was the largest for ED.

3.2.2. Influence of the ECAP Routes on the Mechanical Properties in the Particular Tensile Directions

In order to facilitate the overview of the SFs and the mechanical parameters relations, we plotted the SF dependence of YS and the ultimate elongation in Figures 7 and 8.

ED (Figures 6e, 7 and 8a)—the YS increases with the decreasing SF_{basal} and increasing $SF_{prismatic}$. The higher critical resolved shear stress (CRSS) of the prismatic <a>-slip was in the background of this effect. Further, the prismatic <a> dislocations acted as forest dislocations for the movement of the basal <a> ones [24], which caused hardening. The high $SF_{twinning}$ value was responsible for softening in the route B_c case [25]. The uniform elongation behaved in the opposite way and increased with increasing SF_{basal} and decreased with decreasing prismatic $SF_{prismatic}$. It seems that the twinning was responsible for the similar values of route A and B_c elongations.

TD (Figures 6f, 7, and 8b)—in TD, the values were the opposite compared to that for ED. The strength values for route C exceeded that for the other two routes. The uniform elongations were almost the same for all routes. The reason for this behavior is given by the fact that the SF_{basal} value was the largest and the $SF_{prismatic}$ value was the smallest for route B_c, respectively. Accordingly, the SF dependence of the uniform elongation was the same as for ED.

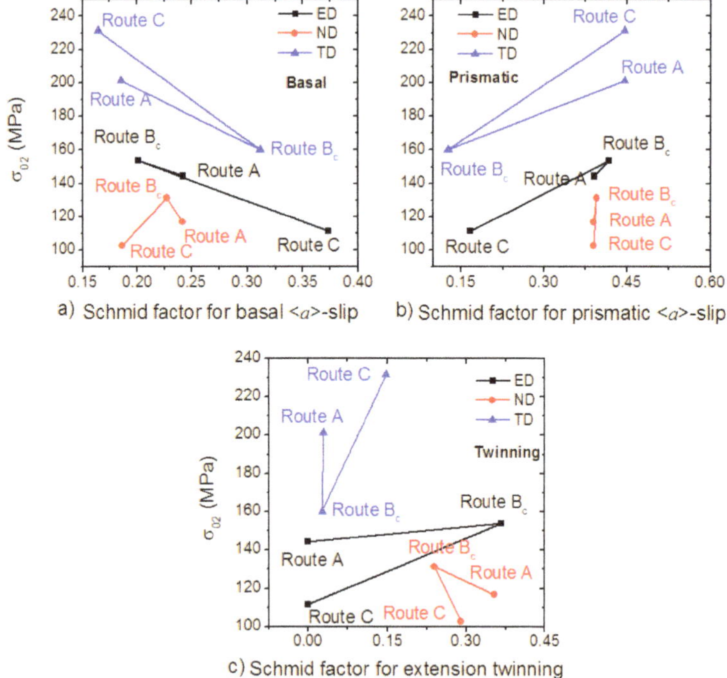

Figure 7. Dependence of the yield stress on the Schmid factor for (**a**) the basal <a>-slip, (**b**) the prismatic <a>-slip; and(**c**) extension twinning.

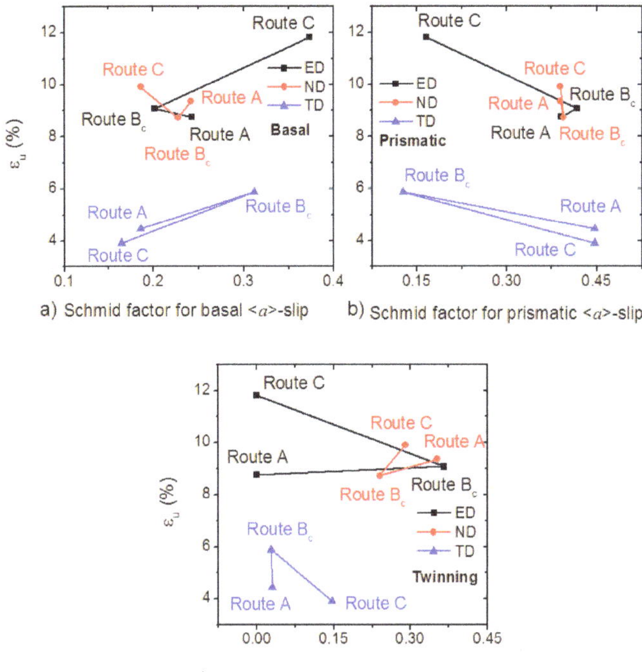

a) Schmid factor for basal <*a*>-slip b) Schmid factor for prismatic <*a*>-slip

c) Schmid factor for extension twinning

Figure 8. Dependence of the uniform elongation on the Schmid factor for (**a**) the basal <*a*>-slip, (**b**) the prismatic <*a*>-slip and (**c**) extension twinning.

ND (Figures 6g, 7, and 8c)—the YS values were similar to ED ($YS_B > YS_A > YS_C$). The UTS was the highest for the route A. The ultimate elongation values are in order route C > route A > route B. In this direction, the SF values were close to each other, which caused the mechanical parameters to be similar.

4. Conclusions

The micro tensile properties of 4x ECAPed AZ31 magnesium alloy were examined as a function of the ECAP processing route and tensile direction. The variation of the mechanical properties for the particular samples was substantiated by the different textures, causing various conditions for the deformation mechanisms. The processing routes and the direction of the tensile tests had a substantial impact on the mechanical properties. The following particular conclusions can be drawn.

Influence of the Processing Routes:

- The Schmid factors for the basal <*a*> slip and extension twinning were foremost responsible for the values of uniform elongation.
- The highest values of yield stress and yield strength were found for samples with the highest Schmid factor for the prismatic <*a*> and pyramidal <*c+a*>-slips and the lowest Schmid factor for basal <*a*> dislocations.

Influence of the Tensile Direction:

- In ED and TD, the yield stress was determined with the ratio of Schmid factors for basal <*a*> and prismatic <*a*> slips. The higher the value, the lower the yield stress. In ND, the twinning activity was significant for all routes.

Author Contributions: K.M. and Z.T. conceived and designed the experiments; K.H. and J.D. prepared the ECAPed samples; M.K. performed the tensile tests; J.S. studied the microstructure of the samples using EBSD; K.M. and Z.T. analyzed the data; K.M. and Z.T. wrote the paper.

Funding: This research was funded by the Czech Science Foundation grant number 14-36566G. K.M. acknowledges the support of the Operational Programme Research, Development and Education, The Ministry of Education, Youth and Sports (OP RDE, MEYS) [CZ.02.1.01/0.0/0.0/16_013/0001794].

Acknowledgments: The authors thank Peter Palček for the provision of Figure 1a,b.

Conflicts of Interest: The authors declare no conflicts of interest.

References

1. Eichenhueller, B.; Egerer, E.; Engel, U. Microforming at elevated temperature—Forming and material behavior. *Int. J. Adv. Manuf. Technol.* **2007**, *33*, 119–124. [CrossRef]
2. Vollertsen, F.; Biermann, D.; Hansen, H.N.; Jawahir, I.S.; Kuzman, K. Size effects in manufacturing of metallic components. *CIRP Ann. Manuf. Technol.* **2009**, *58*, 566–587. [CrossRef]
3. Valiev, R.Z.; Islamgaliev, R.K.; Alexandrov, I.V. Bulk nanostructured materials from severe plastic deformation. *Prog. Mater. Sci.* **2000**, *45*, 103–189. [CrossRef]
4. Kim, W.J.; Sa, Y.K. Micro-extrusion of ECAP processed magnesium alloy for production of high strength magnesium micro-gears. *Scr. Mater.* **2006**, *54*, 1391–1395. [CrossRef]
5. Estrin, Y.; Janeček, M.; Raab, G.I.; Valiev, R.Z.; Zi, E. Severe plastic deformation as a means of producing uItra-fine-grained net-shaped micro electro-mechanical systems parts. *Metall. Mater. Trans. A* **2007**, *38A*, 1906–1909. [CrossRef]
6. Horita, Z.; Fujinami, T.; Langdon, T.G. The potential for scaling ECAP: Effect of sample size on grain refinement and mechanical properties. *Mater. Sci. Eng.* **2001**, *A318*, 34–41. [CrossRef]
7. Xu, C.; Száraz, Z.; Trojanová, Z.; Lukáč, P.; Langdon, T.G. Evaluating plastic anisotropy in two aluminum alloys processed by equal-channel angular pressing. *Mater. Sci. Eng. A* **2008**, *497*, 206–211. [CrossRef]
8. Figueiredo, R.B.; Száraz, Z.; Trojanová, Z.; Lukáč, P.; Langdon, T.G. Significance of twinning in the anisotropic behavior of a magnesium alloy processed by equal-channel angular pressing. *Scr. Mater.* **2010**, *63*, 504–507. [CrossRef]
9. Fu, E.K.Y.; Bellam, H.C.; Qazi, J.I.; Rack, H.J.; Stolyarov, V. Reciprocating-sliding wear of ultra-fine grained Ti-6Al-4V. In *Ultrafine Grained Materials*; Zhu, Y.T., Ed.; TMS: Warrendale, PA, USA, 2004; p. 547.
10. Rund, M.; Procházka, R.; Konopík, P.; Džugan, J.; Folgar, H. Investigation of sample-size influence on tensile test results at different strain rates. *Procedia Eng.* **2015**, *114*, 410–415. [CrossRef]
11. Krajňák, T.; Minárik, P.; Gubicza, J.; Máthis, K.; Kužel, R.; Janeček, M. Influence of equal channel angular pressing routes on texture, microstructure and mechanical properties of extruded AX41 magnesium alloy. *Mater. Charact.* **2017**, *123*, 282–293. [CrossRef]
12. Máthis, K.; Rauch, E.F. Microstructural characterization of a fine-grained ultra low carbon steel. *Mater. Sci. Eng. A* **2007**, *462*, 248–252. [CrossRef]
13. Figueiredo, R.B.; Langdon, T.G. Grain refinement and mechanical behavior of a magnesium alloy processed by ECAP. *J. Mater. Sci.* **2010**, *45*, 4827–4836. [CrossRef]
14. Beausir, B.S.; Biswas, B.S.; Kim, D.I.; Toth, L.S.; Suwas, S. Analysis of microstructure and texture evolution in pure magnesium during symmetric and asymmetric rolling. *Acta Mater.* **2009**, *57*, 5061–5077. [CrossRef]
15. Krajňák, T.; Minárik, P.; Stráská, J.; Gubicza, J.; Máthis, K.; Janeček, M. Influence of equal channel angular pressing temperature on texture, microstructure and mechanical properties of extruded AX41 magnesium. *J. Alloy. Compd.* **2017**, *705*, 273–282. [CrossRef]
16. Lin, H.K.; Huang, J.C.; Langdon, T.G. Relationship between texture and low temperature superplasticity in an extruded AZ31 Mg alloy processed by ECAP. *Mater. Sci. Eng. A* **2005**, *402*, 250–257. [CrossRef]
17. Yamashita, A.; Horita, Z.; Langdon, T.G. Improving the mechanical properties of magnesium and a magnesium alloy through severe plastic deformation. *Mater. Sci. Eng. A* **2001**, *300*, 142–147. [CrossRef]
18. Ding, S.X.; Lee, W.T.; Chang, C.P.; Chang, L.W.; Kao, P.W. Improvement of strength of magnesium alloy processed by equal channel angular extrusion. *Scr. Mater.* **2008**, *59*, 1006–1009. [CrossRef]
19. Chapuis, A.; Driver, J.H. Temperature dependency of slip and twinning in plane strain compressed magnesium single crystals. *Acta Mater.* **2011**, *59*, 1986–1994. [CrossRef]

20. Hutchinson, J.W. Creep and plasticity of hexagonal polycrystals as related to single-crystal slip. *Metall. Mater. Trans. A* **1977**, *8*, 1465–1469. [CrossRef]

21. Yuan, W.; Mishra, R.S. Grain size and texture effects on deformation behavior of AZ31 magnesium alloy. *Mater. Sci. Eng. A* **2012**, *558*, 716–724. [CrossRef]

22. Bohlen, J.; Dobroň, P.; Swiostek, J.; Letzig, D.; Chmelík, F.; Lukáč, P.; Kainer, K.U. Acoustic emission during stress relaxation of pure magnesium and AZ magnesium alloys. *Mater. Sci. Eng. A* **2007**, *462*, 302–306. [CrossRef]

23. Von Mises, R. Mechanik der festen Körper im plastisch deformablen Zustand. *Göttin. Nachr. Math. Phys.* **1913**, *1*, 582–592.

24. Lukáč, P. Hardening and softening during plastic deformation of hexagonal metals. *Czechoslov. J. Phys.* **1985**, *35*, 275–285. [CrossRef]

25. Agnew, S.R.; Duygulu, O. Plastic anisotropy and the role of non-basal slip in magnesium alloy AZ31B. *Int. J. Plast.* **2005**, *21*, 1161–1193. [CrossRef]

Article

Bio-Inspired Functional Surface Fabricated by Electrically Assisted Micro-Embossing of AZ31 Magnesium Alloy

Xinwei Wang [1,2,3], Jie Xu [1], Chunju Wang [1,2,4,*], Antonio J. Sánchez Egea [5], Jianwei Li [1,6], Chen Liu [2,*], Zhenlong Wang [3], Tiejun Zhang [6], Bin Guo [1,2] and Jian Cao [7]

[1] Key Laboratory of Micro-Systems and Micro-Structures Manufacturing, Ministry of Education, Harbin Institute of Technology, Harbin 150080, China; xinweiwang@hit.edu.cn (X.W.); xjhit@hit.edu.cn (J.X.); 13B909069@hit.edu.cn (J.L.); bguo@hit.edu.cn (B.G.)

[2] Laboratory for Space Environment and Physical Sciences, Harbin Institute of Technology, Harbin 150001, China

[3] School of Mechanical Engineering, Harbin Institute of Technology, Harbin 150001, China; wangzl@hit.edu.cn

[4] School of Mechanical and Electrical Engineering, Robotics and Microsystems Center, Soochow University, Suzhou 215131, China

[5] Department of Mechanical Engineering, Universitat Politècnica de Catalunya, Av. Eduard Maristany, 16, 08019 Barcelona, Spain; sanchezegea.antonio@gmail.com

[6] Beijing Hangxing Machinery Manufacture Limited Corporation, Beijing 100013, China; 7w7z@sohu.com

[7] Department of Mechanical Engineering, Northwestern University, 2145 Sheridan Rd, Evanston, IL 60208, USA; jcao@northwestern.edu

* Correspondence: cjwang1978@hit.edu.cn (C.W.); liuchen2016@hit.edu.cn (C.L.)

Received: 23 December 2019; Accepted: 12 January 2020; Published: 16 January 2020

Abstract: Developing bio-inspired functional surfaces on engineering metals is of extreme importance, involving different industrial sectors, like automotive or aeronautics. In particular, micro-embossing is one of the efficient and large-scale processes for manufacturing bio-inspired textures on metallic surfaces. However, this process faces some problems, such as filling defects and die breakage due to size effect, which restrict this technology for some components. Electrically assisted micro-forming has demonstrated the ability of reducing size effects, improving formability and decreasing flow stress, making it a promising hybrid process to control the filling quality of micro-scale features. This research focuses on the use of different current densities to perform embossed micro-channels of 7 μm and sharklet patterns of 10 μm in textured bulk metallic glass dies. These dies are prepared by thermoplastic forming based on the compression of photolithographic silicon molds. The results show that large areas of bio-inspired textures could be fabricated on magnesium alloy when current densities higher than 6 A/mm^2 (threshold) are used. The optimal surface quality scenario is obtained for a current density of 13 A/mm^2. Additionally, filling depth and depth–width ratio nonlinearly increases when higher current densities are used, where the temperature is a key parameter to control, keeping it below the temperature of the glass transition to avoid melting or an early breakage of the die.

Keywords: electrically assisted; micro-embossing; bio-inspired functional surface; bulk metallic glass; photolithography

---

1. Introduction

Nature has developed biological surfaces with periodic multiscale arrangements (macro/micro/nano levels), showing a high state of intelligent functionality. One example of these is the shark skin, which has placoid scales, with a rectangular base implanted in the skin with small spines or bristles that emerge to the surface. This particular scale configuration keeps the skin clean, and also reduces drag [1,2]. In recent years, researchers have fabricated many bio-inspired functional surfaces by using different materials and processes to achieve multiple intelligent performances, such as antibiofouling, super-hydrophobicity and wear-resistance. In particular, Chen et al. [2] made an overview of the bio-inspired sharkskin surface in terms of drag reduction mechanism, fabrication methods and applications. Sharklet AF^{TM} [3] developed silicon wafers by using photolithography that was also coated with a reactive ion etching. This reactive coating transferred defence properties to the elastomers against bacteria [4]. Accordingly, micro-manufacturing processes, like photolithography, micro-machining, laser ablation, direct 3D printing, micro-molding and micro-embossing, are commonly used for bionic microstructure fabrication, although each of these processes presents some difficulties. Photolithography is a complex and a high cost process, which transfers relatively poor mechanical capabilities to silicon. In micro-machining, micro-machining tools are easy to wear during cutting micro-patterns. Laser ablation achieves micro-features with high accuracy, despite being a low efficiency and high cost process in terms of energy consumption. Similar to laser ablation, direct 3D printing could produce very complicated patterns with multiscale features, but is time-consuming and limited in materials [1,2]. Finally, micro-molding is an economic process to obtain bionic microstructures by using organic polymers, although poor mechanical properties and easy aging are denoted in the dies.

Micro-embossing based on plastic deformation is a simple and highly efficient method to mass produce bio-inspired functional surfaces on different types of materials, including metals. This process presents several advantages: it has a high strength, low cost and is environmentally friendly. The most relevant issue to address when using micro-embossing is size effects, which can cause filling defects and breakage of micro-features on the embossed dies. Cao et al. [5] embossed micro-channel patterns with 20 μm depth and 100 μm width on 500 μm thick AA5052 sheets by micro-rolling-based surface texturing. They concluded that the relative velocity between the upper and lower rollers significantly affected the filling ability. Also, Wang et al. [6] investigated the effects of grain size and cavity width on the filling ability of pure nickel during micro-embossing. They observed that the worst filling ability occurred for a width of 50 μm and ratio of cavity of 1.04, which was attributed to the coupling effect of grain size and cavity dimension. Sareh [7] studied a double corrugation surface (origami pattern) with the aim to analyze the flat-foldability depending on the surface characteristics. They found the interrelation of crystallography and computational geometry for designing complex origami structures. Also, Le and Goo [8] analyzed the thermomechanical impact of thermal layer protection based on bio-inspired, corrugated-core sandwich structures. These sandwich structures improved the deflection capabilities and reduced weight up to 65% compared with non-textured structures. Qiao et al. [9] remonstrated the poor filling quality and high damage rate of micro-features of the dies when performing embossed micro-channel of 10 μm in Al-1050 at room temperature. On the contrary, a relatively good quality of micro-channel patterns was obtained when embossing ultra-fine-grained material at elevated temperatures. Consequently, a possible way of reducing the size effect on filling quality is to increase forming temperature and improve plastic flow based on superplasticity, but a relatively longer processing time would be required with traditional thermal-assisted processes. In order to reduce this processing time, researchers [10] found that the passage of electricity during plastic deformation would give rise to multiple alterations of forming property and microstructure. For example, Tang et al. [11] made a comprehensive review of the electrically assisted (EA) forming processes including EA drawing, EA rolling and EA punching, which consistently proved that an electric current could reduce deformation resistance, improve plasticity, simplify processes, increase energy efficiency, lower cost and increase time-effectiveness. However, it should be noted that few works have focused on the EA micro-forming. Recently, Wang et al. [12] found that current-induced softening increases with decreasing grain size but increasing specimen size. Other researchers [13,14] concluded that localized Joule heating at the grain boundaries would affect the mechanical properties, making the Hall–Petch effect smaller as compared to the electrical/thermal decoupling tests. This assumption was demonstrated by Cao et al. [15], where local intergranular cavitation and local grain boundary melting were observed in tensile samples subjected to an electric current. Additionally, Lai et al. [16] fabricated micro-channels on 316 L stainless steel sheets by using EA micro-embossing. The results showed that the channels were deeper and the residual stress was smaller when the process was electrically assisted. Finally, Cao et al. [17] had also induced an electric current during a micro-rolling surface texturing process, in order to increase the depth–width ratio of micro-channels. They found that the channel depths of AA3003-H14 and Ti6Al4V increased by 15% and 200%, respectively. These changes were attributed to the difference in electrical resistivity and, consequently, the Joule heating effect, which are much higher in titanium alloy.

Considering the possible potential of controlling the filling quality and the difficulty of forming bio-inspired functional surfaces with high aspect ratios on metals, especially for difficult-to-form materials with high anisotropy [18], this research focuses on investigating the EA micro-embossing process in AZ31 magnesium alloy. This hybrid process tries to fabricate micro-channels and sharklet patterns in magnesium alloy when using dies manufactured with photolithographic silicon-molding, thermoplastic-forming of bulk metallic glass (BMG) and EA micro-embossing. Then, the EA's capabilities regarding the filling ability of bio-inspired micro-features are investigated. As a result, these EA textured surfaces were well transferred from BMG dies to large AZ31 areas, achieving micro-features down to ~2 μm and a depth–width ratio up to ~1.4.

2. Methodology

2.1. Sample Preparation

The material used in this research was a commercial drawn AZ31 magnesium alloy rod with a diameter of 15 mm and the following chemical composition: 94.8 wt% Mg, 3.5 wt% Al, 1.2 wt% Zn and 0.5 wt% Mn. The as-received materials were annealed at 400 °C for 2 h, which brought equiaxed grains with an average size of 7.1 ± 1.1 μm. Figure 1 shows the material microstructure after the procedures of grinding, polishing and etching. Micro-embossing specimens were cut into cuboid shapes with the height (H) and the base side length (L) of 2.2 and 1.5 mm, respectively. A precision CNC milling machine was used to cut the specimens from the axial cross sections of the annealed rods along the axial direction, as shown in Figure 2. Note that the machining tolerance for all the geometric dimensions was ±0.1 mm. Both end surfaces of the samples were mechanically polished using sandpaper to avoid surface oxidation prior to testing, since bio-inspired micro-features were embossed on the end surfaces.

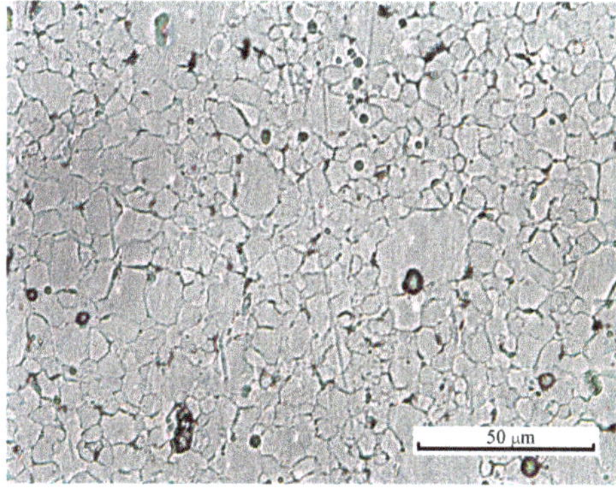

Figure 1. Optical micrograph of the studied AZ31 magnesium alloy to determine the grain size.

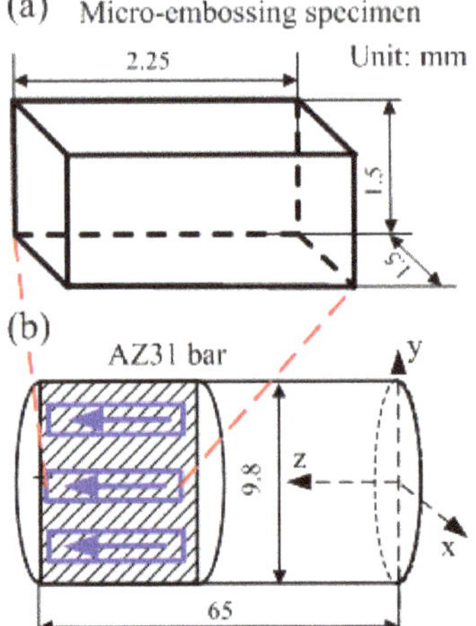

Figure 2. Schematic illustration of specimen preparation by using a CNC milling machine: (**a**) the geometry of the micro-embossing sample; (**b**) the sample cutting section and direction.

2.2. Preparation of Micro-Channel and Sharklet Dies

Micro-channel and sharklet pattern dies were fabricated by integrating the use of the properties of BMGs, photolithography, BMGs' crystallization and thermoplastic forming (TPF) in supercooled liquid region (SCLR). Specifically, silicon dies with micro-channel and sharklet patterns were first fabricated using photolithography. Afterward, we compressed BMGs on the silicon dies with TPF to make our BMG dies. These two operations are thoroughly described below.

(1) Photolithographic Silicon Molding: In this study, the silicon dies with micro-channels and sharklet patterns were prepared at Northwestern University, Evanston, IL, USA. A deep reactive ion etching with an etching rate of ~3.6 µm/min was deployed for 6 min. Figure 3a,b exhibit the photomask designed features of the micro-channel and sharklet patterns, respectively. The white areas in Figure 3 had no photoresist covering, and were etched away to produce depth channels of 15–20 µm. Through a series of photolithographic procedures, such as cleaning, coating, exposure, developing, etching and photoresist removal, multiple silicon molds were made to have large areas of micro-channel patterns, as well as sharklet patterns. Table 1 lists the average depth and width dimensions, measured at five points on the silicon molds by using Alicona Infinite Focus-Optical 3D measurement and inspection. The average depth and width for the sharklet pattern were very close to our designed features. Note that the design of the micro-channel width was 7 µm, although the silicon molds had widths of 8–9 µm. This dimensional dispersion can be associated with the leaking and scattering of the UV light caused by dust between the wafer and mask during the exposure process.

(2) TPF of BMG dies: BMGs are known to have dramatic softening when reheated into the SCLR (between the glass transition temperature T_g and the crystallization temperature T_x), where the glass relaxes into a metastable liquid before its eventual crystallization. This behavior allows to BMGs to have a high viscosity, high formability and be insensitive to heterogeneous influences, which is comparable to plastics [19]. As a result, many TPF-based processes [20] were developed for BMGs in SCLR to achieve high dimensional accuracy in complex geometries, especially in the micro-forming industry [21,22].

$Zr_{35}Ti_{30}Cu_{8.25}Be_{26.75}$ was selected as the as-received material, since it has a small T_g and a large SCLR, i.e., $\Delta T = T_x - T_g = 159$ K, for any known commercial BMG. The thermal, mechanical and rheological properties of $Zr_{35}Ti_{30}Cu_{8.25}Be_{26.75}$ are listed in Table 2. TPF generally occurs at temperatures above T_g and below T_x for glassy material, to apparently decrease flow stress and avoid crystallization. The fabrication of the bio-inspired functional surface on BMGs based on TPF includes pre-compression, polishing and thermoplastic embossing. Pre-compression of small pieces of round BMG specimens was conducted at 673 K for several seconds under 1000 N by using an Instron compression machine. Afterward, each pre-compressed round specimen was polished with 1 µm polishing slurry to obtain a smooth and parallel surface prior to embossing bio-inspired micro-features. The thickness of the pre-compressed BMG samples was around 0.6 mm. Thermoplastic embossing of BMGs when using the 7 µm micro-channel, and 10 µm sharklet pattern of silicon molds was performed at 700 K for 30 s under 3200 and 4300 N, respectively. After that, the BMG dies with bio-inspired functional surfaces were obtained by etching out the silicon set in micro-features in 20% KOH solution at 120 °C. Micro-channels and sharklet features on the fabricated BMG dies were characterized, as shown in Figure 4. The figure shows well-ordered micro-scale patterns over large areas fabricated on $Zr_{35}Ti_{30}Cu_{8.25}Be_{26.75}$ BMG samples using the TPF-based processes. The micro-scale features were measured at least three times, e.g., the micro-channel BMG die has 7.07 ± 0.27 µm of width and 15.28 ± 0.3 µm of depth with a depth–width ratio over two; the micro-sharklet BMG die has 15.62 ± 1.21 µm of width and 8.88 ± 0.33 µm of depth, with a depth–width ratio below one. These results show that it is harder to emboss the sharklet micro-feature, perhaps because the deformation state tends to be more complicated due to the complex sharklet pattern, impeding plastic flow.

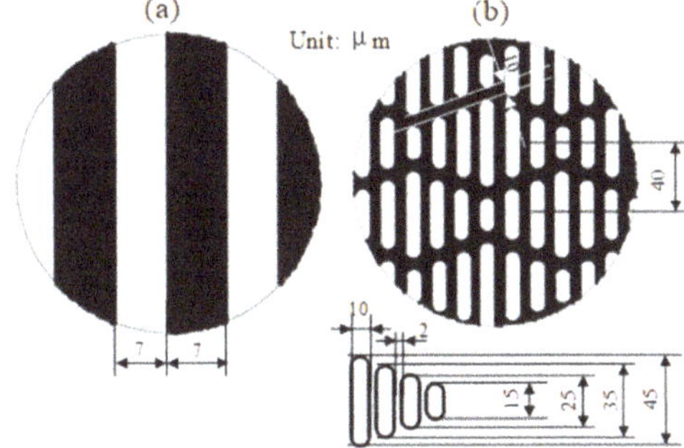

Figure 3. Photomask designs: (**a**) the micro-channel and (**b**) the sharklet patterns.

Table 1. Geometrical measurements of depth and width of the silicon molds.

Silicon Die	Avg. Depth (µm)	Avg. Width (µm)	Stdev. of Width (µm)
Channel pattern	20.13	8.03	0.95
Sharklet pattern	19.94	10.09	1.43

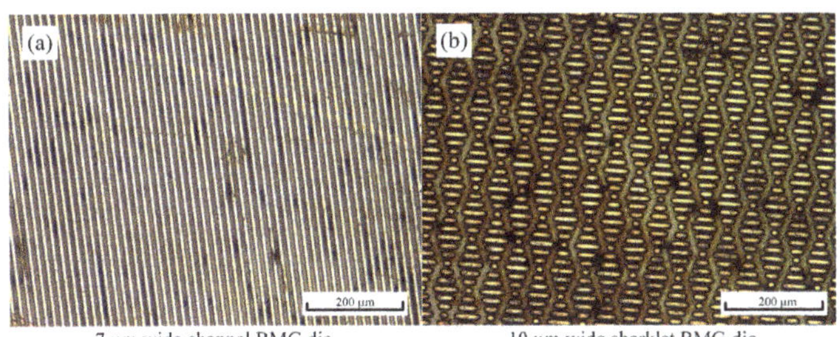

Figure 4. Characterizations of the textured bulk metallic glass (BMG) dies: (**a**) the micro-channels and (**b**) the sharklet patterns.

Table 2. Thermal, mechanical, and rheological properties of $Zr_{35}Ti_{30}Cu_{8.25}Be_{26.75}$.

Glass Transition Temperature T_g (K)	Crystallization Temperature T_x (K)	Liquidus Temperature T_l (K)	Angell Fragility m	Thermal Stability S	Shear Modulus G (GPa)	Poisson Ratio v
578	737	1044	65.6	0.34	31.8	0.37

2.3. EA Micro-Embossing Test

The EA micro-embossing system was schematically shown in Figure 5. The BMG dies were placed on the lower platen of the micro-embossing machine with the bio-inspired functional surface upward. Then, the cuboid specimen was placed on the BMG dies with the polished surface contacting the bio-inspired micro-scale patterns. During micro-embossing, the upper and lower crossbeams were driven by a DC motor to move downward and upward, respectively. The embossing force was measured by a load cell of 1000 N capacity with a resolution of 0.1 N. Similar to in [23], a rectifier-based DC power supply with a maximum output current intensity of 300 A was used to pass a continuous constant current through the specimen, which was insulated from the loading system. The temperature was measured by an infrared camera located on the back side of the sample. A layer of black paint was added on the back surface prior to tests to set the emissivity and reduce the temperature errors with the infrared camera. Furthermore, different current densities were chosen (i.e., 0 A/mm^2, 6 A/mm^2, 10 A/mm^2, 13 A/mm^2) to study the current induced effect on the fillability of bio-inspired micro-features. All the tests were conducted with a fixed strain rate of 0.01 s^{-1} for a duration of ~50 s and stopped after reaching ~300 N. All the workpieces were cleaned with acetone in an ultrasonic cleaning tank after the EA micro-embossing tests.

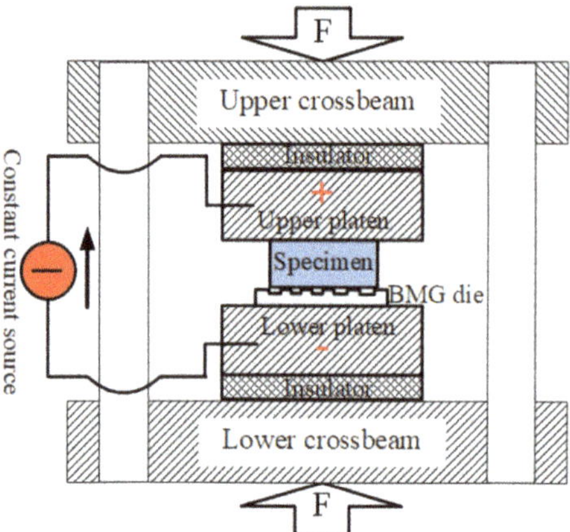

Figure 5. Schematic illustration of the electrically assisted (EA) micro-embossing system.

3. Results and Discussion

Figure 6 shows four embossed areas performed with different current densities in AZ31 magnesium alloy. As can be seen, the channel pattern transferred from the BMG dies is disorderly at 0 A/mm^2, with the micro-feature size probably approximate to the metal cutting error. During EA micro-embossing, it is observed that the geometry integrity of the micro-channel pattern on the sample surfaces is becoming more accurate with the increase in current density. Note that the channel width is not uniform, probably because Joule heating tends to concentrate at a few local areas of the embossing interfaces. Accordingly, the non-uniform distributions of current density, plastic deformation and uneven contact/friction favors local melting areas between channels. The channel depth and width were measured at least three times in each sample for repeatability purposes. The results also exhibit that the channel width is ~7 µm, which is very consistent with that of the BMG dies. Figure 7 shows the relationship of the channel depth, depth–width ratio and current density. This figure denotes the non-linear increase in the channel depth with respect to current density, showing a faster increasing rate at the higher current density. The variation in depth–width ratio with respect to the current density also exhibits a lower convex relationship. An unexpected result for the embossed depth was found, i.e., a depth 20 times higher at 13 A/mm^2 compared to 0 A/mm^2. An area of channel patterns with a depth-width ratio of ~1.4 was obtained (exceeding the common ratio of ~1 in a traditional micro-embossing at a width far below 50 µm [6]). Note that the channel depth does not increase much for current densities lower than 6 A/mm^2, indicating that there also exists a current density threshold [12] during EA micro-embossing. The Joule heating temperature also non-linearly increases with the current density, which was found to be <200 °C (below the glass transition temperature of the BMG without damage of the BMG micro-patterns). Consequently, it resulted in a faster softening and a better filling ability for the embossing samples.

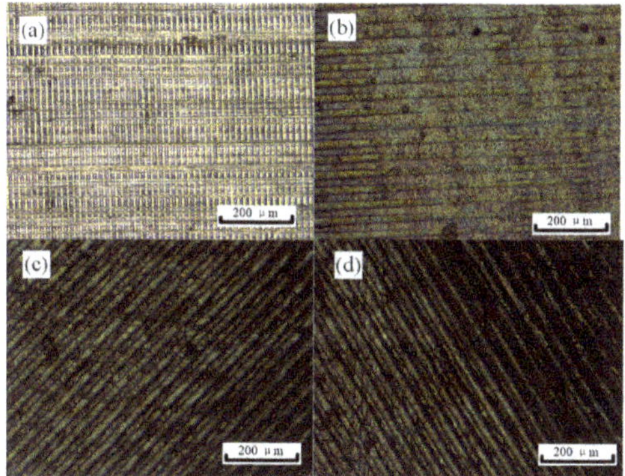

Figure 6. Micrographs of the micro-channel patterns obtained by EA micro-embossing at different current densities: (**a**) 0 A/mm^2, (**b**) 6 A/mm^2, (**c**) 10 A/mm^2, (**d**) 13 A/mm^2.

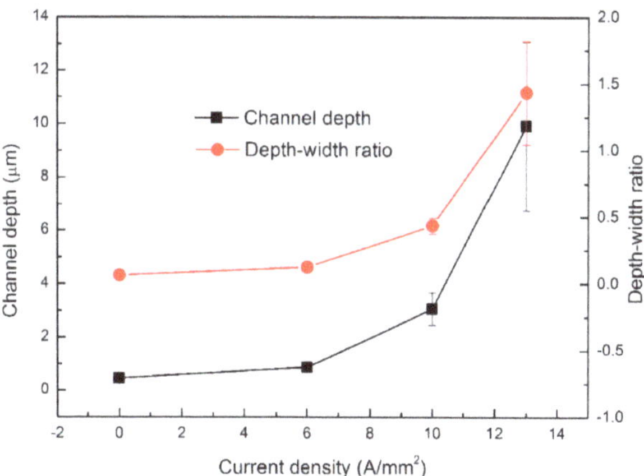

Figure 7. Variations in channel depth and depth–width ratio with respect to different current densities.

Figure 8 exhibits the sharklet patterns on AZ31 sample surfaces transferred from BMG dies for different current densities. Only a few parallel ridges with different lengths were marked into the workpieces during micro-embossing without applying an electric current. Similar to the observations in Figure 6, the passage of an electric current during micro-embossing improves the geometry integrity of the sharklet patterns and fillability, particularly for higher current densities. Also, the sharklet feature copied from BMG dies has a high repeatability over the sample surface when using current densities above 10 A/mm^2 (the threshold was found at 6 A/mm^2). A similar result is observed in Figure 9, where the embossed sharklet depth nonlinearly increases with current density and the width varies in a smaller order. The depth–width ratio in Figure 9 is found to not change with the current density as much, as compared to the micro-channel embossing, e.g., ~0.33 at 13 A/mm^2 vs. 0.06 at 0 A/mm^2 during shark embossing. These differences could be attributed to the enhanced material flow resistance caused by the complex micro-ridges/channels in different directions. The increasing rate of sharklet depth against current density also rises over 6 A/mm^2, considered as the current density threshold where the electric current starts to significantly improve the embossing fillability. The Joule heating temperature is relatively consistent with that in the EA micro-embossing of channel patterns, i.e., below 200 °C (below the glass transition temperature of the BMG without damage of the BMG micro-patterns). An excessive Joule heating would occur above a current density, e.g., ~20 A/mm^2, giving rise to a sharp deterioration in the embossed texture quality due to die breakage/melting.

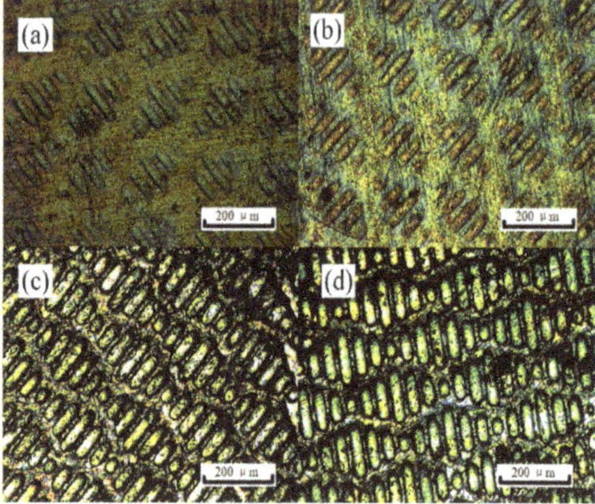

Figure 8. Micrographs of the sharklet patterns obtained by EA micro-embossing at different current densities: (**a**) 0 A/mm^2, (**b**) 6 A/mm^2, (**c**) 10 A/mm^2, (**d**) 13 A/mm^2.

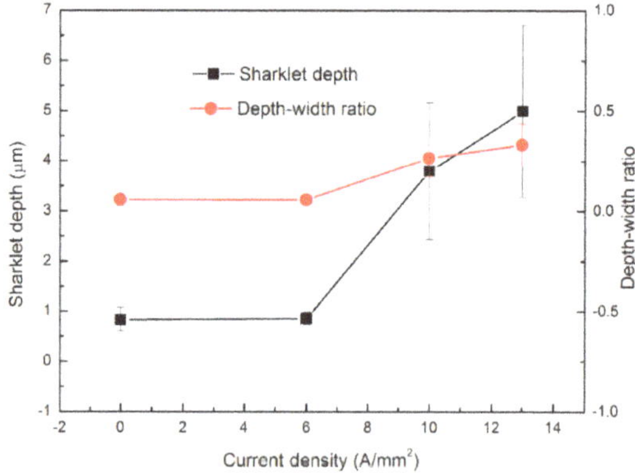

Figure 9. Variations in sharklet depth and depth–width ratio for different current densities.

4. Conclusions

In this research, typical bio-inspired functional surfaces with a micro-channel of 7 µm and sharklet patterns of 10 µm were EA micro-embossed on AZ31 magnesium alloy. To do that, textured BMG dies prepared by TPF-based compression on photolithographic silicon molds were used. We found that both the filling depth and depth–width ratio non-linearly increased with current density. The results also showed that larger and accurate areas of channel and sharklet textures could be fabricated by using EA micro-embossing. The smallest feature, down to ~2 μm, and a depth–width ratio of up to ~1.4 were addressed when using current densities of 13 A/mm^2 and temperatures below the glass transition of the BMG. Finally, the geometry integrity of bio-inspired functional features could not be well obtained below 6 A/mm^2, indicating that the threshold of current density occurs in this EA micro-embossing process.

Author Contributions: Conceptualization, J.C. and B.G.; Data curation, X.W. and C.W.; Formal analysis, X.W. and A.J.S.E.; Funding acquisition, X.W., C.W.; Investigation, X.W., J.L. and C.L.; Methodology, X.W. and J.C.; Resources, J.X., Z.W. and T.Z.; Supervision, B.G., Z.W. and T.Z.; Writing—original draft, X.W.; Writing—review & editing, C.W., C.L. and A.J.S.E. All authors have read and agreed to the published version of the manuscript.

Funding: This work was supported by the National Natural Science Foundation of China (Grant No. 51705101), Key Laboratory of Micro-Systems and Micro-Structures Manufacturing (Harbin Institute of Technology), Ministry of Education (Grant No. 2019KM009), China Postdoctoral Science Foundation (Grant No. 2018T110293), Natural Science Foundation of Jiangsu Province (BK20192007), and Serra Húnter program (Generalitat de Catalunya), grant number UPC-LE-304 (year 2018).

Conflicts of Interest: The authors declare no conflict of interest.

References

1. Malshe, A.; Rajurkar, K.; Samant, A.; Hansen, H.N.; Bapat, S.; Jiang, W. Bio-inspired functional surfaces for advanced applications. *CIRP Ann. Manuf. Technol.* **2013**, *62*, 607–628. [CrossRef]
2. Chen, D.; Liu, Y.; Chen, H.; Zhang, D. Bio-inspired drag reduction surface from sharkskin. *Biosurf. Biotribol.* **2018**, *4*, 39–45. [CrossRef]
3. Dundar Arisoy, F.; Kolewe, K.; Homyak, B.; Kurtz, I.; Schiffman, J.; Watkins, J. Bioinspired Photocatalytic Shark Skin Surfaces with Antibacterial and Antifouling Activity via Nanoimprint Lithography. *ACS Appl. Mater. Interfaces* **2018**, *10*, 20055–20063. [CrossRef] [PubMed]
4. Carman, M.L.; Estes, T.G.; Feinberg, A.W.; Schumacher, J.F.; Wilkerson, W.; Wilson, L.H.; Callow, M.E.; Callow, J.A.; Brennan, A.B. Engineered antifouling microtopographies—Correlating wettability with cell attachment. *Biofouling* **2006**, *22*, 11–21. [CrossRef] [PubMed]
5. Ng, M.-K.; Magargee, J.; Cao, J.; Ehmann, K.F. Microrolling-Based Surface Texturing. In Proceedings of the 10th International Conference on Multi-Material Micro Manufacturing (4M), San Sebastian, Spain, 1 October 2013.
6. Wang, C.; Wang, C.; Jie, X.; Peng, Z.; Shan, D.; Guo, B. Interactive effect of microstructure and cavity dimension on filling behavior in micro coining of pure nickel. *Sci. Rep.* **2016**, *6*, 23895. [CrossRef]
7. Sareh, P. The least symmetric crystallographic derivative of the developable double corrugation surface: Computational design using underlying conic and cubic curves. *Mater. Des.* **2019**, *183*, 108128. [CrossRef]
8. Le, V.T.; Goo, N. Thermomechanical Performance of Bio-Inspired Corrugated-Core Sandwich Structure for a Thermal Protection System Panel. *Appl. Sci.* **2019**, *9*, 5541. [CrossRef]
9. Qiao, X.G.; Gao, N.; Moktadir, Z.; Kraft, M.; Starink, M.J. Fabrication of MEMS components using ultrafine-grained aluminium alloys. *J. Micromech. Microeng.* **2010**, *20*, 045029. [CrossRef]
10. Xue, S.; Wang, C.; Chen, P.; Xu, Z.; Cheng, L.; Guo, B.; Shan, D. Investigation of Electrically-Assisted Rolling Process of Corrugated Surface Microstructure with T2 Copper Foil. *Materials* **2019**, *12*. [CrossRef] [PubMed]
11. Guan, L.; Tang, G.; Chu, P.K. Recent advances and challenges in electroplastic manufacturing processing of metals. *J. Mater. Res.* **2010**, *25*, 1215–1224. [CrossRef]
12. Wang, X.; Xu, J.; Shan, D.; Guo, B.; Cao, J. Effects of specimen and grain size on electrically-induced softening behavior in uniaxial micro-tension of AZ31 magnesium alloy: Experiment and modeling. *Mater. Des.* **2017**, *127*, 134–143. [CrossRef]
13. Wang, X.; Xu, J.; Jiang, Z.; Zhu, W.-L.; Shan, D.; Guo, B.; Cao, J. Size effects on flow stress behavior during electrically-assisted micro-tension in a magnesium alloy AZ31. *Mater. Sci. Eng. A* **2016**, *659*, 215–224. [CrossRef]

14. Sánchez Egea, A.; Jorba Peiró, J.; Signorelli, J.; Rojas, H.; Celentano, D. On the microstructure effects when using electropulsing versus furnace treatments while drawing inox 308L. *J. Mater. Res. Technol.* **2019**, *8*, 2269–2279. [CrossRef]

15. Fan, R.; Magargee, J.; Hu, P.; Cao, J. Influence of grain size and grain boundaries on the thermal and mechanical behavior of 70/30 brass under electrically-assisted deformation. *Mater. Sci. Eng. A* **2013**, *574*, 218–225. [CrossRef]

16. Mai, J.; Peng, L.; Lai, X.; Lin, Z. Electrical-assisted embossing process for fabrication of micro-channels on 316L stainless steel plate. *J. Mater. Process. Technol.* **2013**, *213*, 314–321. [CrossRef]

17. Ng, M.-K.; Fan, Z.; Gao, R.X.; Smith, E.F.; Cao, J. Characterization of electrically-assisted micro-rolling for surface texturing using embedded sensor. *Cirp Ann. Manuf. Technol.* **2014**, *63*, 269–272. [CrossRef]

18. Mathis, K.; Kövér, M.; Stráská, J.; Trojanova, Z.; Džugan, J.; Halmešová, K. Micro-Tensile Behavior of Mg-Al-Zn Alloy Processed by Equal Channel Angular Pressing (ECAP). *Materials* **2018**, *11*, 1644. [CrossRef]

19. Schroers, J.; Paton, N. Amorphous Metal Alloys Form Like Plastics. *Adv. Mater. Process.* **2006**, *164*, 61–63.

20. Schroers, J. Processing of Bulk Metallic Glass. *Adv. Mater.* **2010**, *22*, 1566–1597. [CrossRef]

21. Saotome, Y.; Miwa, S.; Zhang, T.; Inoue, A. The micro-formability of Zr-based amorphous alloys in the supercooled liquid state and their application to micro-dies. *J. Mater. Process. Technol.* **2001**, *113*, 64–69. [CrossRef]

22. Han, G.; Peng, Z.; Xu, L.; Li, N. Ultrasonic Vibration Facilitates the Micro-Formability of a Zr-Based Metallic Glass. *Materials* **2018**, *11*, 2568. [CrossRef] [PubMed]

23. Wang, X.; Sánchez Egea, A.; Xu, J.; Meng, X.; Wang, Z.; Shan, D.; Guo, B.; Cao, J. Current-Induced Ductility Enhancement of a Magnesium Alloy AZ31 in Uniaxial Micro-Tension Below 373 K. *Materials* **2019**, *12*, 111. [CrossRef] [PubMed]

Article

Enhancing the Surface Quality of Micro Titanium Alloy Specimen in WEDM Process by Adopting TGRA-Based Optimization

Muthuramalingam Thangaraj [1], Ramamurthy Annamalai [2], Khaja Moiduddin [3,*], Mohammed Alkindi [4], Sundar Ramalingam [4] and Osama Alghamdi [4]

[1] Department of Mechatronics Engineering, SRM Institute of Science and Technology, Kattankulathur 603203, India; muthurat@srmist.edu.in

[2] Department of Mechanical Engineering, Saveetha Engineering College, Thandalam, Chennai 602105, India; ramapec@gmail.com

[3] Advanced Manufacturing Institute, King Saud University, Riyadh 11421, Saudi Arabia

[4] Department of Oral and Maxillofacial Surgery, College of Dentistry, King Saud University, Riyadh 11545, Saudi Arabia; malkindi@ksu.edu.sa (M.A.); smunusamy@ksu.edu.sa (S.R.); oghamdi@ksu.edu.sa (O.A.)

* Correspondence: khussain1@ksu.edu.sa

Received: 16 February 2020; Accepted: 17 March 2020; Published: 21 March 2020

Abstract: The surface measures of machined titanium alloys as dental materials can be enhanced by adopting a decision-making algorithm in the machining process. The surface quality is normally characterized by more than one quality parameter. Hence, it is very important to establish multi-criteria decision making to compute the optimal process factors. In the present study, Taguchi–Grey analysis-based criteria decision making has been applied to the input process factors in the wire EDM (electric discharge machining) process. The recast layer thickness, wire wear ratio and micro hardness have been chosen to evaluate the quality measures. It was found that the wire electrode selection was the most influential factor on the quality measures in the WEDM process, due to its significance in creating spark energy. The optimal arrangement of the input process parameters has been found using the proposed approach as gap voltage (70 V), discharge current (15 A) and duty factor (0.6). It was proved that the proposed method can enhance the efficacy of the process. Utilizing the computed combination of optimal process parameters in surface quality analysis has significantly contributed to improving the quality of machining surface.

Keywords: EDM; surface; optimization; machining; titanium

---

1. Introduction

Due to its unique physical properties such as higher corrosion resistance and considerable strength, titanium (α-β) alloy (Ti-6Al-4V) is employed in synthesizing dental specimens [1]. As a dental implant material, titanium alloy must possess an adequate surface quality, free from residual stress. It is very difficult to remove the material using traditional machining processes due its high strength, and as such, nontraditional material removal processes such as laser beam machining (LBM), hybrid machining, electro chemical machining (ECM), wire electrical discharge machining (WEDM) and abrasive-water jet machining (AWJM) are utilized. Titanium alloy as dental material should have an optimal surface finish through the machining process. The conventional machining method produces higher residual stress due to vibrations made during the process [2]. The LBM and hybrid machining processes produce a high heat affected zone (HZ) on the machined specimens [3]. The improper selection of laser power results in affecting the machining performance of titanium alloy in the LBM process [4]. The AWJM process causes the titanium alloy specimens to considerably taper [5]. The ECM process may result in the corrosion of the workpiece specimen [6]. For the utilization of titanium alloy as a bio material, the specimen should have an optimal surface finish and performance during the machining process [7]. The quality measures of the machined specimens should be as high as possible in order to manufacture the product with favorable performance measures. The WEDM process is widely used to machine titanium species as it produces relatively lower taperness and kerf widths. The material removal is achieved in this process by applying a pulsed DC supply between the workpiece and wire electrode in an insulated environment. As the WEDM process is of a nonlinear nature, the enhancement of process parameters is required to obtain better performance measures. The surface quality can be effectively controlled by white layer formation in the EDM process [8]. The surface quality performance measures are mostly influenced by enhancing the input process factors in WEDM. The optimization of input process factors in machining methods such as the WEDM process is very tedious due to their unsystematic nature [9]. It is important to establish multi-response optimization techniques to determine the optimal parameter combination in the WEDM process [10,11]. Many multiple performance decision-making techniques such as the assignment of the weight method, genetic algorithms, the Taguchi data envelopment analysis ranking (DEAR) method and the Taguchi–Grey relation analysis (TGRA) that are available can convert multiple response characteristics into a single performance measure in any process. Amongst these, TGRA is widely used as it has higher efficacy and easy adaptability. Nanthakumar et al. made an attempt to introduce the TGRA method as a means of optimizing process parameters in the materials development process. It has been found that the proposed method can significantly improve quality measures [12]. The optimal set of sintering process factors in the grinding process was found using the TGRA method. It has been observed that the TGRA method can determine the optimal combination effectively in any manufacturing process [13]. Pillai et al. effectively applied the TGRA method to optimize the parameters involved in the robotics-assisted machining process [14]. It was inferred that the TGRA method can compute the optimal process parameters, the significance of which determines responses in machining processes [15–17]. The grinding parameters of green manufacturing processes can be optimized using the TGRA method. It has been found that the proposed approach can increase prediction accuracy [18,19]. Product design can be further enhanced by the TGRA method [20]. The detailed survey showed that only multi-criteria decision making (MCDM) can provide better process factors in machining processes. It was also found that little attention was given to optimizing surface quality performance measures such as white layer thickness, wire wear ratio and micro hardness in the WEDM process of machining titanium alloy. In regards to structure, the surface should be of the highest possible quality. MCDM can be utilized in achieving this. In the present study, Taguchi's experiment model and Grey's relational analysis methodology were applied in order to enhance the surface performance measures in cutting titanium alpha-beta (Ti-6Al-4V) alloy with the WEDM process. The following are the primary aims of the investigation on machinability using various process factors:

1. To compute the optimal process factors for obtaining better surface quality measures of titanium alloy specimens using the TGRA method.
2. To evaluate the influence of input factors on surface measures.
3. To investigate the surface quality at optimal levels in the process.

2. Materials and Methods

Titanium(α-β) alloy was chosen as the specimen due to its usability as a dental implant material. Despite possessing a higher corrosion resistance and lighter weight, it is a high strength material [1]. The measurement approaches of quality measures and design of experiments are also discussed in the present subsection. Due to their efficacy in evaluating surface related parameters, pulse-on time (T_{on}), Pulse-off time (T_{off}), servo voltage (SV), wire electrode (WE) and wire tension (WT) were selected as the input factors of the multi criteria optimization in the present study. The selection of process factors is given in Table 1 [17].

The surface quality of machined workpiece specimens, average white layer thickness (AWLT), micro hardness (MH) and wire wear ratio (WWR) were selected as the surface measures in the present study. In the WEDM process, the machining quality of the specimen is considerably characterized by the wire wear ratio due to its importance in evaluating the discharge energy of every pulse cycle. WWR can be calculated using the following Equation (1): [20,21]

$$WWR = \frac{W_I - W_F}{W_I} \tag{1}$$

where W_I-Initial weight of the workpiece specimen; W_F-Final weight of the workpiece specimen after the machining process.

Table 1. Selection of Process factors.

Control Factor	Level I	Level II	Level III	Unit
T_{on}	110	120	130	μs
T_{off}	30	40	50	μs
SV	40	60	80	V
W_b	5	7	9	Kg
WE	Brass Wire Electrode (BWE)	Zinc coated Brass Wire Electrode (ZWE)	Diffused Brass Wire Electrode (DWE)	-
Wire diameter	0.25			mm
Wire feed rate	4			m/min
Dielectric medium	Deionized water			-
Dielectric flow rate	1.2			bar
Peak current	16			A

The weight of workpiece specimens was calculated using electronics balances with an accuracy of 0.001 g [22]. The micro hardness (HV) of the processed workpiece was computed using Vickers-based micro hardness tester in Kg/mm^2. The applied load was considered as 300 g. Due to the divergent width of the AWLT over the machined surface, it must be taken for the purpose of analysis and was calculated using the Equation (2) as follows:

$$AWLT = \frac{\text{Area of recast layer}}{\text{Length of recast layer}} \tag{2}$$

The AWLT area was computed by sketching a polyline along the white layer on the specimen using WEDM [8,23]. Figure 1 illustrates the steps involved in TGRA of the present work.

Selection of orthogonal array (OA)

- OA ≥ DOE
- $DOF = F(P - 1) + Q(P - 1)^2 + 1$

DOF - Degree of freedom
F - Number of independent variables
p - Number their levels
Q - Number of interactions

Grey relational analysis

Calculation of S/N ratio:

- Larger the better characteristics

$$\frac{S}{N} \, ratio = -10 \log \frac{1}{m} \sum \frac{1}{Y_{ni}^2}$$

- Smaller the better characteristics

$$\frac{S}{N} \, ratio = -10 \log \frac{1}{m} \sum Y_{ni}^2$$

m - Number of experimental replication
Y_{ni} - Response of nth trial of ith dependent level.

Calculation of normalized S/N ratio:

- Larger the better characteristics

$$Z_{ni} = \frac{Y_{ni} - \min(Y_{ni})}{\max(Y_{ni}) - \min(Y_{ni})}$$

- Smaller the better characteristics

$$Z_{ni} = \frac{\max(Y_{ni}) - Y_{ni}}{\max(Y_{ni}) - \min(Y_{ni})}$$

Z_{ni} - Normalized of nth trial of ith dependent level.

Computation of grey co-efficient (GC):

$$GC_{ni} = \frac{(\Psi_{min} + \delta\Psi_{max})}{(\Psi_{ni} + \delta\Psi_{max})}$$

Ψ_{ni} - Grey co-efficient for n^{th} trial of ith dependent response
δ - Distinctive co efficient which has value from 0 to 1

Computation of grey relation grade (G_n):

$$G_n = \frac{1}{Q} \sum GC_{ni}$$

q - Number of the performance measures

Figure 1. Steps involved in TGRA computation.

3. Results and Discussion

Titanium alloy specimens were machined using the WEDM method into rectangular specimens in accordance with the Taguchi system. The performance measures of each trial have been measured and tabulated. Figure 2 demonstrates the surface topography of a machined titanium alloy specimen in WEDM. In the EDM process, the surface morphology replicates the tool electrode. The surface patterns caused by the wire electrodes can be clearly viewed in the machined surface as shown in Figure 2. Table 2 shows the L_{27} orthogonal table with input factors and response values in the EDM process. Table 3 illustrates the signal-to-noise (S/N) ratio with their normalized value (N S/N) of the selected performance measures. Micro hardness was chosen as a larger-the-better (LTB) quality, whereas WWR and AWLT were chosen as smaller-the-better (STB) quality level characteristics. As the present study of surface performance measures was completed with both the LTB and STM quality characteristics, the distinguishing coefficient value was selected as 0.5 [6].

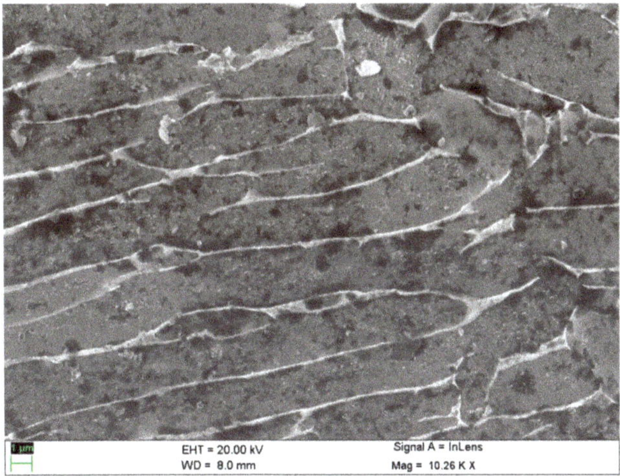

Figure 2. Surface topography of the machined Ti-6Al-4V alloy in WEDM process.

Table 2. OA with performance measures.

Trial	T_{on}	T_{off}	SV	WT	WE	WWR	MH	AWLT
1	110	30	40	5	BWE	0.1666	516.76	5.1
2	110	30	40	5	ZWE	0.0909	465.2	2.11
3	110	30	40	5	DWE	0.0686	494.86	2.11
4	110	40	60	7	BWE	0.1248	518.3	2.72
5	110	40	60	7	ZWE	0.0454	500.33	1.85
6	110	40	60	7	DWE	0.1111	393.9	1.75
7	110	50	80	9	BWE	0.1682	514.2	5.33
8	110	50	80	9	ZWE	0.0919	445.4	3.43
9	110	50	80	9	DWE	0.0258	425.86	2.401
10	120	30	60	9	BWE	0.1686	429.9	3.17
11	120	30	60	9	ZWE	0.0908	466.1	0.55
12	120	30	60	9	DWE	0.0682	663.16	4.28
13	120	40	80	5	BWE	0.125	425.76	2.34

Table 2. *Cont.*

Trial	T_{on}	T_{off}	SV	WT	WE	WWR	MH	AWLT
14	120	40	80	5	ZWE	0.0929	487.1	2.72
15	120	40	80	5	DWE	0.0222	421.73	4.59
16	120	50	40	7	BWE	0.125	460.26	5.67
17	120	50	40	7	ZWE	0.045	557.73	1.88
18	120	50	40	7	DWE	0.1121	543	5.84
19	130	30	80	7	BWE	0.125	401.96	3.73
20	130	30	80	7	ZWE	0.1165	563.26	2.55
21	130	30	80	7	DWE	0.1107	401.3	4.31
22	130	40	40	9	BWE	0.125	512.13	3.36
23	130	40	40	9	ZWE	0.0454	496.13	3.08
24	130	40	40	9	DWE	0.0666	508.96	3.54
25	130	50	60	5	BWE	0.125	366.46	2.77
26	130	50	60	5	ZWE	0.1365	534.8	2.37
27	130	50	60	5	DWE	0.1131	478.73	2.35
			Mean			0.10025556	481.233	3.18152
			Standard deviation			0.04096588	64.0385	1.3149
			Standard error			0.00788412	12.3246	0.25306

Table 3. S/N value with its normalized value.

Trial No.	WWR		MH		AWLT	
	S/N Ratio	N S/N Ratio	S/N Ratio	N S/N Ratio	S/N Ratio	N S/N Ratio
1.	1.55665	0.994114	5.426578	0.579453	−1.41514	0.942649
2.	2.082872	0.695296	5.335279	0.402238	−0.64856	0.569094
3.	2.327352	0.556466	5.388965	0.506444	−0.64856	0.569094
4.	1.807571	0.851627	5.429162	0.58447	−0.86914	0.67658
5.	2.685888	0.352869	5.398513	0.524978	−0.53434	0.513433
6.	1.908572	0.794273	5.190772	0.121741	−0.48608	0.489912
7.	1.548348	0.998829	5.422264	0.57108	−1.45345	0.96132
8.	2.073369	0.700692	5.2975	0.328907	−1.07059	0.774748
9.	3.176761	0.074124	5.258534	0.25327	−0.76078	0.623779
10.	1.546285	1	5.266735	0.269189	−1.00212	0.741382
11.	2.083828	0.694753	5.336958	0.405496	0.519275	0
12.	2.332431	0.553582	5.643237	1	−1.26289	0.868456
13.	1.80618	0.852417	5.25833	0.252874	−0.73843	0.612886
14.	2.063969	0.70603	5.375236	0.479796	−0.86914	0.67658
15.	3.307294	0	5.250069	0.23684	−1.32363	0.898054
16.	1.80618	0.852417	5.326006	0.384238	−1.50717	0.987494
17.	2.693575	0.348504	5.492848	0.708087	−0.54832	0.520242
18.	1.900789	0.798693	5.4696	0.662961	−1.53283	1
19.	1.80618	0.852417	5.208366	0.155891	−1.14342	0.810238

Table 3. *Cont.*

Trial No.	WWR		MH		AWLT	
	S/N Ratio	N S/N Ratio	S/N Ratio	N S/N Ratio	S/N Ratio	N S/N Ratio
20.	1.867348	0.817682	5.501418	0.724721	−0.81308	0.649263
21.	1.911705	0.792494	5.206938	0.153121	−1.26895	0.871412
22.	1.80618	0.852417	5.41876	0.564279	−1.05268	0.76602
23.	2.685888	0.352869	5.391191	0.510765	−0.9771	0.729191
24.	2.353052	0.541873	5.413367	0.55381	−1.09801	0.788109
25.	1.80618	0.852417	5.128053	0	−0.88496	0.68429
26.	1.729735	0.895827	5.456383	0.637306	−0.7495	0.618278
27.	1.893075	0.803073	5.360181	0.450574	−0.74214	0.614691

3.1. Computation of Optimal Process Parameters

The values of Grey Relational (GR) components along with their rank of all trials are given in Table 4. Table 5 shows the average of the GR scale for all the levels of process factors. The average Grey technique value specifies the relationship levels among the comparative values and a reference value. Hence, the optimal assessment of each process factor is the highest average GR value in the process.

Table 4. GR coefficient with its rank.

No.	GR Coefficient			GR Grade
	WWR	MH	AWLT	
1.	0.988365	0.543155	0.897101	0.809541
2.	0.621346	0.455472	0.537111	0.537976
3.	0.529923	0.503243	0.537111	0.523426
4.	0.771161	0.546132	0.607223	0.641505
5.	0.43587	0.512809	0.506808	0.485162
6.	0.708489	0.362776	0.495006	0.522091
7.	0.997662	0.538259	0.928195	0.821372
8.	0.625541	0.426951	0.689415	0.580636
9.	0.350662	0.401049	0.570632	0.440781
10.	1	0.406236	0.659093	0.688443
11.	0.620927	0.456828	0.333333	0.470363
12.	0.528308	1	0.79171	0.773339
13.	0.772102	0.400922	0.563626	0.578883
14.	0.629747	0.490098	0.607223	0.575689
15.	0.333333	0.395833	0.830639	0.519935
16.	0.772102	0.448124	0.975598	0.731941
17.	0.434218	0.631382	0.51033	0.52531
18.	0.712954	0.597343	1	0.770099
19.	0.772102	0.371994	0.724887	0.622994
20.	0.732796	0.644929	0.587726	0.65515

Table 4. *Cont.*

No.	GR Coefficient			GR Grade
	WWR	**MH**	**AWLT**	
21.	0.706708	0.371229	0.795434	0.624457
22.	0.772102	0.534347	0.681218	0.662556
23.	0.43587	0.505441	0.648669	0.529993
24.	0.521851	0.528435	0.702354	0.584214
25.	0.772102	0.333333	0.612963	0.572799
26.	0.827577	0.57958	0.567072	0.658076
27.	0.717436	0.476451	0.564775	0.58622

Table 5. Average GR grade for input factors.

Factor Notation	Average GR Grade			High-Low
	1	**2**	**3**	
T_{on}	0.5958	0.6260	0.6107	0.0302
T_{off}	0.5706	0.5667	0.6319	0.0652
SV	0.6306	0.5998	0.6022	0.0308
WT	0.5958	0.6199	0.6169	0.0240
WE	0.6811	0.5576	0.5938	0.1235

Total mean GR grade = 0.6066.

Figure 3 shows the response graph of average Grey Relational grades. It was observed that the optimal values of parameters are level 2 (T_{on}), level 3 (T_{off}), level 1 (SV), level 2 (WT) and level 1 (WE). The high-low indicates the level of the most dominant process parameter in formulating the performance measures among all the input process parameters in any machining process. It was observed that the wire electrode significantly influences the quality measures such as white layer thickness, micro hardness and WWR in the WEDM process. The crater size produced by the discharge energy is mainly characterized by the electric current conductance of the electrode in WEDM. As the surface quality of the machined workpiece is evaluated using crater size and material removal, the wire electrode possesses a vital role in evaluating the surface performance measures in WEDM [22].

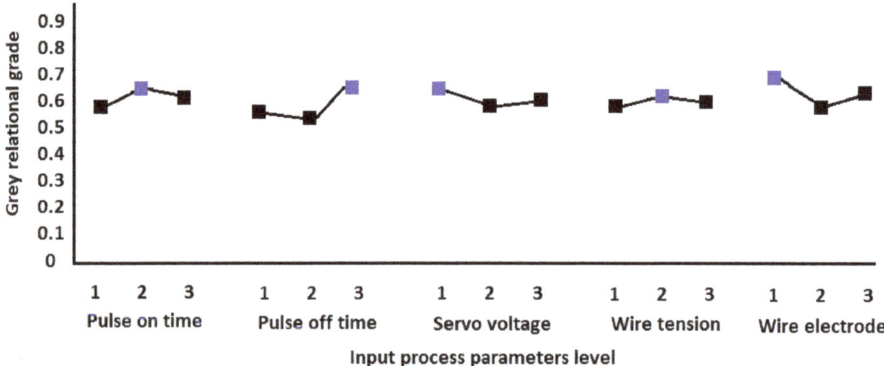

Figure 3. Based optimal process parameters computation.

3.2. Confirmation Experiment

Following the detection of the optimal factor combination, the confirmation test was performed to examine its confidence. In this present test, the experiment was conducted in WEDM under the optimal factor combination [7]. The predicted GR grade (G_a) was computed as per the following Equation (3):

$$G_a = G_b + \sum (G_c - G_b)$$ (3)

where G_b-total average GR grade and G_c-optimal average GR grade. The predicted value was found as 0.762953. The response values were obtained with an optimal factor combination of 0.1675 (WWR), 512.8 (MH) and 4.35 (AWLT). The GR grade was calculated as 0.787367. The GR grade value was improved by 3.2% from the predicted mean value.

The main effect plot was used to examine the significance of the input factors on responses using Minitab software [16]. Figure 4 shows the effects of input factors on Grey relational grades. The surface measures of the machined specimens were considerably characterized by the wire wear ratio due to its significance in examining the discharge energy of every pulse cycle. The micro hardness and white layer thickness are characterized by the amount of resolidification of the workpiece and tool electrode. As the servo voltage contributes mostly to resolidification, it considerably influences the micro hardness and AWLT. The physical characteristics of the wire electrode influence the white layer thickness, as the recast layer consists of melted wire electrode material. The selection of the wire electrode has a vital role in determining the AWLT due to its weight in formulating the recast layer thickness. As the wire electrode has a considerable effect on determining the surface quality related parameters, it has a highly influential role in deciding surface measures in WEDM.

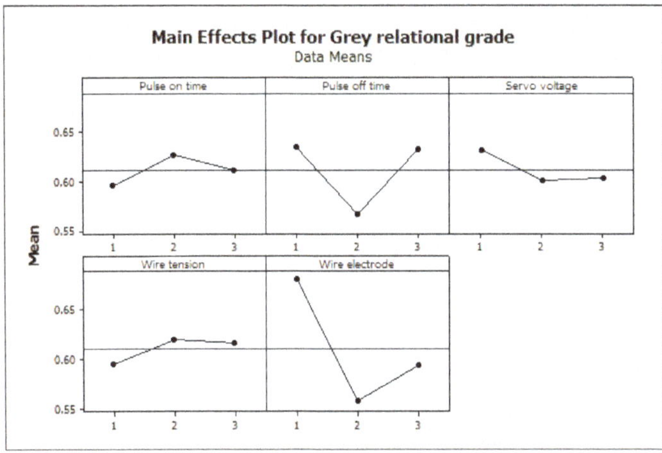

Figure 4. Effects of input process factors on Grey relational grade.

4. Surface Analysis under Optimal Process Parameters Combination

The surface quality measures of the processed specimen under the optimal input factors such as T_{on} (120 µs), T_{off} (50 µs), SV (40 V), WT (7 Kg) and WE (BWE) have been analyzed for surface morphology enhancement.

As surface hardness is the surface layer property, the AWLT has a highly influential role in determining the MH of the specimen. The recast layer of the processed specimen in the WEDM process should have a uniform and minimal thickness in order to enhance the quality measures. While the white layer formation cannot be avoided, its thickness should be as minimal as possible. The specimen processed under the optimal input parameters combination displays a uniform and low AWLT as shown in Figure 5. The electrical pulse energy is proportional to the width of the white layer. The pulse duration can increase the pulse energy in the WEDM process. The pulse energy impacts the resolidification of the melted particles. The pulse energy level as displayed in Figure 5 results in a lower recast layer thickness and uniform distribution.

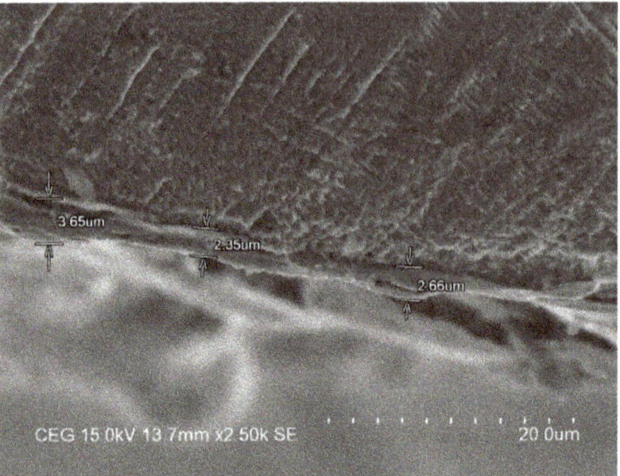

Figure 5. White layer distribution of machined specimen in WEDM process.

Figure 6 shows the minimum white layer zone and a few cracks on the processed surface under the optimal process parameters combination. A higher white layer could increase the residual stress on the machined surface in the EDM process. The delivery of the higher pulse energy creates higher residual stress, which subsequently creates micro cracks. Due to this high residual stress, these cracks then propagate. This could affect the surface performance measures and fatigue life of titanium workpiece specimens in EDM. Dental implants should have lower residual stress to increase the life and quality of the products [1]. The surface quality of the specimens at optimal process parameters was observed to have fewer micro cracks, as shown in Figure 6. Hence, it was proved that the fatigue life of the machined components could be considerably enhanced by adopting the proposed MCDM technique. Dental implants should have considerably lower surface roughness. The surface roughness of the titanium alloy can be effectively modified by the pulse energy during the machining process. The heat affected zone (HAZ) can be viewed as the white region. It was noted that more HAZ was found in the trial with a high discharge energy combination than in the trial using the optimal combination. The surface roughness of machined specimens in the EDM process can be reduced by uniform and tiny craters on the processed surface [17]. In Figure 7, the distance between C and D indicates the evaluation length of the surface roughness measurement, while the distance between A and B specifies the maximum peak value of roughness. The average value of roughness was inferred from the figure itself. It was observed that the surface roughness could be effectively reduced by incorporating the optimal input parameters combination in the EDM process due to the optimal pulse energy as shown in Figure 7.

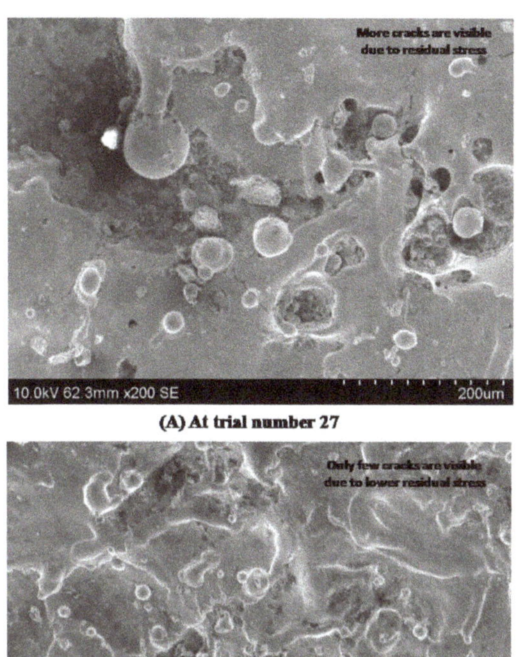

(A) At trial number 27

(B) At optimal process parameters combination

Figure 6. Comparison between surface morphology of machined specimens in WEDM process.

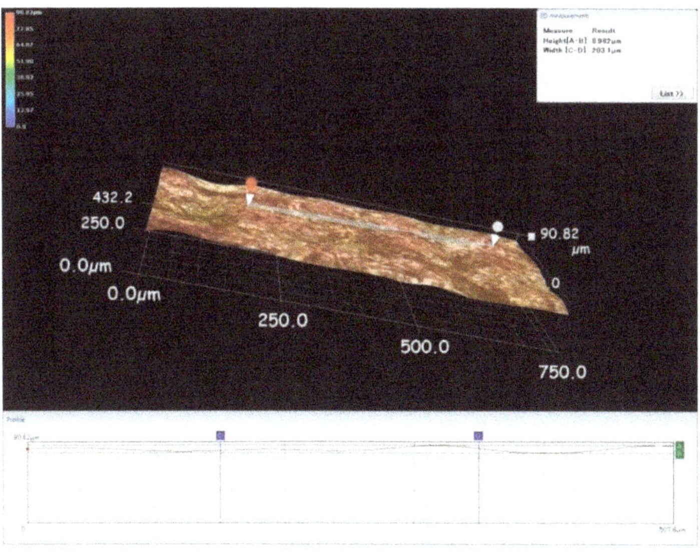

Figure 7. 3D surface analysis of machined specimen in WEDM process.

5. Conclusions

The TGRA method was used to assess the optimal factors combination in obtaining optimal surface measures such as wire wear ratio, MH and AWLT when machining titanium (α-β) alloy with the WEDM process. The following conclusions were made:

- In achieving better quality measures, the optimal electrical factors amongst the existing factor combinations were found to be gap voltage (70 V), discharge current (15 A) and duty factor (0.6).
- The maximum high-low grade value shows that the wire electrode affects the surface measures due to its significance in determining spark energy in WEDM.
- Using a TGRA based MCDM approach, the surface quality analysis has also shown that the optimal input factors combination significantly contributes to improving the quality of the machined surface.

Author Contributions: Conceptualization, M.T. and R.A.; methodology, K.M.; software, M.T.; validation, M.T. and M.A.; formal analysis, K.M. and S.R.; investigation, M.T. and K.M.; resources, O.A.; writing—original draft preparation, M.T. and K.M.; project administration, M.T., S.R. and O.A.; funding acquisition, K.M. and M.A.; All authors have read and agreed to the published version of the manuscript.

Funding: The research was funded by the Deanship of Scientific Research, King Saud University: Research group no. RG-1440-075.

Acknowledgments: The authors extend their appreciation to the Deanship of Scientific Research at King Saud University for funding this work through research group no. (RG-1440-075).

Conflicts of Interest: The authors have no conflict of interest to declare.

References

1. Guehennec, L.L.; Soueidan, A.; Layrolle, P.; Amouriq, Y. Surface treatments of titanium dental implants for rapid osseointegration. *Dent. Mater.* **2007**, *23*, 844–854. [CrossRef] [PubMed]
2. Arrazola, P.J.; Garay, A.; Iriarte, L.M.; Armendia, M.; Marya, S.; Maitre, F.L. Machinability of titanium alloys (Ti6Al4V and Ti555.3). *J. Mater. Process. Technol.* **2009**, *209*, 2223–2230. [CrossRef]
3. Dandekar, C.R.; Shin, Y.C.; Barnes, J. Machinability improvement of titanium alloy (Ti–6Al–4V) via LAM and hybrid machining. *Int. J. Mach. Tool. Manuf.* **2010**, *50*, 174–182. [CrossRef]
4. Rashid, R.A.R.; Sun, S.; Wang, G.; Dargusch, M.S. The effect of laser power on the machinability of the Ti-6Cr-5Mo-5V-4Al beta titanium alloy during laser assisted machining. *Int. J. Mach. Tool. Manuf.* **2012**, *63*, 41–43. [CrossRef]
5. Seo, Y.W.; Ramulu, M.; Kim, D. Machinability of titanium alloy (Ti-6Al-4V) by abrasive waterjets. *Proc. Inst. Mech. Eng. B J. Eng. Manuf.* **2003**, *217*, 1709–1721. [CrossRef]
6. Geethapriyan, T.; Kalaichelvan, K.; Muthuramalingam, T.; Rajadurai, A. Performance analysis of process parameters on machining α-β titanium alloy in electrochemical micromachining process. *Proc. Inst. Mech. Eng. B J. Eng. Manuf.* **2018**, *232*, 1577–1589. [CrossRef]
7. Ohkubo, C.; Watanabe, I.; Ford, J.P.; Nakajima, H.; Hosoi, T.; Okabe, T. The machinability of cast titanium and Ti–6Al–4V. *Biomaterials* **2000**, *21*, 421–428. [CrossRef]
8. Muthuramalingam, T. Measuring the influence of discharge energy on white layer thickness in electrical discharge machining process. *Measurement* **2019**, *131*, 694–700. [CrossRef]
9. Muthuramalingam, T.; Mohan, B. Performance analysis of iso current pulse generator on machining characteristics in EDM process. *Arch. Civil Mech. Eng.* **2014**, *14*, 383–390. [CrossRef]
10. Lin, Y.C.; Cheng, C.H.; Su, B.L.; Hwang, L.R. Machining Characteristics and Optimization of Machining Parameters of SKH 57 High-Speed Steel Using Electrical-Discharge Machining Based on Taguchi Method. *Mater. Manuf. Process.* **2006**, *21*, 922–929. [CrossRef]
11. Lin, Y.C.; Wang, A.C.; Wang, D.A.; Chen, C.C. Machining Performance and Optimizing Machining Parameters of Al$_2$O$_3$–TiC Ceramics Using EDM Based on the Taguchi Method. *Mater. Manuf. Process.* **2006**, *24*, 667–674. [CrossRef]

12. Nanthakumar, P.; Rajadurai, A.; Muthuramalingam, T. Multi Response Optimization on Mechanical Properties of Silica Fly Ash Filled Polyester Composites Using Taguchi-Grey Relational Analysis. *Silicon* **2018**, *10*, 1723–1729. [CrossRef]
13. Jailani, H.S.; Rajadurai, A.; Mohan, B.; Kumar, A.S.; Kumar, T.S. Multi-response optimization of sintering parameters if Al-Si alloy/fly ash composite using Taguchi method and grey relational analysis. *Int. J. Adv. Manuf. Technol.* **2009**, *45*, 362–369. [CrossRef]
14. Pillai, J.U.; Sanghrajka, I.; Shunmugavel, M.; Muthuramalingam, T.; Goldberg, M.; Littlefair, G. Optimisation of multiple response characteristics on end milling of aluminium alloy using Taguchi-Grey relational approach. *Measurement* **2018**, *124*, 291–298. [CrossRef]
15. Manoj, M.; Jinu, G.R.; Muthuramalingam, T. Multi Response Optimization of AWJM Process Parameters on Machining TiB$_2$ Particles Reinforced Al7075 Composite Using Taguchi-DEAR Methodology. *Silicon* **2018**, *10*, 2287–2293. [CrossRef]
16. Nguyen, P.H.; Long, B.T.; Dung, L.Q.; Toan, N.T.; Muthuramalingam, T. Multi-Criteria Decision Making Using Preferential Selection Index in Titanium based Die-Sinking PMEDM. *J. Korean Soc. Precis. Eng.* **2015**, *19*, 577–592. [CrossRef]
17. Ramamurthy, A.; Sivaramakrishnan, R.; Muthuramalingam, T. Taguchi-Grey computation methodology for optimum multiple performance measures on machining titanium alloy in WEDM process. *Indian J. Eng. Mater. Sci.* **2015**, *22*, 181–186.
18. Wang, Z.; Zhang, T.; Yu, T.; Zhao, J. Assessment and optimization of grinding process on AISI 1045 steel in terms of green manufacturing using orthogonal experimental design and grey relational analysis. *J. Clean. Prod.* **2020**, *253*, 119896. [CrossRef]
19. Garg, A.; Lam, J.S.L. Modeling multiple-response environmental and manufacturing characteristics of EDM process. *J. Clean. Prod.* **2016**, *137*, 1588–1601. [CrossRef]
20. Wu, Y.; Zhou, F.; Kong, J. Innovative design approach for product design based on TRIZ, AD, fuzzy and Grey relational analysis. *Comput. Ind. Eng.* **2020**, *140*, 106276. [CrossRef]
21. Ramakrishnan, R.; Karunamoorthy, L. Multi response optimization of wire EDM operations using robust design of experiments. *Int. J. Adv. Manuf. Technol.* **2006**, *29*, 105–112. [CrossRef]
22. Muthuramalingam, T.; Mohan, B.; Rajadurai, A.; Prakash, M.D.A.A. Experimental Investigation of Iso Energy Pulse Generator on Performance Measures in EDM. *Mater. Manuf. Process.* **2013**, *28*, 1137–1142. [CrossRef]
23. Muthuramalingam, T. Effect of diluted dielectric medium on spark energy in green EDM process using TGRA approach. *J. Clean. Prod.* **2019**, *238*, 117894. [CrossRef]

Article

Investigation of Electrically-Assisted Rolling Process of Corrugated Surface Microstructure with T2 Copper Foil

Shaoxi Xue [1], Chunju Wang [1,2,*], Pengyu Chen [1], Zhenhai Xu [1], Lidong Cheng [2], Bin Guo [1,*] and Debin Shan [1]

[1] National Key Laboratory for Precision Hot Processing of Metals, Harbin Institute of Technology, Harbin 150001, China; xueshaoxi@hit.edu.cn (S.X.); chenpengyu66@126.com (P.C.); xzhenhai@163.com (Z.X.); shandb@hit.edu.cn (D.S.)

[2] Robotics and Microsystems Center, Soochow University, Suzhou 215131, China; ldcheng@suda.edu.cn

* Correspondence: cjwang@suda.edu.cn (C.W.); bguo@hit.edu.cn (B.G.); Tel.: +86-451-8641-8183 (B.G.)

Received: 31 October 2019; Accepted: 9 December 2019; Published: 11 December 2019

Abstract: Electrically-assisted (EA) forming is a low-cost and high-efficiency method to enhance the formability of materials. In the study, EAF tensile tests are carried out to study the properties of T2 copper foil in an annealed state, and the effect of the electric current on the forming quality of corrugated foils is further studied in the EA rolling forming process. The result shows that the current reduces the flow stress and the fracture strain, which is different from the result of rolled samples. The joule heating effect on mechanical properties is significant in EA tension, and the softening effect of the surface layer can be observed at tensile strength, due to the grain size effect. Moreover, the current can weaken the grain size effect. In the rolling forming process, the influence of different electrical parameters on the forming height is remarkable, especially for the rolled T2 copper. The appropriate electrical parameters can improve the forming height, while keeping a small thickness thinning. Nevertheless, the high current density will lead to local rupture. This study proves that the current can improve the forming quality of the corrugated foils and is a promising surface texture forming process.

Keywords: surface microstructure; electrically-assisted rolling; current density; T2 copper foil

---

1. Introduction

Surface texture is highly critical in aviation, aerospace, and so forth, due to the special function of physics and chemicals, such as the drag reduction, the hydrophobicity, optics, heat and mass transfer [1]. The fabrication methods of the micro texture include electrochemical micromachining, laser surface texturing, electric discharge texturing, lithography, and micromachining. Electrical chemical micro machining is an advanced process which can fabricate the surface texture in metallic plates. Lee et al. [2] studied the effects of the inter-electrode gap, pulse rate, and electrolytic inflow velocity on the forming accuracy of the micro channel. A laser surface texturing to fabricate the surface texture was conducted by Tang et al [3]. They produced spikes on brass substrate, which had a super-hydrophobic function, using a pulse laser ablation process, and the height of these spikes related to the power of the laser beam. Electric discharge texturing can change the mechanical properties of the sample surface and form the micro structure by a high temperature [4]. Lithography involves multiple steps, like insulation, mask generation, and machining. He et al. [5] conducted lithography and an etching experiment to generate the micro-pillar-based surface textures on a silicon wafer and fabricate the nano pillars on the micro pillars. Hung and Lin [6] studied the micro scale tool piece electrode and the fabricating high-aspect-ratio micro channel on bipolar plates by electrical discharge machining. The surface texture can be easily fabricated by these methods mentioned above. However, laser surface texturing and electric discharge texturing can generate heat-affected zones or thermal residual stresses. Micromachining is not economically viable because of a lot of material waste. Electrochemical micromachining and lithography pose environmental concerns. A method which meets the demands of mass production and environmental protection is needed.

Rolling forming is a promising process for manufacturing surface microstructure sheets because of its low forming force, high efficiency, low cost, and good forming accuracy [7]. Ma et al. [8] found that the grain size and the crystal texture had a significant effect on the springback by rolling forming. Huang et al. [9] observed that the maximum thickness thinning of the micro channels with aspect ratios up to 1.0 was 18.7% by the micro-channel rolling forming test. Zhou et al. [10] improved the forming depth of the micro-structure and flatness of sheets by increasing the relative speed of the upper and lower rollers using a new desktop roll forming tool.

When the ratio of sheet thickness to grain size was small, the formability decreased [11]. The studies show that the current can reduce the flow stress, improve plasticity, and formability, this is called the electroplastic effect [12]. Ross et al. [13] presented the direct current (DC) EA compression experiments of Ti-6Al-4V alloy. The forming load was significantly reduced and the formability was improved. Andrawes et al. [14] found that the forming energy decreased, but the ductility and the elongation also decreased by the DC EA tension of aluminum alloy. However, the pulse current can not only reduce the flow stress but also significantly increase the elongation of aluminum alloy, according to the studies of Salandro et al. [15] and Roh et al. [16]. Jeong et al. [17] reported that the elongation significantly decreased when the pulse current was applied via the tensile test of trip-aided steel. They believed that the increase of the stability of retained austenite inhibited the effect of mechanically induced martensite transformation with the increase of temperature. In the investigation of Gennari et al. [18], the uniform elongation and total elongation were improved greatly in the EA tensile test, while the changes of the yield stress and ultimate tensile stress were inapparent, compared to the thermal test. Perkins et al. [19] conducted a series of studies by the EA compression of different metals, and found that there was a threshold current density in EA compression. Later on, Jones et al. [20] found that when the current density reached 30 A/mm^2, the compressive properties of AZ31 magnesium alloys were remarkably improved without fracture. Siopis et al. [21] reported that with the increase of grain size, the degree of flow stress reduction became smaller, while the threshold value of the current density became larger. The decrease of the flow stress was clearer using a current when the samples underwent pre-plastic deformation [22,23]. Ross et al. [13] indicated that the thermal effect was not the main factor. The decrease of the grain size improved the degree of the joule heat rise and stress decrease, moreover, local intergranular voids and partial grain boundaries melted, according to the study of Fan et al. [24]. The effect of the current on the size effect was studied by Siopis et al. [21]. They believed that when the current was applied, the data dispersion would be reduced. Their studies indicated that the current can weaken the size effect [25,26]. Compared with the tensions at room temperature, oven-heated and air-cooled conditions, the fracture stresses were smallest and the fracture strains were largest in the EA tension [27].

Scholars also studied the EA forming process, such as drawing, rolling, bending, and embossing. Egea et al. [28] studied the current-assisted drawing process of 308L stainless steel. It was found that the material formability and energy efficiency increased by 11.9% and 7.6%, respectively. Li et al. [29] improved the toughness and strength of zirconium by EA rolling and subsequent low temperature annealing. Khal et al. [30] reported that with the increase of the current density, pulse frequency, and pulse duration, the springback can be further reduced by EA bending. Mai et al. [31] conducted the EA micro-channel embossing process of SS316L. When the current density increased to 50 A/mm^2, the micro-channel depth increased to 26%. Cao et al. [32,33] studied the effects of the rolling force and the joule heat and friction on the forming quality of the microstructure by the EA roll forming process. The forming depth and width of the micro-channels increased with the increase of the temperature, rolling force, and friction force. It was proved that the joule heat was the main reason for the increase in the forming depth.

To conclude, the EA tension tensile and compression mechanical property have been sufficiently investigated. However, there are few studies on the EA process, especially on the thin sheet metal with microstructures. The influence of electrical parameters on the forming quality of the microstructures is not clear. In this paper, the EA uniaxial tension of T2 copper is studied experimentally. The effects of different current parameters on the flow stress, fracture strain, and grain size effect are analyzed. Furthermore, the EA rolling forming process is investigated, and the results of different electrical parameters on the forming quality are analyzed, which verify the feasibility of the EA rolling forming process.

2. Materials and Methods

2.1. Experimental Materials

In this work, commercial T2 copper foils with the thickness of 100 μm in the rolled state and annealed state (annealed at 310 °C and air cooled) were chosen as the experimental materials. To study the grain size effect on the deformation behavior, T2 copper foils in the rolled state were annealed at the temperature of 350, 450, 550, and 650 °C for 1 h to obtain different grain sizes. The microstructures were examined using optical microscope (OM) after grinding, polishing, and etching. The microstructure and obtained grain size are given in Figure A1 and Table 1, respectively. The T2 copper foil was machined to tensile test samples by the electrical discharge machining (EDM) method, as shown in Figure A2.

Table 1. Grain sizes under different temperatures.

Temperature (°C)	350	450	550	650
Grain size d (10^{-6}m)	12.31	17.73	27.87	44.49

2.2. Experimental Set-Up

The EA micro-tension device was developed on the basis of a tension testing machine (CMT8502). Two insulation blocks were inserted between the EA tension grips and tension testing machine, as illustrated in Figure A3a. As shown in Figure A3b, the EA rolling system and corrugated surface microstructure were designed to conduct the EA rolling experiment. The power supply (JX-HC DC pulse power, Lanzhou, China) had a maximum voltage of 30 V, which could output a high frequency square wave pulse current, as shown in Table 2. The temperature of specimens was measured by a FLIR infrared camera.

Table 2. DC power parameters.

Voltage/V	Frequency/Hz	Pulse Duration/10^{-6} s
0~30	10~1000	3~500

3. Results and Discussion

3.1. EA Uniaxial Tensile Test

EA uniaxial tensile tests using samples of annealed state were carried out under a strain rate of 10^{-3}s^{-1} at room temperature. The current started and continued 30 s when the tensile test was conducted for 100 s. It is found that the flow stress decreases nearly instantly as soon as the electric current is present, and the flow stress decreases with the increase of voltage, as shown in Figure 1. The flow stress increases after closing the power, but compared with the uniaxial tension of no current, the flow stress is still lower, especially at the great voltage.

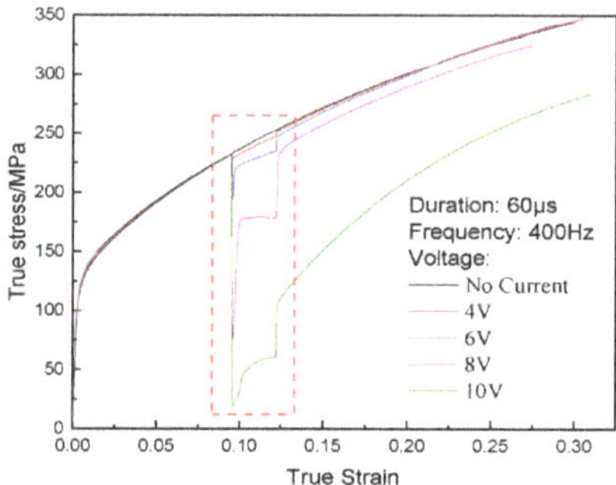

Figure 1. True stress–strain curves of annealed T2 copper.

To further study the effect of the current on mechanical properties, the current was present during the tension test. The influences of voltage, frequency, and pulse duration on the mechanical properties and temperature are shown in Figure 2. It is seen from Figure 2a that the flow stress and the fracture strain decrease with the increase of the voltage in the EA tension. The result is similar to that of Zhang et al. [27]. They conclude that the effect of the joule heating on the mechanical properties can be neglected. Figure 2b presents the maximum temperature of the specimen in the gauge section. The temperature increases instantly by joule heating and then changes slowly for the balance of the joule heating and air cooling until the fracture of samples. Furthermore, as the the voltage increases, the temperature increases because of the Joule heating effect. Mean temperatures are 60.35 °C at 4 V, 138.25 °C at 6 V, 301.45 °C at 8 V, and 447.00 °C at 10 V, respectively. When the voltage increases from 4 to 10 V, the temperature increases from 60.35 to 447.00 °C, the reduction rate of the tensile strength increases from 15.68% to 77.24%. The maximum reduction rate (77.24%) is much larger than that (23%) in the reference [27]. It needs to be noted that the temperature rise is significant in this study, and maybe the joule heating effect is the main reason for the change of mechanical properties.

Figure 2c,d depicts the variations in the true stress–strain for various pulse durations and frequencies, respectively. It can be seen that the flow stress and fracture strain decrease as the pulse duration or frequency increases. Note that the flow stress continues to decrease, but a change of the fracture strain is not apparent, which indicates that the high frequency is conducive to the plastic deformation.

From Figure 2, it is found that when the temperature is high, such as 447 °C, the flow stress reduction remarkably increases, which indicates that the current density (related to the voltage, frequency, and pulse duration) threshold is mainly temperature-independent.

It is interesting that the results are different from that of rolled samples. The flow stress drop is more remarkable, moreover, the fracture strain increases significantly in the EA tension test of T2 copper foil (rolled state) [34]. The current promotes dislocation movement, makes dislocation easier to overcome obstacles, and restrains the slip band generation. Hence, it can be understood, from the results, that the current has a more significant influence on the samples of higher dislocation density.

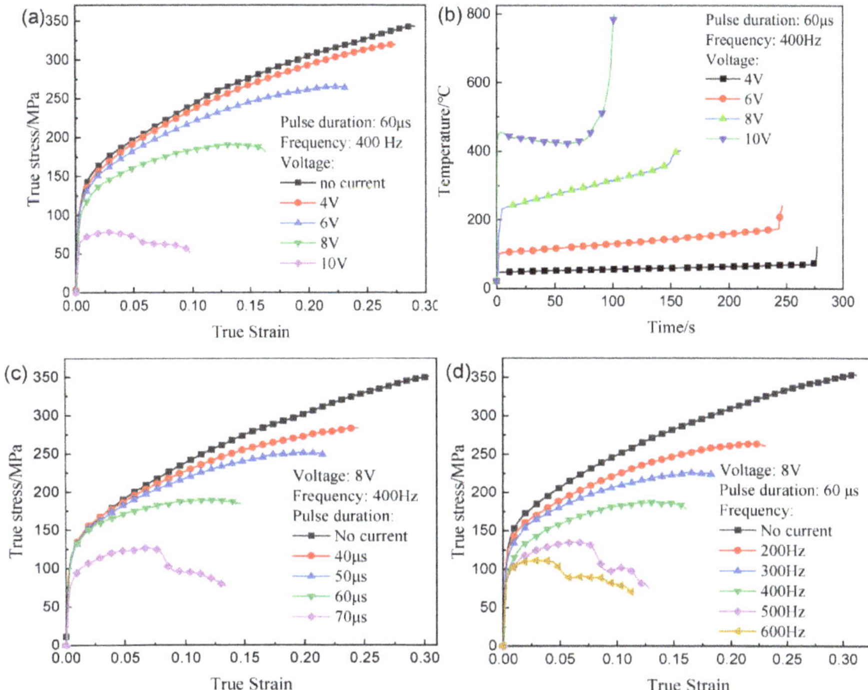

Figure 2. EA tensile test results: (**a**) different voltages; (**b**) temperature of gauge; (**c**) different durations; (**d**) different frequencies.

In order to investigate the grain size effect on the EA tension, the EA tensile tests of various grain sizes and N values were carried out, as shown in Figure 3. Figure 3a shows that the flow stress and fracture strain decrease in the EA tension, due to the increase in the grain sizes, which corresponds to results in Figure 2a. The relationship between the tensile strength and grain sizes at a voltage of 6 V is depicted in Figure 3b. This cannot be interpreted by the Hall–Patch relation, due to the softening effect of the free surface at d = 44.49 µm [35]. This is due to the interactive effect of specimen and grain sizes. However, the softening effect is weakened when the current is present. This result indicates that the current can weaken the grain size effect, which corresponds to the result of Wang et al [25].

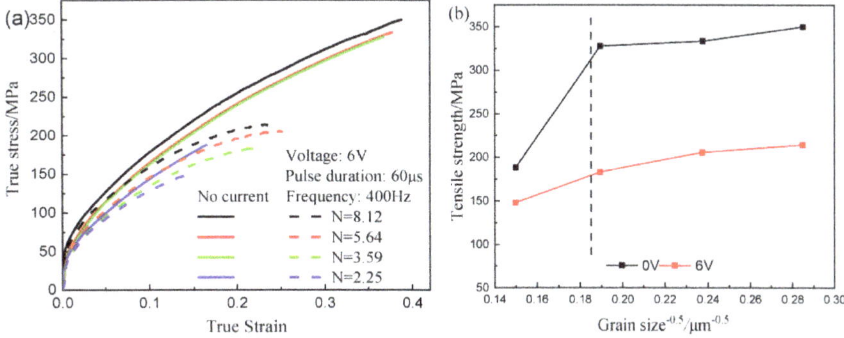

Figure 3. Tensile test results with different N value: (**a**) EA tensile test; (**b**) flow stress and fracture strain.

3.2. Rolling Forming of Corrugated Surface Microstructure

The rolling tests were conducted to investigate the effect of clearance between the upper roller and lower roller on the formability of the surface microstructure. The clearance is relative to the initial position in Figure 4. Figures 4 and 5 show the forming height and the rolling load by the load sensor (Interface MSC-130KN-375, Shenzhen, China) under different clearances. The rolling load and the forming height increase with the decrease of clearance.

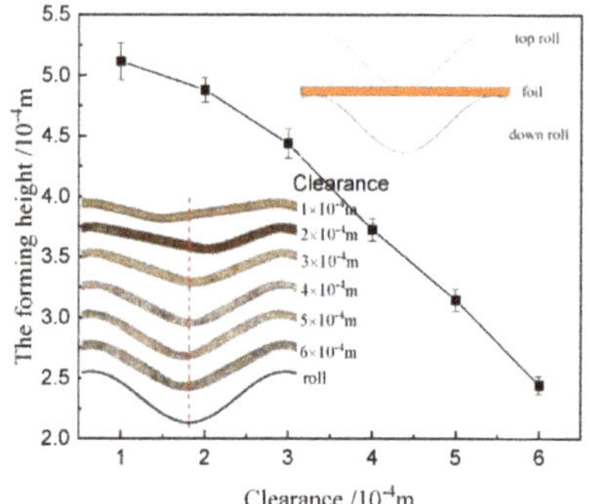

Figure 4. Forming height with different clearances.

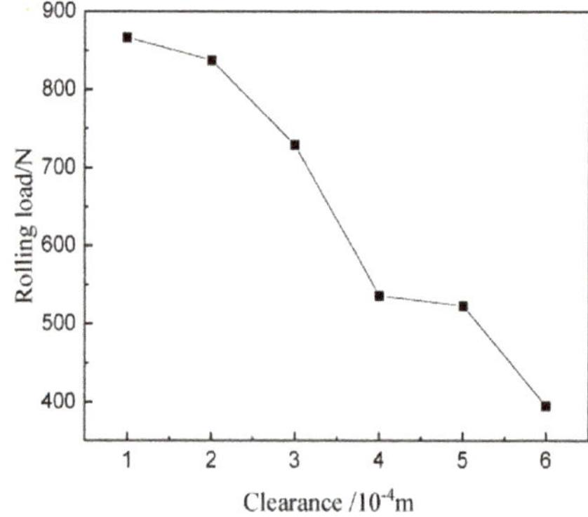

Figure 5. Rolling load with different clearances.

As the clearance decreases, the maximum thickness thinning rate and data dispersion increase, as shown in Figure 6a. The thicknesses of the wave crest and trough are at a minimum (in Figure 6b). Figure 7a presents the scheme of mechanical analysis. The bending and tensile deformation occur at part A, but the tensile deformation is only at part B, moreover, the tensile stresses are similar at part A and part B. But part A has a different bending deformation that results in the large deformation and thickness thinning (Figure 7b).

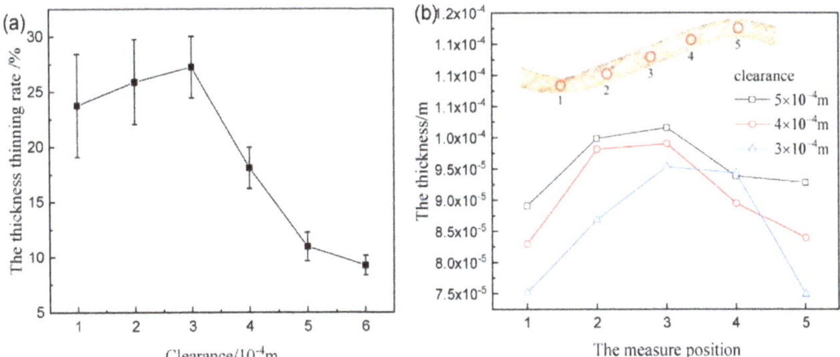

Figure 6. Thickness thinning measure results: (**a**) maximum thickness thinning rate; (**b**) distribution of thickness.

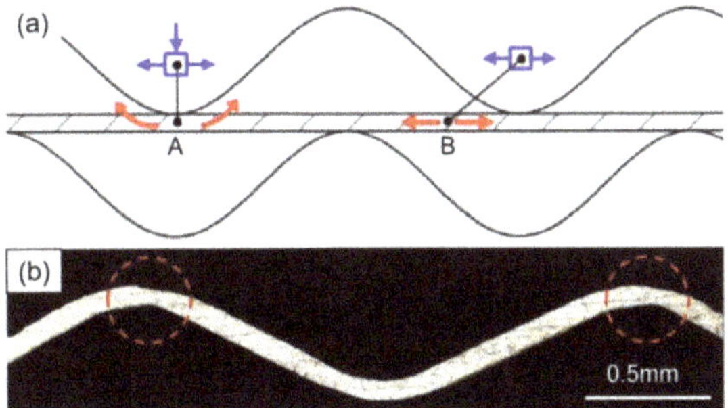

Figure 7. Mechanical analysis in rolling process: (**a**) scheme of mechanical analysis; (**b**) thickness thinning.

3.3. EA Rolling Forming

The EA rolling tests were conducted at a roll speed of 0.75 r/min and a clearance of 0.4 mm at room temperature. The influences of various voltages, frequencies, and pulse durations on the forming height and the thickness thinning rate are demonstrated in Figure 8, Figure 9, and Figure 10, respectively. It can be seen from Figure 8a that the height has a significant improvement at 15 V for two kinds of materials, especially the rolled samples, which is attributed to the flow stress drop due to the joule heating effect.

The thickness thinning rate of annealed samples is significantly lower than that of rolled samples, as shown in Figure 8b. But the difference between various voltages is insignificant, whither in a rolled or annealed state. A small part of the wave crest and trough crack for the rolled samples is due to the higher temperature when the voltage increases to 20 V, but this is not the case for the annealed samples.

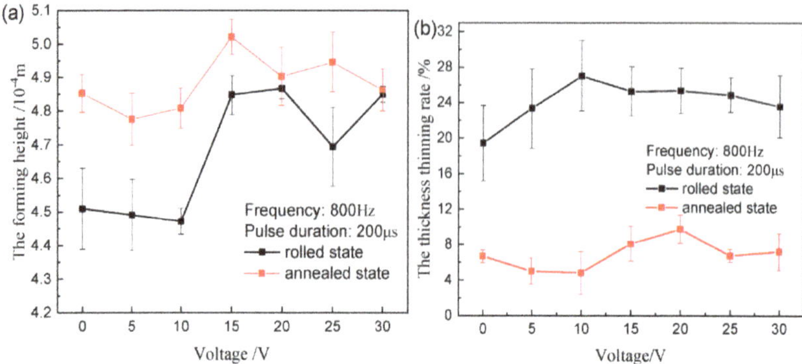

Figure 8. Experimental results of different voltage: (**a**) forming height; (**b**) maximum thickness thinning rate.

It is observed from Figure 9a that the height remarkably increases at 800 Hz for two kinds of materials. The difference of the thickness thinning rate between various frequencies is also not significant. However, the higher frequency can not only increase the forming height, but also reduces the thickness thinning, as shown in Figure 9b.

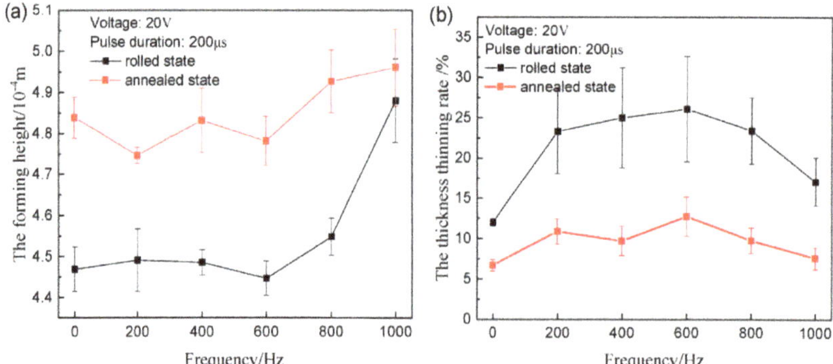

Figure 9. Experimental results of different frequency: (**a**) forming height; (**b**) maximum thickness thinning rate.

It is observed from Figure 10a that as the pulse duration increases, the forming height increases for two materials. Compared to the effect of the voltage and frequency, the change of height is smaller. Meanwhile, the thickness thinning rates also differ slightly (Figure 10b). For the high pulse duration (300 µs), some cracks also are observed for two state samples.

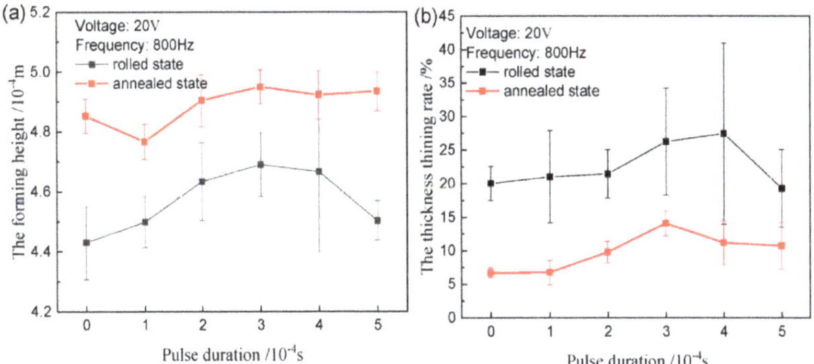

Figure 10. Experimental results of different duration: (**a**) forming height; (**b**) maximum thickness thinning rate.

To summarize, when the current density threshold is reached, the joule heating temperature is sufficiently high to reduce the flow stress remarkably, there is a range of current density, which can significantly improve the forming height and maintain a smaller thickness thinning rate. Nevertheless, a larger current density can lead to a local fracture.

4. Conclusions

In this work, the effects of voltage, frequency, and pulse duration on tensile properties and rolling forming process of T2 copper are investigated by the EA tension and EA rolling process. The influence of the current on the grain size effect is studied by EA tension with T2 copper foil of various grain sizes. The following results can be concluded:

(1) The flow stress and fracture strain decrease as the current density increases, which is mainly attributed to the joule heating effect.
(2) The softening effect of the surface layer is significant, and this is due to the grain size effect at coarse grains. However, it can be weakened in the EA tension.
(3) With the decrease of clearance, the forming load and the forming height of the microstructures increase gradually, and the thinning of the wave peaks and troughs is more serious than that of other parts.
(4) The forming height significantly increases by the EA rolling forming process, moreover, the wall thickness change slightly under proper current parameters. Thus, it is conducive to the formation of corrugated foils.

Author Contributions: Conceptualization, B.G., and C.W.; methodology, C.W., and Z.X.; validation, D.S., and B.G.; formal analysis, Z.X., and L.C.; investigation, S.X., and P.C.; resources, C.W.; data curation, L.C., and Z.X.; writing—original draft preparation, S.X.; writing—review and editing, C.W.; supervision, D.S., and B.G.; project administration, L.C., and Z.X.; funding acquisition, C.W.

Funding: This research was funded by the Natural Science Foundation of Jiangsu Province (BK20192007), National Science Foundation of China (51635005, 51875128), and the Six Talent Peaks in Jiangsu Province (GDZB-069).

Conflicts of Interest: The authors declare no conflict of interest.

Appendix A

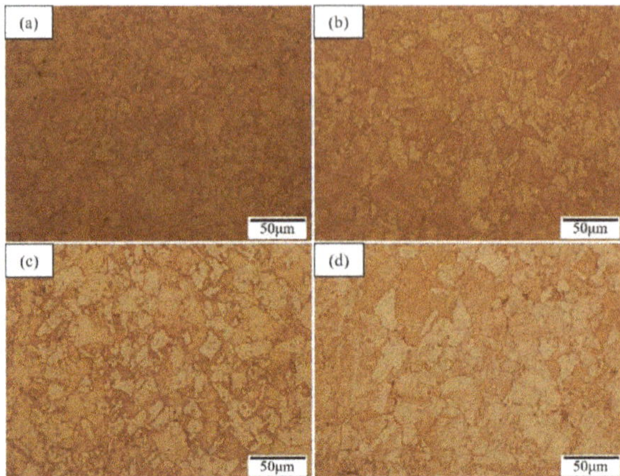

Figure A1. Microstructure of T2 copper foil along the thickness direction after annealing treatment:(**a**) 350 °C; (**b**) 450 °C; (**c**) 550 °C; (**d**) 650 °C.

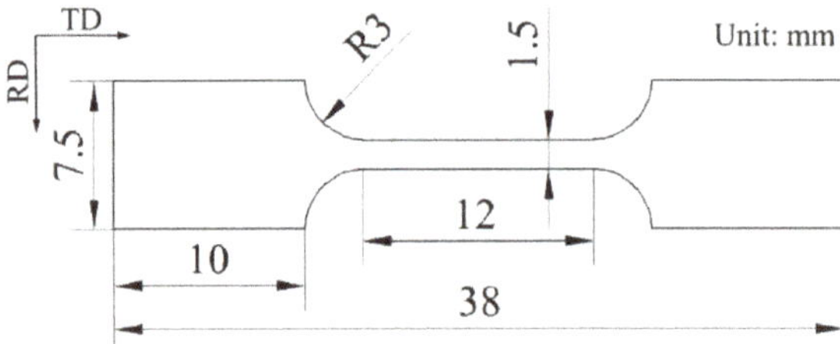

Figure A2. Geometry of tensile samples.

Figure A3. EA device: (**a**) tensile insulated grip; (**b**) EA rolling device.

References

1. Patel, D.; Jain, V.K.; Ramkumar, J. Micro texturing on metallic surfaces: State of the art. *Proc. Inst. Mech. Eng. Part B-J. Eng. Manuf.* **2018**, *232*, 941–964. [CrossRef]
2. Lee, S.J.; Lee, C.Y.; Yang, K.T.; Kuan, F.H.; Lai, P.H. Simulation and fabrication of micro-scaled flow channels for metallic bipolar plates by the electrochemical micro-machining process. *J. Power Sources* **2008**, *185*, 1115–1121. [CrossRef]
3. Tang, M.; Shim, V.; Pan, Z.Y.; Choo, Y.S.; Hong, M.H. Laser ablation of metal substrates for super-hydrophobic effect. *J. Laser Micro Nanoeng.* **2011**, *6*, 6–9. [CrossRef]

4. Aspinwall, D.K.; Wise, M.L.H.; Stout, K.J.; Goh, T.H.A.; Zhao, F.L.; El-Menshawy, M.F. Electrical discharge texturing. *Int. J. Mach. Tools Manuf.* **1992**, *32*, 183–193. [CrossRef]

5. He, Y.; Jiang, C.; Cao, X.; Chen, J.; Tian, W.; Yuan, W. Reducing ice adhesion by hierarchical micro-nano-pillars. *Appl. Surf. Sci.* **2014**, *305*, 589–595. [CrossRef]

6. Hung, J.C.; Chang, D.H.; Chuang, Y. The fabrication of high-aspect-ratio micro-flow channels on metallic bipolar plates using die-sinking micro-electrical discharge machining. *J. Power Sources.* **2012**, *198*, 158–163. [CrossRef]

7. Zhi, Y.; Wang, X.; Wang, S.; Liu, X. A review on the rolling technology of shape flat products. *Int. J. Adv. Manuf. Technol.* **2018**, *94*, 1–12. [CrossRef]

8. Ma, Z.W.; Tong, G.Q.; Chen, F.; Wang, Q.; Wang, S.C. Grain size effect on springback behavior in bending of Ti-2. *5Al-1.5Mn foils. J. Mater. Process. Technol.* **2015**, *224*, 11–17. [CrossRef]

9. Huang, J.; Deng, Y.; Yi, P.Y.; Peng, L.F. Experimental and numerical investigation on thin sheet metal roll forming process of micro channels with high aspect ratio. *Int. J. Adv. Manuf. Technol.* **2018**, *100*, 117–129. [CrossRef]

10. Zhou, R.; Cao, J.; Ehmann, K.; Xu, C. An Investigation On Deformation-Based Surface Texturing. *J. Manuf. Sci. Eng. -Trans. ASME* **2011**, *133*, 061017. [CrossRef]

11. Gau, J.T.; Principe, C.; Wang, J. An experimental study on size effects on flow stress and formability of aluminm and brass for microforming. *J. Mater. Process. Technol.* **2007**, *184*, 42–46. [CrossRef]

12. Ruszkiewicz, B.J.; Grimm, T.; Ragai, I.; Mears, L.; Roth, J.T. A review of electrically-assisted manufacturing with emphasis on modeling and understanding of the electroplastic effect. *J. Manuf. Sci. Eng. -Trans. ASME* **2017**, *139*, 110801. [CrossRef]

13. Ross, C.D.; Kronenberger, T.J.; Roth, J.T. Effect of dc on the Formability of Ti–6Al–4V. *J. Eng. Mater. Technol.-Trans. ASME* **2009**, *131*, 031004. [CrossRef]

14. Andrawes, J.S.; Kronenberger, T.J.; Perkins, T.A.; Roth, J.T.; Warley, R.L. Effects of DC current on the mechanical behavior of AlMg1SiCu. *Mater. Manuf. Process.* **2007**, *22*, 91–101. [CrossRef]

15. Salandro, W.A.; Jones, J.J.; McNeal, T.A.; Roth, J.T.; Hong, S.T.; Smith, M.T. Formability of Al 5xxx sheet metals using pulsed current for various heat treatments. *J. Manuf. Sci. Eng. -Trans. ASME* **2010**, *132*, 051016. [CrossRef]

16. Roh, J.H.; Seo, J.J.; Hong, S.T.; Kim, M.J.; Han, H.N.; Roth, J.T. The mechanical behavior of 5052-H32 aluminum alloys under a pulsed electric current. *Int. J. Plast.* **2014**, *58*, 84–99. [CrossRef]

17. Jeong, H.J.; Park, J.; Jeong, K.J.; Hwang, N.M.; Hong, S.T.; Han, H.N. Effect of pulsed electric current on TRIP-aided steel. *Int. J. Precis Eng Manuf-Green Technol.* **2018**, *6*, 315–327. [CrossRef]

18. Gennari, C.; Pezzato, L.; Simonetto, E.; Gobbo, R.; Forzan, M.; Calliari, I. Investigation of Electroplastic Effect on Four Grades of Duplex Stainless Steels. *Materials* **2019**, *12*, 1911. [CrossRef]

19. Perkins, T.A.; Kronenberger, T.J.; Roth, J.T. Metallic forging using electrical flow as an alternative to warm/hot working. *J. Manuf. Sci. Eng. -Trans. ASME* **2007**, *129*, 84–94. [CrossRef]

20. Jones, J.J.; Mears, L.; Roth, J.T. Electrically-assisted forming of magnesium AZ31: Effect of current magnitude and deformation rate on forgeability. *J. Manuf. Sci. Eng. -Trans. ASME* **2012**, *134*, 034504. [CrossRef]

21. Siopis, M.S.; Kinsey, B.L. Experimental investigation of grain and specimen size effects during electrical-assisted forming. *J. Manuf. Sci. Eng. -Trans. ASME* **2010**, *132*, 021004. [CrossRef]

22. Siopis, M.S.; Kinsey, B.L.; Kota, N.; Ozdoganlar, O.B. Effect of severe prior deformation on electrical-assisted compression of copper specimens. *J. Manuf. Sci. Eng. -Trans. ASME* **2011**, *133*, 064502. [CrossRef]

23. Wu, W.C.; Xu, C.; Si, C.R.; Xue, T. Influence of Dislocation Density and Solute Atoms Concentration on the Electroplastic Effect of Al-Cu Alloy. In Proceedings of the ASME 12th International Manufacturing Science and Engineering Conference, Los Angeles, CA, USA, 4–8 June 2017; AMER SOC Mechanical Engineers: New York, NY, USA, 2017.

24. Fan, R.; Magargee, J.; Hu, P.; Cao, J. Influence of grain size and grain boundaries on the thermal and mechanical behavior of 70/30 brass under electrically-assisted deformation. *Mater. Sci. Eng. A-Struct. Mater. Prop. Microstruct. Process.* **2013**, *574*, 218–225. [CrossRef]

25. Wang, X.W.; Xu, J.; Jiang, Z.L.; Zhu, W.L.; Shan, D.B.; Guo, B.; Cao, J. Size effects on flow stress behavior during electrically-assisted micro-tension in a magnesium alloy AZ31. *Mater. Sci. Eng. A-Struct. Mater. Prop. Microstruct. Process.* **2016**, *659*, 215–224. [CrossRef]

26. Zhang, S.J.; Lu, Y.C.; Gong, X.L.; Shen, Z.H. Investigation of Current Parameters and Size Effects on Mechanical Properties During Pulsed Electrically Assisted Uniaxial Tension in T2 Red Copper Sheets. *J. Mater. Eng. Perform.* **2018**, *27*, 6493–6504. [CrossRef]

27. Wang, X.W.; Egea, A.J.S.; Xu, J.; Meng, X.Y.; Wang, Z.L.; Shan, D.B.; Guo, B.; Cao, J. Current-induced ductility enhancement of a magnesium alloy az31 in uniaxial micro-tension below 373 k. *Materials* **2018**, *12*, 111. [CrossRef]

28. Egea, A.J.S.; Rojas, H.A.G.; Celentano, D.J.; Peiro, J.J. Mechanical and metallurgical changes on 308L wires drawn by electropulses. *Mater. Des.* **2016**, *90*, 1159–1169. [CrossRef]

29. Li, M.; Guo, D.F.; Li, J.T.; Zhu, S.M.; Xu, C.; Li, K.F.; Zhao, Y.; Wei, B.N.; Zhang, Q.; Zhang, X.Y. Achieving heterogeneous structure in hcp Zr via electroplastic rolling. *Mater. Sci. Eng. A-Struct. Mater. Prop. Microstruct. Process.* **2018**, *722*, 93–98. [CrossRef]

30. Khal, A.; Ruszkiewicz, B.J.; Mears, L. Springback evaluation of 304 stainless steel in an electrically assisted air bending operation. In Proceedings of the ASME 11th International Manufacturing Science and Engineering Conference, Blacksburg, VA, USA, 27 June–1 July 2016; AMER SOC Mechanical Engineers: New York, NY, USA, 2016.

31. Mai, J.M.; Peng, L.F.; Lai, X.M.; Lin, Z.Q. Electrical-assisted embossing process for fabrication of micro-channels on 316L stainless steel plate. *J. Mater. Process. Technol.* **2013**, *213*, 314–321. [CrossRef]

32. Fan, R.; Ng, M.K.; Xu, D.K.; Hu, P.; Cao, J. Experimental and numerical study of electrically-assisted micro-rolling. *AIP Conf. Proc.* **2013**, *1532*, 1038–1043.

33. Ng, M.K.; Fan, Z.Y.; Gao, R.X.; Smith, E.F.; Cao, J. Characterization of electrically-assisted micro-rolling for surface texturing using embedded sensor. *CIRP Ann.-Manuf. Technol.* **2014**, *63*, 269–272. [CrossRef]

34. Chen, P.Y.; Cheng, L.D.; Wang, C.J.; Wang, X.W.; Shan, D.B.; Guo, B.; Li, M. Electrically-assisted rolling process of curved thin foil micro-features. *J. Plast. Eng.* **2019**, *26*, 79–84.

35. Wang, C.J.; Xue, S.X.; Chen, G.; Zhang, P. Constitutive model based on dislocation density and ductile fracture of Monel 400 thin sheet under tension. *Met. Mater. Int.* **2017**, *23*, 264–271. [CrossRef]

Article

Adaptive Spiral Tool Path Generation for Diamond Turning of Large Aperture Freeform Optics

Dongfang Wang [1,2,*], Yongxin Sui [1,3], Huaijiang Yang [1,3] and Duo Li [4]

[1] Changchun Institute of Optics, Fine Mechanics and Physics, Chinese Academy of Sciences, Changchun 130033, China; suiyx@sklao.ac.cn (Y.S.); yanghj@sklao.ac.cn (H.Y.)
[2] University of Chinese Academy of Sciences, Beijing 100049, China
[3] Changchun National Extreme Precision Optics Co., Ltd., Changchun 130033, China
[4] Centre for Precision Engineering, Harbin Institute of Technology, Harbin 150006, China; duo.kevin.li@gmail.com
[*] Correspondence: wdfszf@163.com; Tel.: +86-431-8670-8173

Received: 22 January 2019; Accepted: 5 March 2019; Published: 8 March 2019

Abstract: Slow tool servo (STS) diamond turning is a well-developed technique for freeform optics machining. Due to low machining efficiency, fluctuations in side-feeding motion and redundant control points for large aperture optics, this paper reports a novel adaptive tool path generation (ATPG) for STS diamond turning. In ATPG, the sampling intervals both in feeding and cutting direction are independently controlled according to interpolation error and cutting residual tolerance. A smooth curve is approximated to the side-feeding motion for reducing the fluctuations in feeding direction. Comparison of surface generation of typical freeform surfaces with ATPG and commercial software DiffSys is conducted both theoretically and experimentally. The result demonstrates that the ATPG can effectively reduce the volume of control points, decrease the vibration of side-feeding motion and improve machining efficiency while surface quality is well maintained for large aperture freeform optics.

Keywords: freeform optics; slow tool servo; tool path generation; large aperture optics

---

1. Introduction

Most currently conventional optical devices from 2D optical design methods are usually rotationally symmetric. While systems with such optics perform well in numerous applications, the requirement of high energy efficiency or aberration correction often cannot be satisfied with the same optics because of its inherent symmetry limitations on geometry. In order to meet those kinds of demand, 3D optical design methods have been developed, such as the simultaneous multiple surface (SMS) 3D method [1], supporting the quadric method [2]. In general, the results of a 3D design method are nonrotational, nonlinear asymmetric surfaces, known as freeform surfaces. With the development of 3D optical design methods, large aperture freeform optics are increasingly used in both non-imaging and imaging optical systems due to its capacity of improving optical performance [3–6]. There are three ultra-precision machining processes, namely fast tool servo (FTS), slow tool servo (STS) and diamond milling frequently used to produce optical freeform surfaces. STS machining has the advantages of fast setting-up and a large degree of freedom in processing, especially for freeform surfaces with large deviation [7,8]. However, the tool path generation and tool shape compensation must be conducted carefully for efficiently producing contoured surface with submicron form accuracy and nano-scale roughness or less.

Over the past decades, much research work has been conducted on tool path generation for diamond turning of freeform optics, including angle sampling strategy, spiral path and tool radius compensation. Fang et al. [9] used non-uniform rational basis spline (NURBS) surface to represent a freeform surface, and carried out the compensation and optimized values for tool geometry. Gong et al. [10] calculated the non-zero rake angle tool position directly from symbolic computation. In addition, the authors [11] projected space Archimedean spiral onto freeform surface along the normal direction of the base surface to get the diamond tool path. Wang et al. [12] machined a toric surface by generating a spiral tool path with the constant-angle method, and analyzed the different entrance parameters algorithm on Position–Velocity–Time (PVT) interpolation. Zhu et al. [13] proposed the adaptive tool servo (ATS) mode that the sampling angle and feedrate were actively adjusted at any cutting point to adapt shape variation of the desired surface.

During the conventional multi-axis computerized numerical control (CNC) milling of freeform surfaces, it is crucial to adapt the side-feeding motion and forward sampling of tool motions to surface shape variations to achieve uniform surface quality with maximum efficiency [13]. Koren et al. [14] used a non-constant offset of the previous tool path to guarantee the cutter moving in an un-machined area and without redundant machining. However, this method is not suitable for STS because the non-constant offset of the spiral path will lead to additional motion of the lateral feeding axis (x-axis) when it follows the c-axis. Due to the limitation of the dynamic response of the servo axis, this will lead to the following error and vibration.

Generally, the following problems exist in tool path generation algorithm for slow tool servo diamond turning of large aperture freeform optics:

Current angle sampling strategies are still constant-angle and constant arc-length methods [15,16], which are aimed at path generation of axisymmetric surfaces. However, the curvature of a freeform surface is not uniform on the whole part and the strategy of constant-angle or constant arc-length cannot adapt to the change of freeform surface shape. Therefore, the uniform sampling will inevitably lead to an uneven distribution of surface accuracy [11,17]. Especially for large aperture optical surfaces, when the constant-angle method is applied, the surface quality of central region is better than that of the outer region due to the fact that the point density of outer region is relatively sparser than those on the central region. Therefore, in order to ensure the allowable interpolation error on the outer region, a large number of sampling points are needed on the outer region of the surface, which results in redundant control points on the center region. Moreover, the volume of points generated by this method probably exceeds the storage capacity of CNC system.

The cutting residual is caused by the tool radius envelope along the feeding direction, which is regarded as surface roughness. Kwok et al. [18] and Yu et al. [19] developed the model for predicting the residual error in STS, indicating that the residual error is not only dominated by the tool nose radius and feedrate, but also highly dependent on local surface curvature. Currently, the commonly adopted constant feed in STS leads to a heterogeneous distribution of the cutting residual error, especially for freeform optics with large curvature variation. To guarantee the cutting residual within a tolerance level, machining of the entire whole surface is required to maintain the minimum feedrate, which is not necessary for other regions and significantly reduces the machining efficiency.

Current tool path generation methods do not consider the dynamic performance of machine tools. In traditional machining of rotational symmetrical surfaces, the feed axis and spindle are independent, so the dynamic performance of machine tool is not considered. However, when a freeform surface is machined by STS diamond turning, the feed axis and the spindle should maintain a strict spatial position relationship. Therefore, an optimal tool path is very important. Although the ATS method proposed by Zhu [13] can produce control points with a uniform distribution of surface profile errors, and much CAM software used in milling which could be ported to turning can output a constant load or constant scallop height tool path [20,21]. However, because they produce irregular spiral paths, this will bring additional fluctuations to the feed axis, reducing the machining surface quality.

In this paper, a novel path generation of STS diamond turning for large aperture freeform optics; namely, adaptive tool path generation (ATPG) is proposed. Firstly, the principle of the novel generation method is introduced and the path generation algorithm is presented in detail with analysis of the motion characteristics of machine tools. The results show that, on the premise of satisfying the machining accuracy, the proposed algorithm can reduce the volume of control points, and improve the dynamic characteristics of machine tools and machining efficiency. Finally, both theoretical analysis and experimental comparison of the algorithm are conducted.

2. Basic Principle of Adaptive Tool Path Generation

The principle of diamond tool turning freeform optics is that the diamond tool contours the complex surface along the spiral path as shown in Figure 1a. The stroke of the diamond tool is synchronized to the angle and radial position of freeform surface on the machine's spindle. The forward and reverse motion is achieved by z-axes of machine tools (STS).

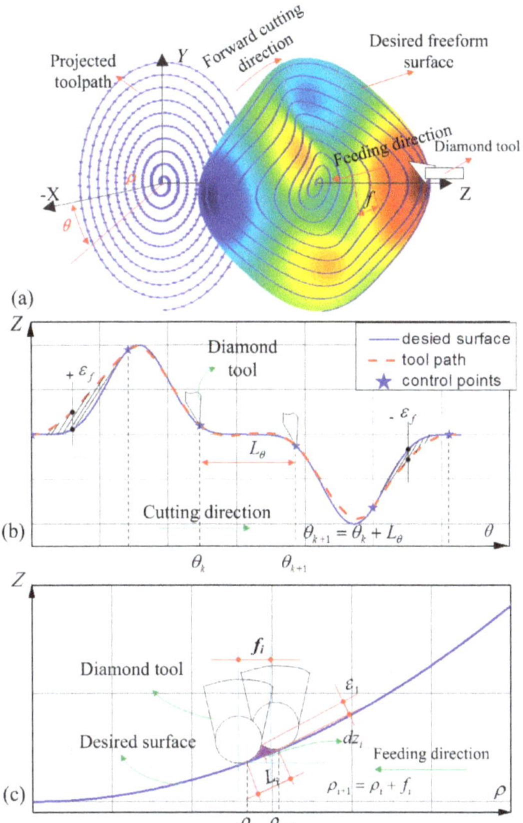

Figure 1. Schematic of principle of slow tool servo (STS), (**a**) the relative motions; (**b**) error in forward cutting direction and (**c**) error in feeding direction.

There are two intrinsic errors in STS machining, namely interpolation error and cutting residual as shown in Figure 1b,c [22]. In the forward cutting direction, the deviation between the desired surface and the tool trajectory leads to the interpolation error and the error is highly dependent on the distance L_θ between the two consecutive control points as well as the local surface profile. Generally, most diamond machine tools offer PVT spline interpolation for STS, which creates a path known as a cubic Hermite Spline. Therefore, for a given interpolation tolerance ε_f, the algorithm adaptively adjusts the angle interval L_θ according to the error of Hermite interpolation to ensure that the error is within the range $[-\varepsilon_f, +\varepsilon_f]$, so that the error is homogeneously distributed in the whole cutting direction.

In the feeding direction, in order to improve the machining efficiency, the feedrate should be adjusted with surface curvature. Because the curvature of a freeform surface not only changes with radius, but also changes with polar angles, as shown in Figure 2. The curvature of a freeform surface at the polar angle θ_1 and θ_2 of the same radius is different, so the calculated feedrate f_1 and f_2 on the same radius are different, which leads to the position of diamond tool changing with spindle angle in the process of feeding. As a result, micro-fluctuations are generated in the process of machining, which would deteriorate the surface quality of machining, just as shown in Figure 3a. The dash line denotes the feedrate calculated by the surface shape (In fact, it is the feedrate in ATS). In contrast, ATPG sets the feedrate to a constant equal to the minimum value of this revolution, which can restrain the fluctuations of feedrate with angles. Therefore, the final feedrate only changes with radial position. Furthermore, if the feedrate varies dramatically along the radius, the feedrate will be re-fitted with a smooth curve to suppress the vibration in the feeding direction, just as shown by the red solid line in Figure 3b. It should be noted that the blue dash dot line donates the feedrate in the traditional machining mode, which maintains the minimum feedrate to guarantee the cutting residual on the whole surface.

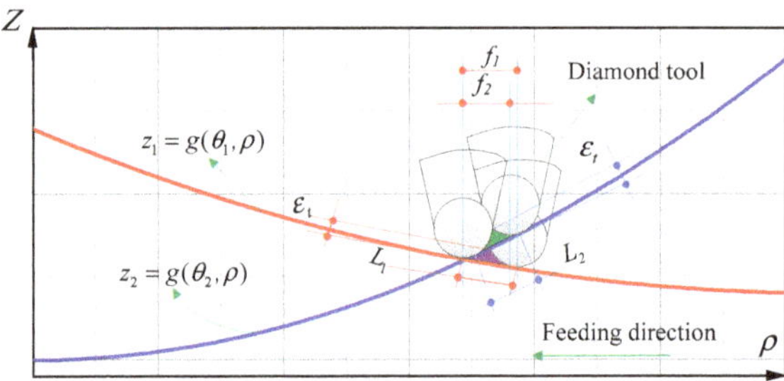

Figure 2. Schematic of feed variation with angle on the same radius for freeform surface.

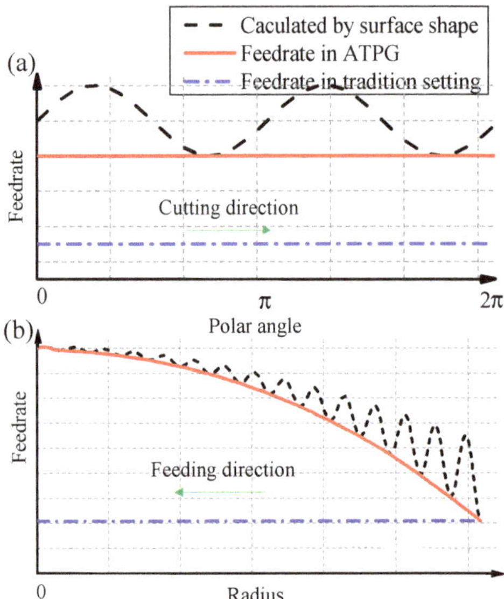

Figure 3. Schematic of feedrate in different methods. (**a**) Feedrate in one revolution of spiral path; (**b**) Feedrate along the radius on the whole surface.

Although the feedrate in ATPG is slightly smaller than ATS, it is still larger than the traditional settings. The advantages of this setting are that it not only improves the machining efficiency, but also suppresses the micro-fluctuation caused by surface shape. Although this will result in a slight heterogeneous distribution of interpolation error at different angles on the same radius, it is almost negligible, and the advantage is to increase the stability of the machine tool.

3. Tool Path Generation for STS Diamond Turning

In the traditional STS diamond turning, the machined surfaces have to be evaluated after the machining process which gives a high risk of the machined surface failing to meet the contour accuracy. The surface generation algorithm is not only the generation of a diamond tool path, but also a predictive method to analyze the contour accuracy by using different cutting strategies.

Figure 4 shows the general process of ATPG, including five steps to generate the final control points: (1) Tool interference check; (2) Determination of the feedrate for every revolution of the spiral path; (3) Determination of smooth curve fitting the relationship between feedrate and radius; (4) Determination of control points in the cutting direction with the constraint of interpolation error; (5) Tool radius compensation.

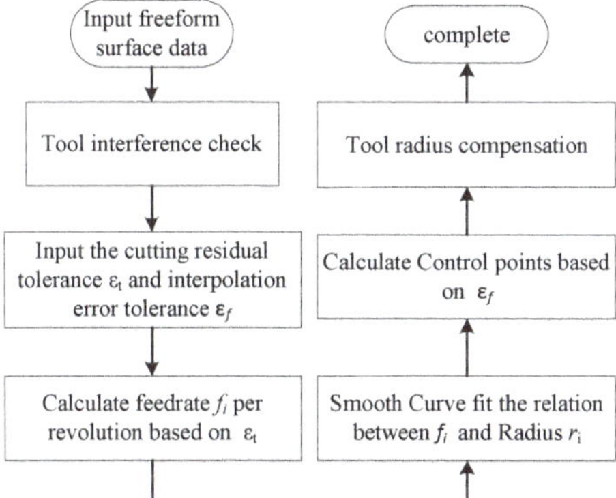

Figure 4. The general process of ATPG for STS.

3.1. Tool Interference Check

Due to the geometrical complexity of freeform surfaces, it is necessary to check tool interference, and the sectional curve method [9] is used to check the tool parameters before machining, which decomposes the whole surface into two-dimensional sectional curves and integrated the tool parameters from curves.

3.2. Determination of the Feedrate for Every Revolution of the Spiral Path

The coordinate system of the workpiece for STS machining is shown in Figure 1a. The desired surface can be expressed as $z = f(x,y)$, which can be also expressed as $z = g(\theta,\rho)$ by coordinate transformation.

Let's suppose that the radius corresponding to the N_i-th revolution of the spiral path is ρ_i at the angle $\theta_i = 2\pi i$, so the corresponding curvature at this revolution can be defined as [23]:

$$K_i = \min\left\{ \left.\frac{\partial^2 g/\partial\rho^2}{[1+\partial g/\partial\rho]^{\frac{3}{2}}}\right|_{\rho=\rho_i}, \forall\theta \in \left[0°,360°\right) \right\} \tag{1}$$

where K_i means the minimum curvature of the cross-sectional profile passing through the rotation center over the whole revolution.

By approximating the local profile as a segment of arc, the distance L_i (Figure 1c) between the two consecutive revolutions N_i and N_{i+1} on the desired surface with respect to a given cutting residual error tolerance ε_t (Figure 1c) can be estimated by [14]:

$$L_i = \sqrt{\frac{8R_t\varepsilon_t}{1 - R_t K_i}} \tag{2}$$

where R_t denotes the nose radius of a diamond tool.

At the same time, we define the slope of the cross-sectional profile dz_i (Figure 1c) at the radius ρ_i as:

$$dz_i = \min\left\{ \left.\frac{\partial g}{\partial \rho}\right|_{\rho=\rho_i}, \forall \theta \in \left[0°, 360°\right) \right\} \tag{3}$$

As shown in Figure 1c, the feedrate f_i for the N_i-th revolution in the projected tool path and radius ρ_{i+1} corresponding to the N_{i+1}-th revolution in the projected tool path can be approximately expressed as:

$$f_i = \frac{L_i}{\sqrt{1+dz_i^2}} \tag{4}$$

$$\rho_{i+1} = \rho_i + f_i \tag{5}$$

The first revolution can be set as:

$$N_1 = 0, \rho_1 = d/2 \tag{6}$$

where d is the inner diameter of the desired surface, and if the desired surface is started from the center, $d = 0$.

Since there is only one parameter ρ_{i+1} unknown, it can be calculated by numerical iteration solving Equations (1), (2) and (6). Thus, we can sequentially obtain series of number pairs (N_i,ρ_i).

3.3. Determination of Smooth Curve about the Relationship between Feedrate and Radius

Obviously, the curve of series of number pairs (N_i,ρ_i) is related to the characteristics of side-feeding motion of machine tool, especially for the machine tool whose spindle is mounted on the x-axis. The drastic fluctuations in the curve may cause the side-feeding acceleration of the machine tool to exceed the allowable range, which may result in the failure of the machining process. Therefore, some drastic fluctuations of the number pairs (N_i,ρ_i) should be limited, if necessary. Hence, we use a continuous and smooth curve to approximate the curve of number pairs (N_i,ρ_i) and the feedrate obtained by fitting should not be greater than number pairs (N_i,ρ_i), just as shown by the red solid line in Figure 3b. After that, we can obtain the functional relationship between the radius ρ and revolution number N, which can be expressed as:

$$\rho = \ell(N) \tag{7}$$

where N denotes revolution of spiral, and it is not necessary to be an integer.

3.4. Determination of the Control Points in Cutting Direction

Let us assume that the k-th cutting point $P_k(\theta_k,\rho_k,z_k)$ in the N_k-th revolution of the spiral path, so the corresponding angle θ_k can be expressed by

$$\theta_k = 2\pi(N_k - [N_k]) \tag{8}$$

where $[\cdot]$ denotes rounding operation.

Because the spiral radius increment and angle increment are very small between the two consecutive control points in the actual tool path, an arc can approximate the segment spiral path.

Hence, the tool spiral path that passes through the point $P_k(\theta_k,\rho_k,z_k)$ can be approximated by the curve $z = g(\theta,\rho_k)$, and the parameter ρ_k is already known, so the curve is a function of one variable, which can be expressed approximately as:

$$z = \phi(\theta) = g(\theta, \rho_k), \ \theta \in (\theta_k - \delta, \theta_k + \delta) \tag{9}$$

where δ is a small number, and $\phi(\cdot)$ denotes the operator notation of one variable function.

PVT mode provides an excellent contouring capability because it takes the interpolated commanded path exactly through the control points. It generates a path known as a Hermite spline. Hence, according to the error of piecewise cubic Hermite interpolation, the relationship between the interpolation error tolerance ε_f and angle increment L_θ between the two consecutive control points $P_k(\theta_k,\rho_k,z_k)$ and $P_{k+1}(\theta_{k+1},\rho_{k+1},z_{k+1})$ can be constructed by [24]:

$$\varepsilon_f \le \frac{L_\theta^4}{384} \max_{\theta_k \le \theta \le \theta_{k+1}} |\phi^{(4)}(\theta)| \tag{10}$$

where the curve $z = \phi(\theta)$ is supposed $\phi(\theta) \in C^4[\theta_k - \delta, \theta_k + \delta]$, and $\phi^{(4)}(\cdot)$ denotes the 4th derivative.

The following relation is also established as:

$$\theta_{k+1} = \theta_k + L_\theta \tag{11}$$

so the unknown parameters L_θ and θ_{k+1} be derived from Equations (10) and (11) by numerical calculation.

The corresponding revolution N_{k+1} for P_{k+1} can be obtained by:

$$N_{k+1} = N_k + L_\theta/2\pi \tag{12}$$

The other parameters ρ for P_{k+1} can be obtained by:

$$\rho_{k+1} = \ell(N_{k+1}) \tag{13}$$

$$z_{k+1} = g(\theta_{k+1}, \rho_{k+1}) \tag{14}$$

Thus, the coordinate of point $P_{k+1}(\theta_{k+1},\rho_{k+1},z_{k+1})$ can be obtained by combining the equations from Equation (8) to Equation (14). By conducting the iterative steps with respect to $N_1 = 0$, the whole toolpath can be adaptively resampled.

3.5. Tool Radius Compensation

Traditionally, there are two methods of tool nose radius compensation for diamond turning, xz-direction compensation [9,25] and z-direction compensation [10]. Generally, the xz-direction tool compensation is accomplished by the normal vector of the machined surface and cutting plane; however, the direction of normal vector of the freeform in the cutting plane periodically changes with angle, resulting in a slight periodic displacement of the tool in side-feeding motion. Especially for the machine tool whose spindle is mounted on the x-axis, the stability of side-feeding motion is particularly important to surface quality. Thus, we adopt the z-direction compensation in ATPG as shown in Figure 5.

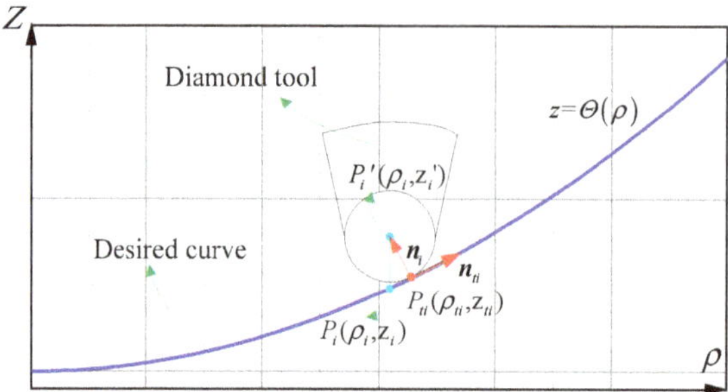

Figure 5. Schematic of tool radius compensation in the z-direction.

Without loss of generality, $P_i(\theta_i,\rho_i,z_i)$ is the control point, and $P_i'(\theta_i,\rho_i,z_i')$ is the corresponding coordinate of tool edge center, where z_i' is the only unknown parameter. The curve that is a cross-sectional profile passing through the rotation center O and P_i can be expressed as:

$$z = g(\theta, \rho)|_{\theta=\theta_i} = \Theta(\rho) \tag{15}$$

$P_{ti}(\theta_i,\rho_{ti},z_{ti})$ is assumed as the corresponding foot point on the curve $z = \Theta(\rho)$ of the $P_i'(\theta_i,\rho_i,z_i')$, where ρ_{ti}, z_{ti} are the unknown parameters as shown in Figure 5. According to the geometric constraint, assuming that a round edged diamond tool with a zero rake angle is adopted, the following relations can be obtained:

$$\begin{cases} \|\overline{P_i'P_{ti}}\| = R_t \\ n_i \cdot n_{ti} = 0 \\ z_{ti} = g(\theta_i,\rho_{ti}) \end{cases} \tag{16}$$

where $\|\cdot\|$ denotes the Euler distance between any two points, and n_i is the normal vector of the curve $z = \Theta(\rho)$ at the point $P_{ti}(\theta_i,\rho_{ti},z_{ti})$, which can be expressed as:

$$n_i = \left(\rho_i - \rho_{ti}, z_i' - z_{ti}\right) \tag{17}$$

where n_{ti} is the tangent vector of the curve $z = \Theta(\rho)$ at the point $P_{ti}(\theta_i,\rho_{ti},z_{ti})$, which can be expressed as:

$$n_{ti} = \left.\left(1, \frac{\partial\Theta}{\partial\rho}\right)\right|_{\rho=\rho_{ti}} \tag{18}$$

Thus, according to Equations (15), (17) and (18), Equation (16) can be solved by Newton's iteration method.

After the tool radius compensation, the whole tool path control points are generated.

4. Theoretical Investigation

4.1. Demonstration of Surface Generation in ATPG

To demonstrate the process of this novel method, a typical astigmatic surface was employed for investigation as shown in Figure 6, which can be mathematically described as $z = K\rho^2\sin(2\theta)$. To clearly show the difference between the ADPT and conventional method, the aperture of the workpiece was set as 5 mm. and the amplitude was set as 1mm. The tool nose radius was set as $R_t = 1.0$ mm with a zero degree rake angle, and the tool parameters have been checked by tool interference check.

Tolerances of interpolation error and cutting residual were set as $\varepsilon_f = 100$ nm and $\varepsilon_t = 100$ nm, respectively, based on the optical requirements.

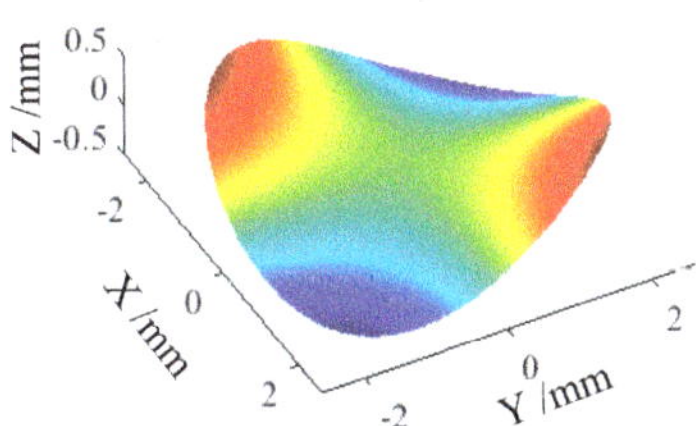

Figure 6. Schematic of desired surface (astigmatic surface).

We firstly calculated the feedrate for each revolution, and Figure 7 showed the calculation process of feedrate for radius ρ = 2 mm (reference radius). Firstly, according to the curvature of the cross-sectional profile passing through the rotation center and the equations mentioned above, the feedrate at different angles was calculated and shown in the blue dash line. It can be seen that the feedrate is different at different angles. This means that the motion of the x-axis was different with the position of a c-axis at this radius, which would cause micro-fluctuations of x-axis. In order to suppress these fluctuations and guarantee the accuracy of cutting residual, we set the feedrate equal to the minimum feedrate, as shown in the red line, ensuring that the x-axis remained uniform motion in all positions of c-axis on this radius. As in so many other revolutions, we can get f_i and N_i along the feeding direction on the whole surface.

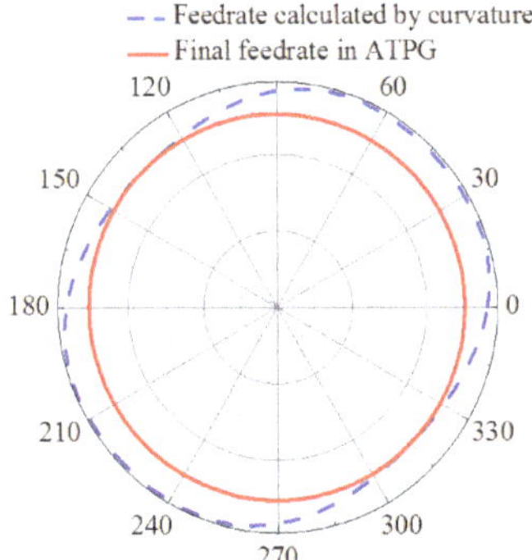

Figure 7. Schematic of feed calculation process at ρ = 2 mm.

After obtaining series of number pairs (N_i,ρ_i), which are shown in Figure 8 with the red dotted line, in order to suppress drastic fluctuations in the side-feeding motion of machine tool, we used a continuous and smooth curve to approximate these number pairs (N_i,ρ_i) by a numerical curve fitting method. This curve model can be any mathematical function. For this case, after comparing the fitting residuals of various functions, cubic polynomials were suitable because its residuals are the smallest, and it can be expressed by:

$$\rho = \ell(N) = aN^3 + bN^2 + cN + d \tag{19}$$

where polynomial coefficient $a = -4.9469 \times 10^{-8}$, $b = -8.774 \times 10^{-7}$, $c = 0.0263$, and $d = -1.141 \times 10^{-4}$. The sum of squares error (SSE) is 2.082×10^{-7}; this means that goodness of fit is very high.

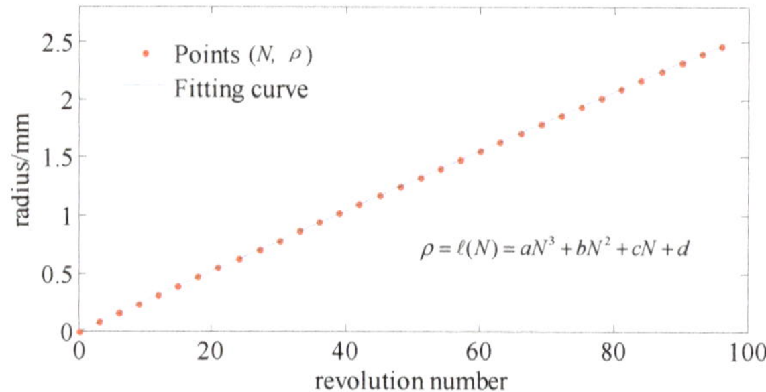

Figure 8. Curve fitting for (N_i, ρ_i).

Thus, we can obtain the whole surface control point coordinates by combining the equations from Equations (8) to (14).

After the tool radius compensation with solving Equation (16), the whole tool path control points were finished.

4.2. Characterization of Motion in ATPG

To characterize the motions of the machine tool, the sampling intervals for the x-axis and z-axis were extracted. The spindle speed was set as 500 rpm. The velocity for each axis is as shown in Figure 9. It is noted that the sampling strategy in the region near the center of rotation follows the constant angle strategy due to the too large interval for the given interpolation error.

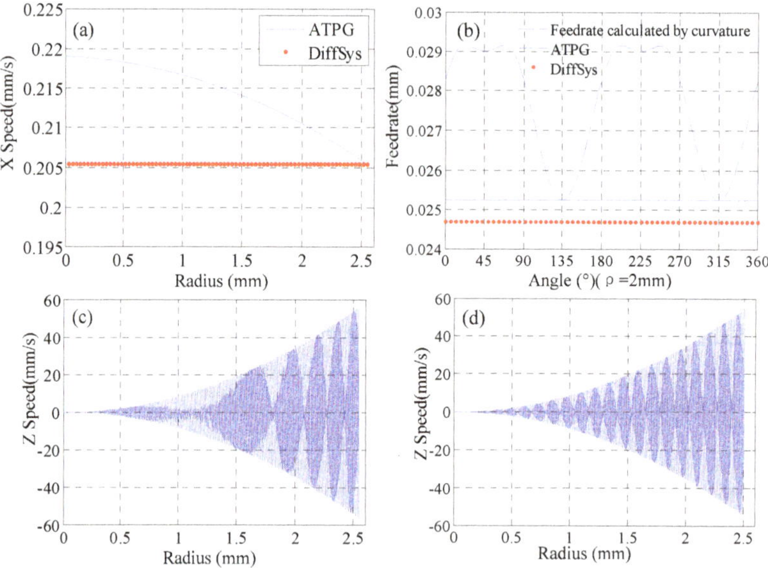

Figure 9. Features of the motion (**a**) side-feeding motion in the feeding direction; (**b**) feedrate with angles at ρ = 2mm in the cutting direction, translational motion along the z-axis in (**c**) ATPG and (**d**) DiffSys.

To have a comparison, the tool path was generated by both ATPG and conventional method, which is calculated by DiffSys (a commercial and professional CAM software for ultra-precision turning, version 3.98, WESTERN ISLE, North Wales, GB). DiffSys follows the constant-angle strategy which was based on the interpolation tolerance and the minimum radius of curvature of the part. It is clear that in the feeding direction the *x*-axis speed varies smoothly according to surface shape in ATPG from Figure 9a, and the speed does not present micro fluctuations during the whole process. This is due to the smooth curve fitting and no compensation on the *x*-axis direction in ATPG. Because the feedrate in DiffSys is defined by the minimum radius of curvature of the part, the feedrate of ATPG is greater than that of DiffSys on the whole part. Meanwhile, although in order to suppress the micro fluctuations of motion in the cutting direction, ATPG takes the minimum feedrate calculated according to the surface shape in every spiral path, it is still higher than that of DiffSys. Through Figure 9a,b, it is clear that the machining efficiency is improved in ATPG. Figure 9c,d show the speeds of the two methods in the *z*-axis, and there is no obvious difference in the *z*-direction.

We computed the intermediate "way-points" for each axis for each point along the spline path by Hermite spline interpolation. After subtracting the desired surface, error maps of the theoretically generated surface for ATPG and DiffSys are illustrated in Figure 10. It can be seen that all interpolation errors are restricted in $[-\varepsilon_f, +\varepsilon_f]$, and the interpolation errors are homogeneously distributed over the entire surface. For DiffSys which adopts the constant-angle methods, the surface quality of central region is better than that of the outer region, and in order to ensure the allowable interpolation error on the outer region, a large number of sampling points are needed on the outer region of the surface, which results in redundant control points on the center region. In this case, the number of control points in ATPG (3,920) was only about 64% of that in DiffSys (6061). Therefore, the machining time in DiffSys would be much longer than that in ATPG to have similar machining accuracy.

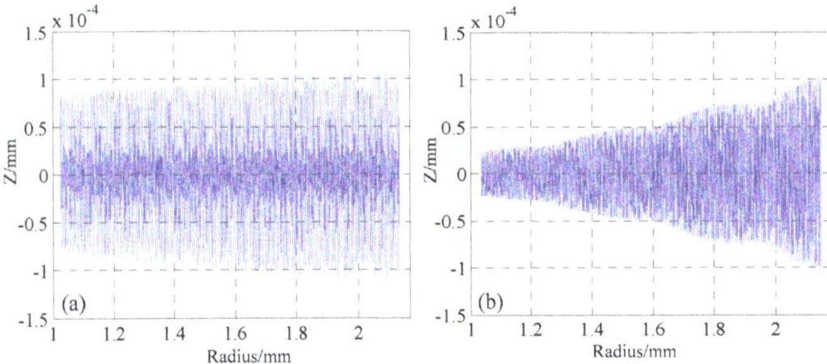

Figure 10. Characteristics of the generated surface error maps along the toolpath in (**a**) ATPG and (**b**) DiffSys.

5. Experimental Results and Discussion

5.1. Experimental Setup

The astigmatic (AST) surface was machined for investigating the performances. Considering the testing ability of the interferometer, the aperture of the workpiece was 50 mm, the amplitude was set as 4.8 μm, respectively. To have a comparison, the tool path was generated by both ATPG and DiffSys, and DiffSys follows the constant-angle strategy that was based on the interpolation tolerance and the minimum radius of curvature of the part. The feedrate in DiffSys was set as $f = 8.9$ μm/rev, which was based on the surface finish specification and the maximum slope of the part, and the tool radius compensation in DiffSys was set along the *z*-direction as well.

In order to validate the performance of ATPG for large aperture optical surfaces, an off-axis paraboloid (OAP) with an aperture of 213 mm was also machined, where the vertex curvature radius R was 915 mm and off-axis magnitude was 915 mm. We compared the programs generated both by ATPG and Diffsys for the same setting rule.

The cutting experiments were performed on a CNC ultra-precision lathe (Precitech Nanoform® 250 ultra, Keene, NH, USA). The configurations of this experiment were shown in Figures 11 and 12a.

A natural single crystal diamond tool with a nose radius of 1.04 mm and a zero rake angle (Contour Fine Tooling, Hertfordshire, UK) was used in this experiment. Before machining, the diamond tool parameters have been checked according to the sectional curve method. The workpiece materials are brass and oxygen-free high-conductivity copper (OFHC), respectively.

Figure 11. Hardware configurations of AST surface machining, where (1) diamond tool; (2) workpiece; (3) fixture; (4) spindle.

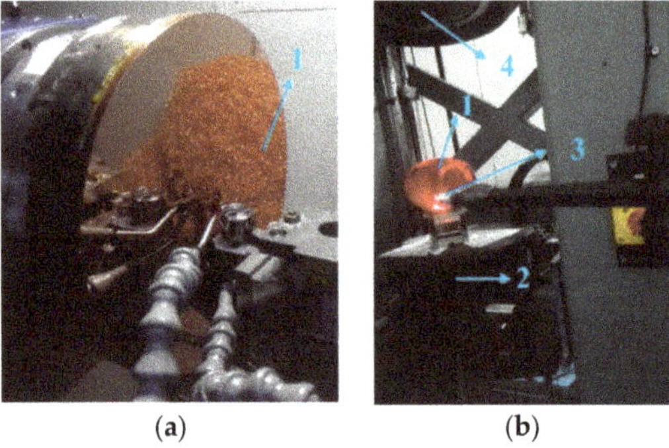

(a) (b)

Figure 12. (**a**) Hardware configurations of off-axis paraboloid (OAP) machining and (**b**) form test, where (1) workpiece; (2) adjustment mechanism; (3) standard ball; (4) planar interferometer.

After turning, a laser Interferometer (Zygo, MST, Middlefield, CT, USA) was employed to measure the machined surface form error. The AST surface was tested directly by the interferometer and the OAP was tested though optical properties of quadric surfaces as shown in Figure 12b. A white light interferometer (Bruker, ContourGT-K1, Madison, WI, USA) was employed to measure surface roughness.

5.2. Results and Discussion

After subtracting the design surface, the form errors for the two machined surfaces (AST) are respectively shown in Figure 13a,b. The roughness was measured by a white light interferometer and shown in Figure 13e,f, respectively. More details of the experimental results are summarized in Table 1. The total volume of control points for ATPG was 23,581, while that for DiffSys was 35,053. As shown in Table 1, it can be seen that the surface machined by ATPG is slightly better, but considering the fabrication and metrology uncertainty, there is no significant difference between the two methods for surface quality. Nevertheless, the number of control points generated by ATPG is 32.7% less than by DiffSys in this case. The machining efficiency is also improved, which is particularly important for large-aperture optical surfaces.

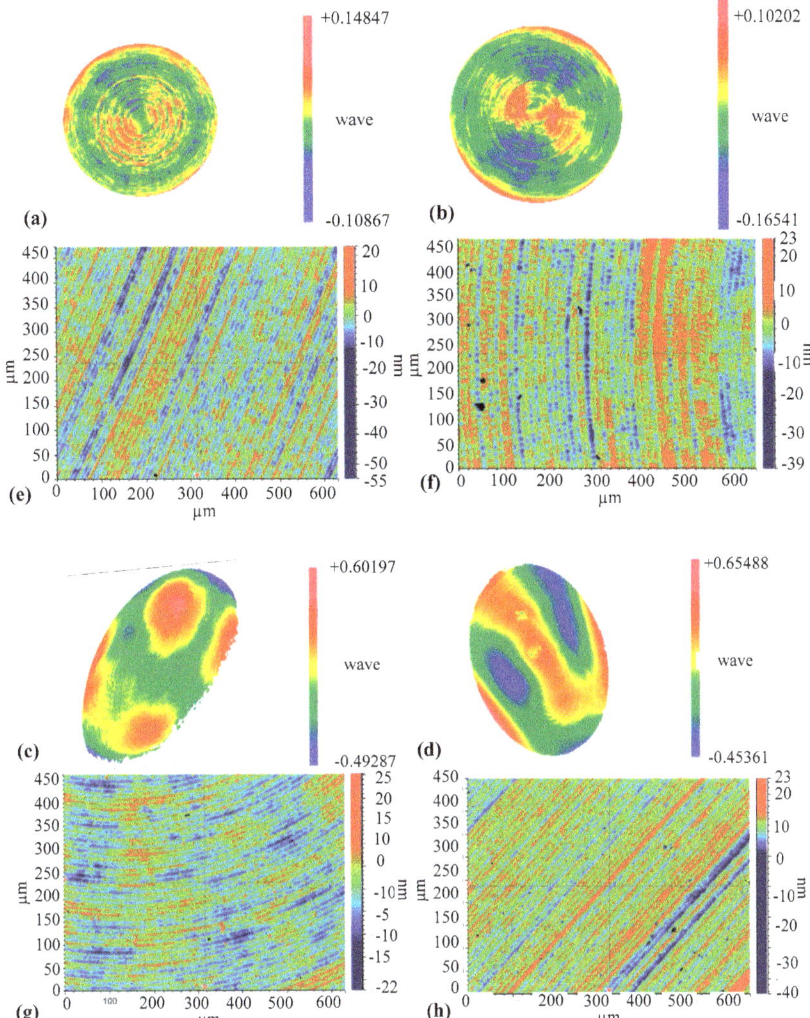

Figure 13. The form error obtained by (**a**) ATPG (AST), (**b**) DiffSys (AST), (**c**) DiffSys (OAP), (**d**) ATPG (OAP); and the roughness of the turned surfaces obtained by (**e**) ATPG (AST); (**f**) DiffSys (AST); (**g**) DiffSys (OAP); (**h**) ATPG (OAP).

Table 1. Result of the experiments.

Item	AST		OAP	
	ATPG	**DiffSys**	**ATPG**	**DiffSys**
Number of control points	23,581	35,053	281,141	592,541
Cutting time	5 min 30 s	8 min	1 h	2.2 h
PV error (nm)	163	169	701	693
Roughness R_a (nm)	4.8	4.9	4.6	4.5

The machined OAP surfaces were measured through optical properties of quadric surfaces as shown in Figure 12b and further shown in Figure 13c,d; the roughness was shown in Figure 13g,h.

Comparing the data in Table 1, the number of control points generated by ATPG is 52.5% less than by DiffSys in this case, and the roughness is still keeping the same level; this is very significant for large aperture optics to reduce the volume of NC program. Although the form errors exceed the preset value, it is mainly caused by low frequency components from the test data, which should be caused by clamping deformation. This is not the subject of this paper.

Generally speaking, more control points mean higher interpolation accuracy. However, as the number of points increases, the number of points is no longer the bottleneck for machining accuracy. Furthermore, because the new tool path makes the axis speed variation very small, the machine tool can adapt to a higher spindle speed. Hence, the comparison verifies the high machining efficiency of the ATPG for achieving the same surface quality. It is noted that the PV value of the form error and roughness is higher than the preset theoretical one. The reason is that, besides the accuracy of interpolation, factors that determine the surface form and roughness still include machine tool vibration, tool wear, lubrication and clamping deformation, etc.

6. Conclusions

This research proposes a novel adaptive tool path generation (ATPG) for freeform optics to improve machining efficiency. The main conclusions are as follows:

- With consideration of machine tool motion characteristics, an adaptive tool path generation is investigated, including establishing the functional relationship between spiral radius and revolution number with respect to cutting residual, smooth curve fitting and solving cutter contact points in the cutting direction with the consideration of interpolation error and tool radius compensation.
- A theoretical investigation for ATPG is conducted. The results indicate that the smooth curve produced by the spiral radius and revolution number can effectively suppress vibration of side-feeding motion, which can make the machining operated at a higher spindle speed.
- A comparison investigation on surface generation in ATPG and DiffSys was experimentally conducted on typical freeform surfaces. By adopting ATPG, the volume of tool path control points and machining time are effectively reduced, while surface form error as well as roughness is maintained. Therefore, the proposed ATPG significantly increases the machining efficiency for large aperture freeform optics.

Author Contributions: Conceptualization, Y.S. and H.Y.; Methodology, D.W.; Investigation, D.W.; Writing—Original Draft Preparation, D.W. and D.L.; Writing—Review & Editing, D.W. and D.L.

Funding: This research was funded by National Science and Technology Major Project of the Ministry of Science and Technology of China (2009ZX02205).

Acknowledgments: The authors would like to express their sincere thanks to the colleagues Dongqi Su and Shijun Peng for the test work of the experiments.

Conflicts of Interest: The authors declare no conflict of interest.

References

1. Benitez, P.; Minano, J.C.; Blen, J.; Mohedano, R.; Chaves, J.; Dross, O.; Hernandez, M.; Hernandez, W. Simultaneous multiple surface optical design method in three dimensions. *Opt. Eng.* **2004**, *43*, 1489–1502. [CrossRef]
2. Oliker, V.; Rubinstein, J.; Wolansky, G. Supporting quadric method in optical design of freeform lenses for illumination control of a collimated light. *Adv. Appl. Math.* **2015**, *62*, 160–183. [CrossRef]
3. Thompson, K.P.; Rolland, J.P. Freeform optical surfaces: A revolution in imaging optical design. *Opt. Photonics News* **2012**, *23*, 5–30. [CrossRef]
4. Ye, J.; Chen, L.; Li, X. Review of optical freeform surface representation technique and its application. *Opt. Eng.* **2017**, *56*, 110901. [CrossRef]

5. Challita, Z.; Agocs, T.; Hugot, E.; Jasko, A.; Kroes, G.; Taylor, W.D.; Miller, C.; Schnetler, H.; Venema, L.; Mosoni, L.; et al. Design and development of a freeform active mirror for an astronomy application. *Opt. Eng.* **2014**, *53*, 031311. [CrossRef]

6. Yang, T.; Cheng, D.; Wang, Y. Aberration analysis for freeform surface terms overlay on general decentered and tilted optical surfaces. *Optics Express* **2018**, *26*, 7751–7770. [CrossRef] [PubMed]

7. Matthias, B.; Johannes, H.; Thomas, P.; Christoph, D.; Andreas, G.; Sebastian, S.; Daniela, S.; Stefan, D.Z.; Ramona, E.; Andreas, T. Development, fabrication, and testing of an anamorphic imaging snap-together freeform telescope. *Appl. Opt.* **2015**, *54*, 3530–3542.

8. To, S.; Zhu, Z.; Wang, H. Virtual spindle based tool servo diamond turning of discontinuously structured micro-optics arrays. *CIRP Annals* **2016**, *65*, 475–478. [CrossRef]

9. Fang, F.; Zhang, X.; Hu, X. Cylindrical coordinate machining of optical freeform surfaces. *Opt. Express* **2008**, *16*, 7323–7329. [CrossRef]

10. Gong, H.; Fang, F.; Hu, X. Accurate spiral tool path generation of ultraprecision three-axis turning for non-zero rake angle using symbolic computation. *Int. J. Adv. Manuf. Technol.* **2012**, *58*, 841–847. [CrossRef]

11. Gong, H.; Wang, Y.; Song, L.; Fang, F.Z. Spiral tool path generation for diamond turning optical freeform surfaces of quasi-revolution. *Comput. Aided Des.* **2015**, *59*, 15–22. [CrossRef]

12. Wang, X.S.; Fu, X.Q.; Li, C.L.; Min, K. Tool path generation for slow tool servo turning of complex optical surfaces. *Int. J. Adv. Manuf. Technol.* **2015**, *79*, 437–448. [CrossRef]

13. Zhu, Z.W.; To, S. Adaptive tool servo diamond turning for enhancing machining efficiency and surface quality of freeform optics. *Opt. Express* **2015**, *23*, 20234–20248. [CrossRef] [PubMed]

14. Lin, R.S.; Koren, Y. Efficient tool-path planning for machining free-form surfaces. *J. Manuf. Sci. Eng.* **1996**, *118*, 20–28. [CrossRef]

15. Zhang, L.; Zhou, W.; Naples, J.N.; Yi, A.Y. Fabrication of an infrared Shack–Hartmann sensor by combining high-speed single-point diamond milling and precision compression molding processes. *Appl. Opt.* **2018**, *57*, 3598–3605. [CrossRef]

16. Li, Z.X.; Fang, F.Z.; Zhang, X.D.; Liu, X.L.; Gao, H.M. Highly efficient machining of non-circular freeform optics using fast tool servo assisted ultra-precision turning. *Opt. Express* **2017**, *25*, 25243–25256. [CrossRef]

17. Li, Z.X.; Fang, F.Z.; Chen, J.J.; Zhang, X.D. Machining approach of freeform optics on infrared materials via ultra-precision turning. *Opt. Express* **2017**, *25*, 2051–2062. [CrossRef]

18. Kwok, T.C.; Cheung, C.F.; Kong, L.; To, S.; Lee, W. Analysis of surface generationin ultra-precision machining with a fast tool servo. *Proc. Inst. Mech. Eng. Part B J. Eng. Manuf.* **2010**, *224*, 1351–1367. [CrossRef]

19. Yu, D.; Wong, Y.; Hong, G. Optimal selection of machining parameters for fast tool servo diamond turning. *Int. J. Adv. Manuf. Technol.* **2011**, *57*, 85–99. [CrossRef]

20. Lasemi, A.; Xue, D.; Gu, P. Recent development in CNC machining of freeform surfaces: A state-of-the-art review. *Comput. Aided Des.* **2010**, *42*, 641–654. [CrossRef]

21. Lin, Z.; Fu, J.; Shen, H.; Gan, W. A generic uniform scallop tool path generation method for five-axis machining of freeform surface. *Comput. Aided Des.* **2014**, *56*, 120–132. [CrossRef]

22. Neo, D.W.K.; Kumar, A.S.; Rahman, M. A novel surface analytical model for cutting linearization error in fast tool/slow slide servo diamond turning. *Precis. Eng.* **2014**, *38*, 849–860. [CrossRef]

23. Gottschling, J.; Saleem, M. *Mathematics for Engineers Part II (ISE)*; University Duisburg-Essen Press: Nordrhein-Westfalen, Germany, 2007; pp. 1, 11–15.

24. Gautschi, W. *Numerical Analysis*, 2nd ed.; Springer Science Business Media: New York, NY, USA, 2012; pp. 107–109.

25. Zhang, X.; Fang, F.; Wang, H.; Wei, G.; Hu, X. Ultra-precision machining of sinusoidal surfaces using the cylindrical coordinate method. *J. Micromech. Microeng.* **2009**, *19*, 054004. [CrossRef]

Technical Note

Experimental Investigation on Form Error for Slow Tool Servo Diamond Turning of Micro Lens Arrays on the Roller Mold

Yutao Liu, Zheng Qiao, Da Qu, Yangong Wu, Jiadai Xue, Duo Li * and Bo Wang *

Centre for Precision Engineering, Harbin Institute of Technology, 92 West Dazhi Street, Nan Gang District, Harbin 150001, China; 16B908095@stu.hit.edu.cn (Y.L.); qiaozhengyunlong@126.com (Z.Q.); bennyqu007@yahoo.com (D.Q.); yuguanzi@163.com (Y.W.); brucexjd@hit.edu.cn (J.X.)
* Correspondence: duo.kevin.li@gmail.com (D.L.); bradywang@hit.edu.cn (B.W.)

Received: 27 August 2018; Accepted: 21 September 2018; Published: 25 September 2018

Abstract: Slow tool servo (STS) assisted ultra-precision diamond turning is considered as a promising machining process with high accuracy and low cost to generate the large-area micro lens arrays (MLAs) on the roller mold. However, the chatter mark is obvious at the cut-in part of every machined micro lens along the cutting direction, which is a common problem for the generation of MLAs using STS. In this study, a novel forming approach based on STS is presented to fabricate MLAs on the aluminum alloy (6061) roller mold, which is a high-efficiency machining approach in comparison to a traditional method based on STS. Based on the different distribution patterns of the discrete point of micro lens, the equal-arc method and the equal-angle method are also proposed to generate the tool path. According to a kinematic analysis of the cutting axis, the chatter mark results from the overlarge instantaneous acceleration oscillations of the cutting axis during STS diamond turning process of MLAs. Cutting parameters including the number of discrete points and cutting time of every discrete point have been experimentally investigated to reduce the chatter mark. Finally, typical MLAs (20.52-μm height and 700-μm aperture) is successfully machined with the optimal cutting parameters. The results are acquired with a fine surface quality, i.e., form error of micro lenses is 0.632 μm, which validate the feasibility of the new machining method.

Keywords: slow tool servo; ultra-precision diamond turning; micro lens arrays (MLAs); chatter mark; forming method

---

1. Introduction

In recent years, microstructure surfaces, such as micro lens arrays (MLAs) with spherical shape and accurate position, have been extensively used in a series of areas, including liquid crystal display (LCD), illumination, data storage, sensor devices, and so forth, due to their excellent optical features [1–4]. Therefore, many researchers have paid attention to exploring more controllable and efficient methods to fabricate MLAs. Investigations showed that the roll-to-roll (RTR) replication technique [5–8], which is a typical continuous forming process, is regarded as a promising approach with high accuracy, high efficiency, and low cost, to generate various microstructure surfaces in comparison with plane replication technology. The cylindrical roller mold containing millions of micro lenses plays a key role in the RTR technique. Although various fabrication methods have already been proposed, including chemical etching [9], laser lithography [10], diamond broaching [11], and end-fly-cutting [12], the methods mentioned are not suitable for fabricating MLAs on the cylindrical roller mold due to their limitations. Micro-milling is also used for fabricating various microstructures on flat or roll molds [13–16], which can cut nonferrous metal and tool steel. But machining efficiency of the micro-milling process is too low to fabricate large-area micro lens arrays.

Up to now, ultra-precision diamond turning using a slow/fast tool servo is regarded as a state-of-the-art method to generate MLAs on the cylindrical roller mold [17–20]. To fabricate MLAs with a high aspect ratio (AR), the clearance angle of the cutting tool has been increased via lowering the position of the cutting tool tip relative to the rotation center of the workpiece in the vertical direction [21]. Based on this novel machining method, the interference phenomenon between the flank surface of the diamond tool and the machined surface can be eliminated when MLAs with high AR are processed using slow tool servo (STS) diamond turning on the roller mold. As a consequence, the AR of MLAs can be increased to about five times larger than previous fabrication results with the method of traditional diamond turning using the same diamond tool. In addition, to solve the problem of cutting tool wear during large-area MLAs machining over a roller mold, Gao et al. [22,23] have developed a fabrication method of tool replacement stitching. The contact force between the cutting tool tip and the roll surface was utilized to scan the interrupted point on the roll surface by a force sensor integrated on the fast tool servo (FS-FTS) system. A sub-micrometer cutting tool positioning for the new replaced cutting tool can be realized derived from the proposed method. Kong et al. [24] have presented a novel orthogonal slow tool servo (OSTS) assisted diamond turning approach. In addition, a tool path generator based on OSTS has been developed for manufacturing different wavy microstructures. In such a manner, the various wavy patterns with fine surface finish can be machined stably on a roller surface. To fabricate the complex microwave pattern of optical films on a Ni-coated steel roller mold, Lee et al. [25] have developed a right/left-horizontal swing fast tool servo (HFTS) with different hinge structures. The simulated analysis and experimental results have shown that HFTS with a single fixed hinge is more appropriate for precision manufacturing of the optical microstructure than HFTS with double fixed hinges because the former can avoid over-constraint conditions. Both the OSTS and the HFTS machining approaches provide feasibility for the fabrication of various wave prism patterns.

Although sufficient research has been dedicated to the generation of microstructures using STS/FTS assisted diamond turning, most of these aimed to manufacture MLAs on the planar mold rather than the roller mold. And the trajectory method is always used for processing MLAs on the planar or cylindrical mold [18,26,27]. The productivity of STS diamond turning is restricted by the preceding tool path. What is more, the obvious chatter mark occurs at the cut-in part of every processed micro lens along the cutting direction, which is a general problem for machining MLAs using STS [21,28]. This defect reduces the surface quality of micro lens and affects its optical properties. But the problem has received relatively little attention in previous research.

In this study, a new machining method based on STS diamond turning is proposed to generate MLAs on the roller mold. Two distribution strategies of the discrete point are defined to generate the tool path. Furthermore, the effects of cutting parameters, such as the number of discrete points and cutting time of every discrete point on the chatter mark of micro lens surface, are analyzed in detail. Finally, a series of cutting experiments are carried out to validate the effectiveness of the proposed methodology and analysis.

2. A New Fabrication Approach of MLAs

2.1. Principle of Machining Method

In the conventional STS diamond turning process, trajectory method is usually adopted to fabricate MLAs on the planar or cylindrical mold. In other words, every micro lens is determined by multiple cutting to generate the complete profile at the axial and circumferential direction, as shown in Figure 1. Therefore, the application of the STS method is already restricted by the relatively low production efficiency of this machining method in the research field of optical manufacturing. To overcome the drawback associated with the traditional trajectory method, a novel and high-efficiency machining approach of MLAs on the roller mold is presented in this study. As illustrated in Figure 2, the forming method based on STS diamond turning was used for the generation of MLAs on the roller mold. Single cutting at the axial direction can process an entire micro lens, so it decreases the period of the machining process. The production efficiency of MLAs can be improved dozens of times in comparison with the conventional trajectory method based on STS diamond turning and micro-milling [14,27].

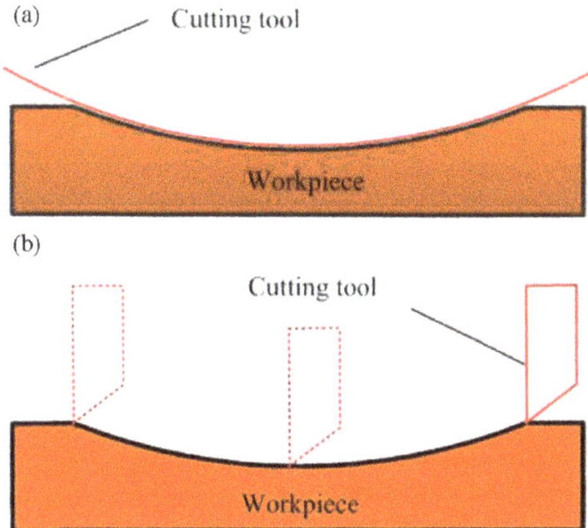

Figure 1. Schematic of the trajectory method. (**a**) The axial direction; (**b**) the circumferential direction.

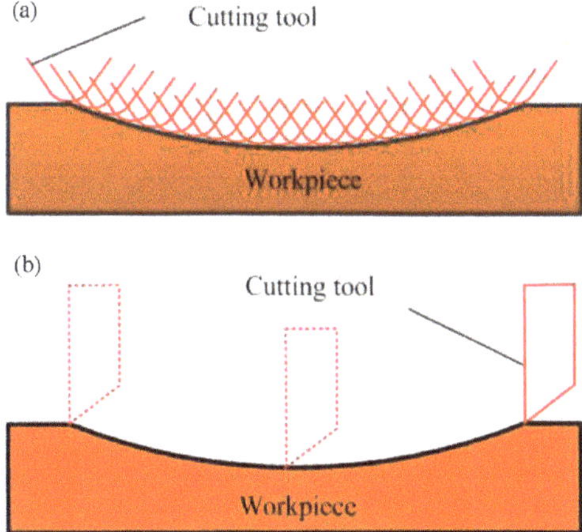

Figure 2. Schematic of the forming method. (**a**) The axial direction; (**b**) the circumferential direction.

2.2. Tool Path Selection

To generate the tool path for the machining of MLAs on the roller mold, the equal-arc method and the equal-angle method are presented in this work, which is based on the various distribution patterns of the discrete point of micro lens. As for the equal-arc method, the arc length of micro lens between any two adjacent discrete points is defined to be equal. While the rotation angle of the C-axis from one discrete point to the next discrete point is different, i.e., $\alpha_1 \neq \alpha_2$ and $\alpha_3 \neq \alpha_4$, as shown in Figure 3. And cutting time of every discrete point is set as a constant value, which is determined by the machining parameters. Therefore, the velocity of the C-axis of the motion system is also varied between the different discrete points during processing. So, this motion control strategy will result in frequent acceleration and deceleration of the spindle during machining. Additionally, because the weight of the roller mold is normally hundreds of kilogram or even a ton, frequently changing speed would greatly affect the dynamic behavior and accuracy of the motion control in the case of the heavy load.

To overcome the deficiency of the equal-arc method, the equal-angle method is developed in the subsequent study. In this motion control strategy, the rotation angle of the C-axis for adjacent discrete points is the identical value, i.e., $\beta_1 = \beta_2$ and $\beta_3 = \beta_4$, as shown in Figure 4. Therefore, the spindle has uniform motion between the different discrete points during processing. In terms of the control and motor system, such a strategy is considered more reasonable and easier to implement. So, the equal-angle method is utilized to fabricate MLAs in all experiments.

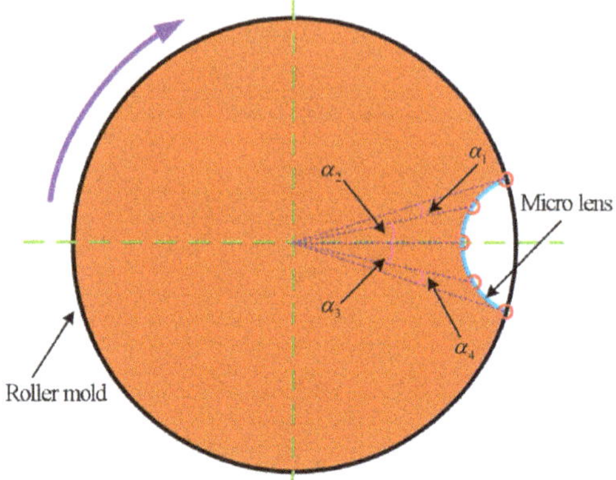

Figure 3. The equal-arc method.

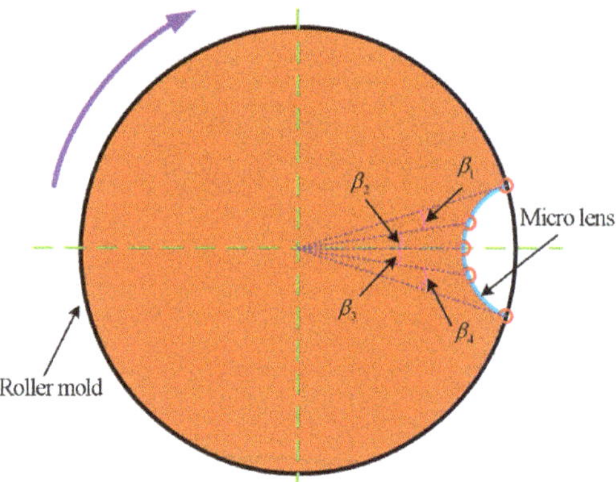

Figure 4. The equal-angle method.

3. Experimental Setup

To validate the effectiveness of the proposed new STS assisted ultra-precision diamond turning method, MLAs were generated using the in-house developed ultra-precision horizontal drum roll lathe (including the linear X-, Z-, and the rotatory C-axes) in this study, as shown in Figure 5. The specification of the ultra-precision drum roll lathe is summarized in Table 1. The machining principle of STS assisted ultra-precision diamond turning of MLAs is shown in Figure 6. Both ends of the roller mold are fixed on the headstock and tailstock spindle through four-jaw independent chuck, respectively. Among all those axes, the X-axis of the machine system is used as the cutting axis, which controls the cutting depth and aperture of a single micro lens. The pitch between two adjacent micro lenses in the radial direction is determined by the rotary motion of the C-axis. The pitch between two adjacent micro lenses in the axial direction, which is parallel to the centerline of the roller mold, is controlled by linear motion of the Z-axis. In addition, the X-axis and C-axis are synchronized to form the designed micro lens via the Universal Motion and Automation Controller (UMAC) in the circumferential direction. Therefore, the spindle speed is related to the diameter of roller mold, number of discrete points and cutting time of every discrete point. When a circle of micro lenses is finished on the roller mold, the Z-axis moves the distance of an axial pitch. Then the next circle of micro lenses can be processed sequentially.

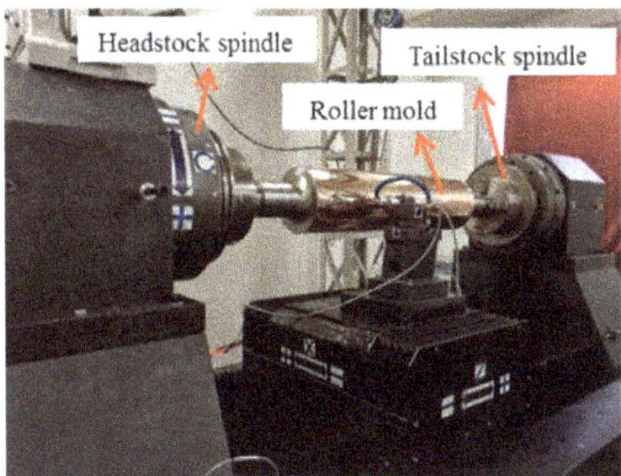

Figure 5. Experimental setup for the machining process.

Table 1. Specification of the machine tool.

The Machine Tool	Values
Positioning accuracy	C-axis: ±3 arc s (compensated) X-axis: 0.73 μm/200 mm (compensated) Z-axis: 0.95 μm/1100 mm (compensated)
Repetitive positioning accuracy	C-axis: ±2 arc s (compensated) X-axis: 0.63 μm/200 mm (compensated) Z-axis: 0.88 μm/1100 mm (compensated)

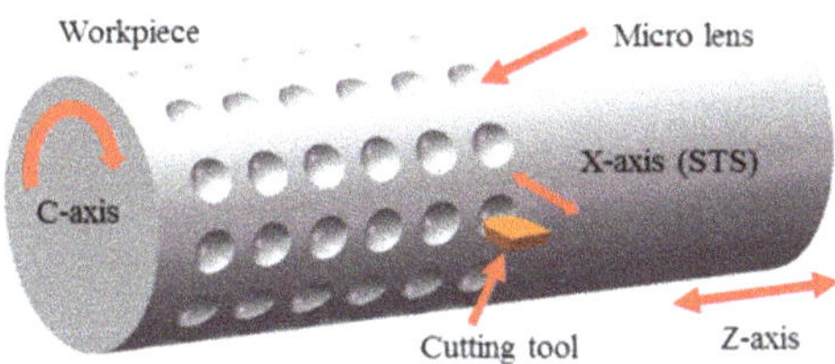

Figure 6. Schematic of slow tool servo (STS) diamond turning of micro lens arrays (MLAs).

Before the fabrication of MLAs, the surface of the roller mold must be machined to eliminate both imbalances from itself and installation error. Generally, this process consists of two parts, including rough machining and finish machining. In the first stage of rough machining, a cutting depth of 15 μm and a feed rate of 20 μm/rev are adopted to remove materials rapidly using a round-shaped polycrystalline diamond (PCD) tool. In the second stage for finish machining, a cutting depth of 5 μm and a feed rate of 5 μm/rev are applied to generate a mirror-like surface with a round-shaped natural diamond (ND) tool. Finally, MLAs are machined using the above-mentioned ND tool, which was used for the mirror finishing. The purpose of this operation is to avoid the error which results from repetitive tool setting. The machining accuracy could be affected by the changes of the cutting depth and the aperture of every micro lens, which are derived from the above tool setting error.

In terms of the MLAs experiments, the experimental conditions including the dimension of the cutting tool, cutting parameters and size of micro lens are listed in Table 2. In general, the size of the roller mold was too large to directly measure MLAs on its surface using existing measuring instrument in the laboratory. To solve this problem, MLAs with the same aperture were processed using the same cutting parameters by the above machine tool on a small roller mold, which could be directly inspected by commercial ultra-precision 3-D metrology devices, as shown in Figure 7.

Table 2. Experimental conditions for generation of micro lens arrays (MLAs).

The Cutting Tool	Values
Tool material	Single-crystal diamond
Tool nose radius r_t	2.995 mm
Tool rake angle α_t	0°
Tool clearance angle γ_t	8°
The Cutting Parameters	
Number of discrete points n	100, 200, 400, 600 and 800
Cutting time of every discrete point t	0.5, 1, 1.5, 2 and 2.5 ms
Workpiece material	Aluminum alloy (6061)
Lubricant	ISOPAR H
The Size of Micro Lens	
Height of the micro lens	20.52 μm
Aperture of micro lens	700 μm
The pitch of radial direction	1 mm
The pitch of axial direction	1 mm

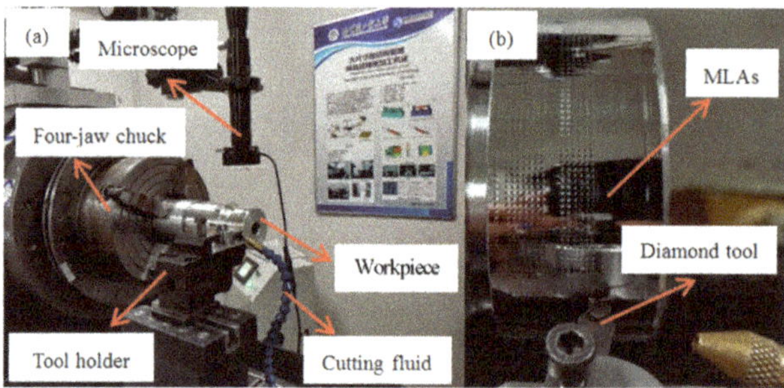

Figure 7. MLAs on the small workpiece. (**a**) The machining system; (**b**) partial enlarged view.

4. Results and Discussion

4.1. Form Error of Micro Lens

As illustrated in Figure 8a, the three-dimensional topography of the machined micro lenses was measured using a white light interferometer (Zygo Newview 8200 from Zygo Corp., Middlefield, CT, USA). Wavy distortions can be clearly observed at the cut-in part along the cutting direction. The form error between the machined micro lens surface and an ideal spherical surface ($R = 2.995$ mm) was derived using a post-process software (Metropro, from Zygo Corp., Middlefield, CT, USA) and revealed in Figure 8b, appearing to obvious chatter mark in the same part. Owing to the location of both the distortion and the chatter mark which are a one-to-one correspondence, it is expected that the distortions of micro lens surface are caused by the chatter mark. It is a general problem with the fabrication of MLAs using trajectory method or forming method based on STS diamond turning.

To investigate the generation of the chatter mark, the real-time motion information of the cutting axis (X-axis) was collected synchronously via the UMAC during processing, including actual acceleration, follow error, actual position, and command position, as revealed in Figure 9. In addition, the green lines as highlighted in Figure 9b represent the surface of the workpiece, i.e., the initial position of the cutting process for every micro lens. As observed in Figure 9a, there are a few overlarge acceleration oscillations occurring for the X-axis in the MLAs processing. Further, the fluctuations of the follow error are accordingly caused by the acceleration oscillations. It can be seen that the maximum amplitude of follow error is up to several micrometers, but the oscillation amplitude will decrease over time. Even so, once the large fluctuations of follow error occur in the cutting area, i.e., the upper part of the green line in Figure 9b, both the actual position and the command position of the cutting axis (X-axis) are obviously misaligned, as highlighted with a black dotted circle in Figure 9b. This phenomenon will result in the distortion of the profile for machined micro lenses. Thus, it can be determined that excessive follow error is reflected to the surface of micro lens in the form of a chatter mark. Further, the chatter mark can be attributed to an overlarge instantaneous acceleration and deceleration of the cutting axis (X-axis) of the machine tool during processing.

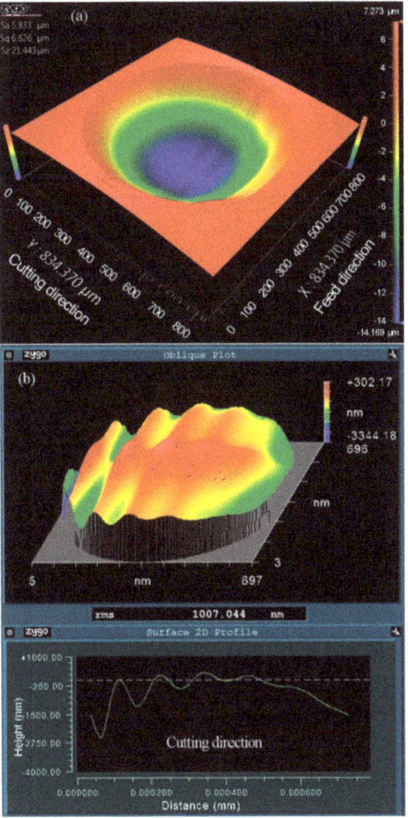

Figure 8. Zygo photographs of micro lens. (**a**) 3-D topography of machined micro lens; (**b**) The form error of machined micro lens surface from a designed sphere ($n = 100$, $t = 1$ ms).

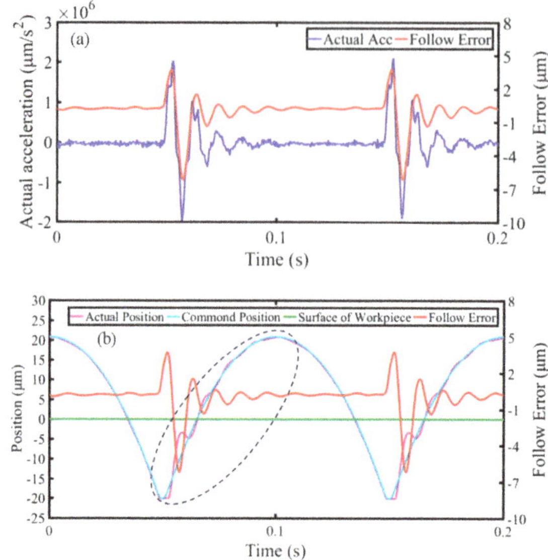

Figure 9. The collected real-time kinematic information of the cutting axis. (**a**) Actual acceleration and follow error; (**b**) Follow error, actual, and command position.

4.2. Effect of the Number of Discrete Points

In this section, the effects of the number of discrete points for the form error of micro lens will be investigated in detail. The corresponding experiments were carried out on the aluminum alloy (6061) roller mold using the STS approach based on the self-developed ultra-precision horizontal drum roll lathe. The number of discrete points is 100, 200, 400, 600, and 800 for a single micro lens, and the cutting time of every discrete point is 1 ms. The size of machined micro lenses is shown in Table 2. The plot of form error of machined micro lenses at a various number of discrete points is shown in Figure 10. In this study, Root-Mean-Square (RMS) is used for characterizing the form error of machined micro lenses. In addition, the measured data were processed using a Gaussian low-pass filter provided by Metropro before the calculation of RMS. The mathematical definition and the digital implementation of RMS can be formulated as:

$$Z_{RMS} = \sqrt{\frac{1}{N}\sum_{n=1}^{N}|z_n|^2} \tag{1}$$

where N is the number of the measurement point of every micro lens and z_n the form error of the profile at point number n.

As it can be seen from Figure 10, when the number of discrete points is 100, RMS is more than 1 µm. The shape of every micro lens distorts at the cut-in part due to the generation of the overlarge chatter mark, as shown in Figure 8b. With the increasing of the number of discrete points, RMS decreased obviously. Furthermore, MLAs have a fine surface quality with 0.724 µm RMS when the discrete point is 400, i.e., RMS is decreased by about 28.10%, as shown in Figure 11. No chatter mark appears at the cut-in part along the cutting direction. But the surface quality of MLAs could not be improved significantly when the discrete point is 800, i.e., RMS is reduced by about 31.88% (RMS = 0.686 µm).

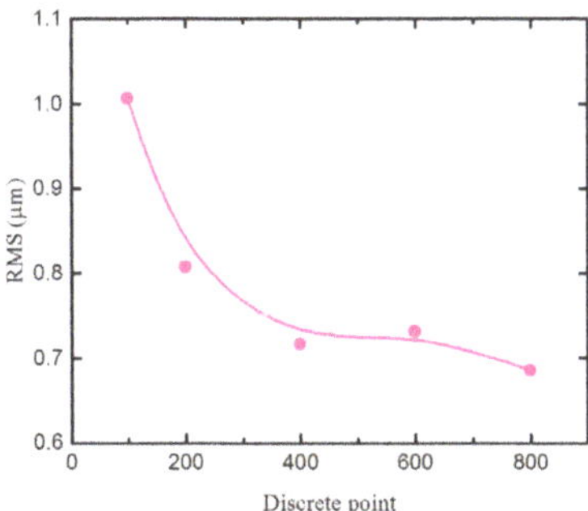

Figure 10. Form error of micro lenses machined at various discrete points.

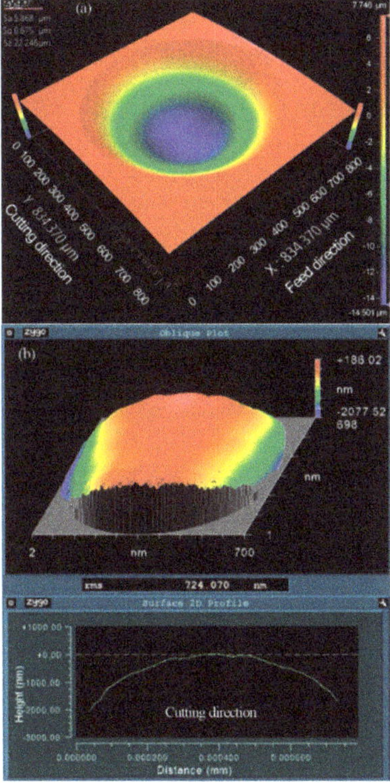

Figure 11. Zygo photographs of micro lens. (**a**) 3-D topography of machined micro lens; (**b**) The form error of machined micro lens surface from a designed sphere ($n = 400$, $t = 1$ ms).

According to the above experimental results, the surface quality can be improved effectively by increasing the number of discrete points of every micro lens. This is because as the number of discrete points increases, the cutting depth at each discrete point decreases accordingly. So, when the cutting time of every discrete point is a constant value, acceleration and deceleration of the cutting axis (X-axis) also reduce during processing. Then above overlarge acceleration oscillations of the cutting axis can be eliminated, and so do fluctuations of its follow error, when the number of discrete points is sufficient. Consequently, the micro lens with no surface chatter mark can be fabricated. It is noted that too many discrete points do not always reduce form error significantly. After the chatter mark disappears on the surface of micro lens, the number of discrete points is no longer the main factor affecting surface quality.

4.3. Effect of Cutting Time of Every Discrete Point

The cutting time of every discrete point is another major cutting parameter for MLAs machining. Cutting time of every discrete point and the number of discrete points together determine the machining speed of the micro lens. To analyze the influences of the cutting time of every discrete point on form error of micro lens, some cutting experiments were conducted at various cutting times when the number of discrete points was a constant value. The form error is not significantly improved when the number of discrete point is greater than 400. Taking the machining efficiency into account, the number of discrete points is 400 for a single micro lens. The cutting time of every discrete point is 0.5, 1, 1.5, 2, and 2.5 ms. The plot of form error of machined micro lens with different cutting times of every discrete point is shown in Figure 12. It can be seen that increasing the cutting time has a certain impact on improving form error of micro lens. When the cutting time of every discrete point is 0.5 ms, RMS is 0.746 µm. As the cutting time of every discrete point increases, RMS decreases slightly. Furthermore, when the cutting time of every discrete point is 1.5, 2, and 2.5 ms, RMS are 0.632, 0.651, and 0.626 µm, respectively.

This is because increasing the cutting time of every discrete point can further reduce acceleration and deceleration of the cutting axis when the cutting depth is the same value. In addition, fluctuations of follow error of the cutting axis also further decrease correspondingly. Thus, the surface quality of micro lens can be improved further. Similarly, too much cutting time of every discrete point do not decrease form error significantly.

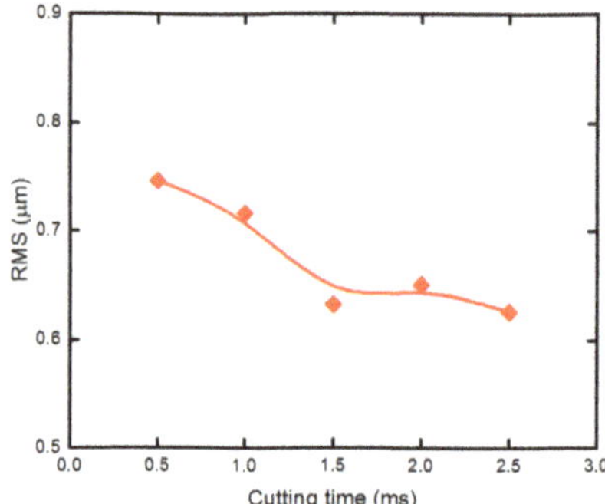

Figure 12. Form error of micro lenses machined at different cutting time of every discrete point.

4.4. Effect of Cutting Depth

The amplitude of the form error is greater at the cut-in part and cut-off part in comparison to other cutting areas along the cutting direction. It is an obvious feature in the 2D sectional profile of the form error of machined micro lens surface, as shown in Figures 8b and 11b. This is because cutting depth of every discrete point is different when the equal-angle method is used during processing, as shown in Figure 13. The maximum value of cutting depth occurs at the first and last discrete point; the minimum value of cutting depth appears at the bottom of every micro lens. With the increasing cutting depth, the cutting force is also bigger. Certainly, the cutting force will determine the elastic recovery of materials and relieving amount, i.e., bigger cutting force generates greater elastic recovery and relieving amount after cutting [29]. The form accuracy of micro lens will degrade due to the existence of elastic recovery and relieving amount. Therefore, the form error at the cut-in part and cut-off part is larger under the influence of bigger elastic recovery and relieving amount.

Based on the discussion above, the optimal cutting parameters were adopted for generating typical MLAs (20.52-µm height and 700-µm aperture), which took into account form accuracy, as well as machining efficiency. The number of discrete points was 400, and the cutting time of every discrete point was 1.5 ms. The measurement result is shown in Figure 14 with 0.632 μm RMS and no chatter mark was observed. Compared with the previous research, both aperture and height of the machined micro lens vary from different researchers. So, RMS cannot be directly compared to evaluate the form error of micro lens. In terms of chatter mark, the acquired surface quality is similar to or even better than the conventional trajectory method based on FTS/STS diamond turning of MLAs [21,28]. Therefore, the experimental result proves the effectiveness of the novel forming method based on STS diamond turning. The extendable fabrication of similar microstructure surfaces can be also performed in this way.

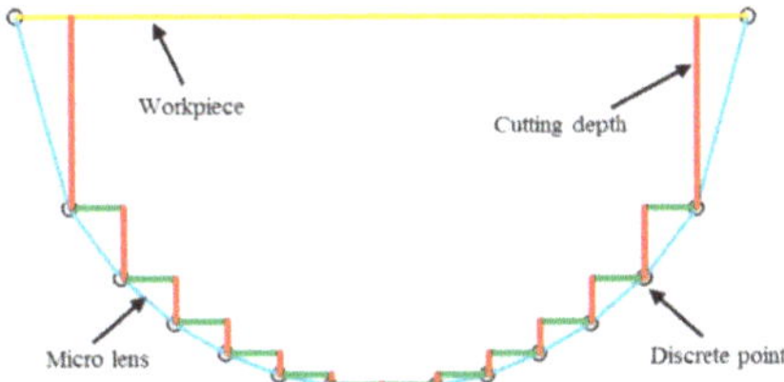

Figure 13. Schematic of the change trend of cutting depth for the equal-angle method.

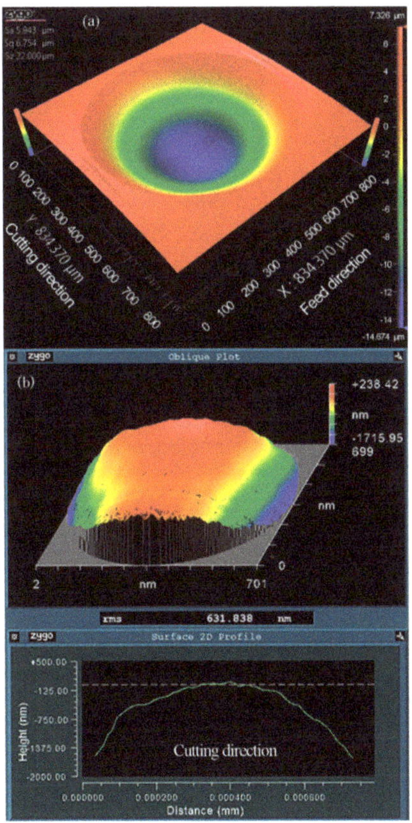

Figure 14. Zygo photographs of micro lens. (**a**) 3-D topography of machined micro lens; (**b**) The form error of machined micro lens surface from a designed sphere ($n = 400$, $t = 1.5$ ms).

5. Conclusions

A novel forming method based on STS diamond turning is presented to generate MLAs on the roller mold, which addresses the inherent drawback of relatively low production efficiency associated with traditional trajectory method. The MLAs (20.52-μm height and 700-μm aperture) is successfully machined without a chatter mark using the optimal cutting parameters. The proposed method can be extensively applied to other microstructures. According to the above experimental results, the conclusions can be presented as follows:

1. Taking into account the capacity of the control and motor system, the equal-angle method is recommended during processing. In this tool path, there is no frequent speed changes of the C-axis in the case of the heavy load during processing;
2. According to the kinematic analysis of the cutting axis, the chatter mark can be attributed to the overlarge instantaneous acceleration oscillations of the cutting axis of the machine tool during STS diamond turning of MLAs;
3. Increasing the number of discrete points of every micro lens will reduce the form error effectively. Furthermore, when the number of discrete points is greater than 400, MLAs have a fine surface quality without the chatter mark. And when the discrete point is 800, the form error of machined micro lens is reduced by about 31.88%, i.e., RMS is 0.686 μm;
4. Increasing the cutting time of every discrete point also has a certain impact on improving the surface quality of micro lens. When the cutting time of every discrete point is 2.5 ms, the form error of machined micro lens is reduced by about 37.84%, i.e., RMS is 0.626 μm. But too much cutting time of every discrete point does not decrease form error significantly.

6. Future Work

In the current study, although the forming method can be used for generating the MLAs with no chatter mark efficiently. As for the form accuracy of the machined micro lens, the effect of elastic recovery and relieving amount for the form accuracy cannot be neglected. If both factors can be studied and compensated, the form accuracy is likely to be improved further. In the future research, compensation strategy of form accuracy will become the focus based on the forming method.

Author Contributions: Investigation, Y.L., D.L., Z.Q. and D.Q.; Methodology, Y.W. and J.X.; Validation, B.W.; Writing for original draft, Y.L.; Writing for review & editing, Y.L. and D.L.

Funding: This work was supported by National Science and Technology Major Project of High-end CNC Machine Tools and Basic Manufacturing Equipment of China (grant number 2011ZX04004–021).

Acknowledgments: The authors would like to thank the fund from National Science and Technology Major Project of High-end CNC Machine Tools and Basic Manufacturing Equipment of China (grant number 2011ZX04004-021).

References

1. Evans, C.J.; Bryan, J.B. "Structured", "textured" or "engineered" surfaces. *CIRP Ann.* **1999**, *48*, 541–556. [CrossRef]
2. Bruzzone, A.A.G.; Costa, H.L.; Lonardo, P.M.; Lucca, D.A. Advances in engineered surfaces for functional performance. *CIRP Ann.* **2008**, *57*, 750–769. [CrossRef]
3. Fang, F.Z.; Zhang, X.D.; Weckenmann, A.; Zhang, G.X.; Evans, C. Manufacturing and measurement of freeform optics. *CIRP Ann.* **2013**, *62*, 823–846. [CrossRef]
4. Yuan, W.; Li, L.-H.; Lee, W.-B.; Chan, C.-Y. Fabrication of microlens array and its application: A review. *Chin. J. Mech. Eng.* **2018**, *31*, 16. [CrossRef]
5. Lee, X.H.; Moreno, I.; Sun, C.C. High-performance led street lighting using microlens arrays. *Opt. Express* **2013**, *21*, 10612–10621. [CrossRef] [PubMed]
6. Oh, J.S.; Song, C.K.; Hwang, J.; Shim, J.Y.; Park, C.H. An ultra-precision lathe for large-area micro-structured roll molds. *J. Korean Soc. Precis. Eng.* **2013**, *30*, 1303–1312. [CrossRef]

7. Wang, M.W.; Tseng, C.C. Analysis and fabrication of a prism film with roll-to-roll fabrication process. *Opt. Express* **2009**, *17*, 4718–4725. [CrossRef] [PubMed]

8. Nieslony, P.; Krolczyk, G.M.; Wojciechowski, S.; Chudy, R.; Zak, K.; Maruda, R.W. Surface quality and topographic inspection of variable compliance part after precise turning. *Appl. Surf. Sci.* **2018**, *434*, 91–101. [CrossRef]

9. Qu, P.; Chen, F.; Liu, H.; Yang, Q.; Lu, J.; Si, J.; Wang, Y.; Hou, X. A simple route to fabricate artificial compound eye structures. *Opt. Express* **2012**, *20*, 5775–5782. [CrossRef] [PubMed]

10. Radtke, D.; Zeitner, U.D. Laser-lithography on non-planar surfaces. *Opt. Express* **2007**, *15*, 1167–1174. [CrossRef] [PubMed]

11. Li, L.; Yi, A.Y. Design and fabrication of a freeform microlens array for a compact large-field-of-view compound-eye camera. *Appl. Opt.* **2012**, *51*, 1843–1852. [CrossRef] [PubMed]

12. Zhu, Z.; To, S.; Zhang, S. Large-scale fabrication of micro-lens array by novel end-fly-cutting-servo diamond machining. *Opt. Express* **2015**, *23*, 20593–20604. [CrossRef] [PubMed]

13. Zhao, Z.; Fu, Y.; Liu, X.; Xu, J.; Wang, J.; Mao, S. Measurement-based geometric reconstruction for milling turbine blade using free-form deformation. *Measurement* **2017**, *101*, 19–27. [CrossRef]

14. Scheiding, S.; Yi, A.Y.; Gebhardt, A.; Loose, R.; Li, L.; Eberhardt, R.; Tünnermann, A. Diamond milling or turning for the fabrication of micro lens arrays: Comparing different diamond machining technologies. In Proceedings of the Advanced Fabrication Technologies for Micro/Nano Optics and Photonics IV, San Francisco, CA, USA, 22–27 January 2011; Volume 7927, p. 79270N.

15. Gao, P.; Liang, Z.; Wang, X.; Zhou, T.; Xie, J.; Li, S.; Shen, W. Fabrication of a micro-lens array mold by micro ball end-milling and its hot embossing. *Micromachines* **2018**, *9*, 96. [CrossRef]

16. Wojciechowski, S.; Wiackiewicz, M.; Krolczyk, G.M. Study on metrological relations between instant tool displacements and surface roughness during precise ball end milling. *Measurement* **2018**, *129*, 686–694. [CrossRef]

17. Fan, Z.; Dai, Y.; Guan, C.; Tie, G.; Zhu, D. Ultra-precision turning processing technique of microstructure array on cylindrical surface. *Nanotechnol. Precis. Eng.* **2014**, *12*, 358–364.

18. Zhu, W.-L.; Duan, F.; Zhang, X.; Zhu, Z.; Ju, B.-F. A new diamond machining approach for extendable fabrication of micro-freeform lens array. *Int. J. Mach. Tools Manuf.* **2018**, *124*, 134–148. [CrossRef]

19. Zhu, L.; Li, Z.; Fang, F.; Huang, S.; Zhang, X. Review on fast tool servo machining of optical freeform surfaces. *Int. J. Adv. Manuf. Technol.* **2017**, *95*, 2071–2092. [CrossRef]

20. Yu, D.P.; Hong, G.S.; Wong, Y.S. Profile error compensation in fast tool servo diamond turning of micro-structured surfaces. *Int. J. Mach. Tools Manuf.* **2012**, *52*, 13–23. [CrossRef]

21. Huang, P.; To, S.; Zhu, Z. Diamond turning of micro-lens array on the roller featuring high aspect ratio. *Int. J. Adv. Manuf. Technol.* **2018**, *96*, 2463–2469. [CrossRef]

22. Gao, W.; Chen, Y.-L.; Lee, K.-W.; Noh, Y.-J.; Shimizu, Y.; Ito, S. Precision tool setting for fabrication of a microstructure array. *CIRP Ann.* **2013**, *62*, 523–526. [CrossRef]

23. Chen, Y.-L.; Gao, W.; Ju, B.-F.; Shimizu, Y.; Ito, S. A measurement method of cutting tool position for relay fabrication of microstructured surface. *Meas. Sci. Technol.* **2014**, *25*, 064018. [CrossRef]

24. Kong, L.B.; Cheung, C.F.; Lee, W.B. A theoretical and experimental investigation of orthogonal slow tool servo machining of wavy microstructured patterns on precision rollers. *Precis. Eng.* **2016**, *43*, 315–327. [CrossRef]

25. Kim, M.; Lee, D.W.; Lee, S.; Kim, Y.; Jung, Y. Effects of hinge design of horizontal-swing fast tool servo (hfts) for micro-patterning on a roll. *Int. J. Adv. Manuf. Technol.* **2017**, *95*, 233–241. [CrossRef]

26. Mukaida, M.; Yan, J. Ductile machining of single-crystal silicon for microlens arrays by ultraprecision diamond turning using a slow tool servo. *Int. J. Mach. Tools Manuf.* **2017**, *115*, 2–14. [CrossRef]

27. Zhang, X.; Fang, F.; Yu, L.; Jiang, L.; Guo, Y. Slow slide servo turning of compound eye lens. *Opt. Eng.* **2013**, *52*, 023401. [CrossRef]

28. Kwok, T.C.; Cheung, C.F.; Kong, L.B.; To, S.; Lee, W.B. Analysis of surface generation in ultra-precision machining with a fast tool servo. *Proc. Inst. Mech. Eng. Part B J. Eng. Manuf.* **2010**, *224*, 1351–1367. [CrossRef]

29. Zong, W.J.; Huang, Y.H.; Zhang, Y.L.; Sun, T. Conservation law of surface roughness in single point diamond turning. *Int. J. Mach. Tools Manuf.* **2014**, *84*, 58–63. [CrossRef]

 *materials* 

Article

Investigation on Microsheet Metal Deformation Behaviors in Ultrasonic-Vibration-Assisted Uniaxial Tension with Aluminum Alloy 5052

Chunju Wang [1,2,*], Weiwei Zhang [3], Lidong Cheng [1], Changqiong Zhu [2], Xinwei Wang [4,*], Haibo Han [5], Haidong He [1] and Risheng Hua [1]

[1] School of Mechanical and Electrical Engineering, Robotics and Microsystems Center, Soochow University, Suzhou 215131, China; ldcheng@suda.edu.cn (L.C.); hdhe@suda.edu.cn (H.H.); 20195229040@stu.suda.edu.cn (R.H.)

[2] School of Materials Science and Engineering, Harbin Institute of Technology, Harbin 150001, China; 13182781268@163.com

[3] Institute of Electronic Engineering, China Academy of Engineering Physics, Mianyang 621999, China; zhangweiwei0509103@163.com

[4] Laboratory for Space Environment and Physical Sciences, Harbin Institute of Technology, Harbin 150080, China

[5] Shanghai Institute of Spacecraft Equipment, Shanghai 200240, China; haibosmile@163.com

* Correspondence: cjwang@suda.edu.cn (C.W.); xinweiwang@hit.edu.cn (X.W.); Tel./Fax: +86-451-86418640 (C.W.)

Received: 22 December 2019; Accepted: 29 January 2020; Published: 31 January 2020

Abstract: Ultrasonic vibration (UV) is widely used in the forming, joining, machining process, etc. for the acoustic softening effect. For parts with small dimensions, UV with limited output energy is very suitable for the microforming process and has been gaininf more and more attention. In this investigation, UV-assisted uniaxial tensile experiments were carried out utilizing GB 5052 thin sheets of different thicknesses and grain sizes, respectively. The coupling effects of UV and the specimen dimension on the properties of the material were analyzed from the viewpoint of acoustic energy in activating dislocations. A reduction of flow stress was found for the existing acoustic softening effects of UV. Additionally, the residual effects of UV were demonstrated when UV was turned off. The uniform deformation ability of thin sheet could be improved by increasing the hardening exponent with UV. The experimental results indicate that UV is very helpful in improving the forming limit in microsheet forming, e.g., microbulging and deep drawing processes.

Keywords: ultrasonic vibration; acoustic softening; residual effect; microthin sheet; forming limit

---

1. Introduction

As demands on miniature/micrometal products increase significantly, for example, metallic parts with microchannels fabricated through ultra-thin sheets are widely used in mass and heat transfer areas, e.g., heat exchangers and proton exchange membrane fuel cells (PEMFCs). Due to its advantageous characteristics for mass production, microforming has become an attractive option in the manufacture of these products [1]. However, the geometrical size of the workpiece, the microstructural length scale of deforming materials, and their interaction significantly affect the deformation response of microscale objects, e.g., flow stress, springback, forming limit. [2–4]. Several kinds of microsheet forming processes, e.g., rigid stamping, hydroforming, and rubber pad forming, have been carried out to manufacture microchannels in recent years [5,6]. Microrigid stamping was a commonly used technology for microchannels due to its advantages of low cost and high productivity. A significant thinning easily occurred at the corner regions of the formed microchannel in rigid stamping, which was associated with the tensile force applied at the corner and the friction, resulting in the occurrence of localized strain [7,8]. Then, forming limit diagrams were used to determine the safe limit of the metallic bipolar plates [9]. Since hydroforming has obvious merits, e.g., flexibility, higher drawing ratio, good surface quality, less springback, and the low cost of mold, it has attracted more attention [10,11]. However, the channel height of a hydroformed workpiece was a function of the pressure, and high pressure and the seal became a big problem [5,12]. For rubber pad forming, only half of a single rigid tool is needed, and the punch-cavity misalignment problem can be avoided. A metallic bipolar plate was successfully fabricated using stainless steel SS304 [13] and titanium with a TiN layer [14]. To improve the filling percentage, thinning percentage, and dimensional accuracy, semi-stamp rubber forming was carried out instead of convectional rubber forming [15]. However, the main drawback of the rubber pad forming process is that the life of the rubber pad is not so long and should be replaced after the production of only about 100 plates. Previous investigations have indicated that the existing process cannot be used for the manufacture of microchannels due to the obvious drawbacks mentioned above.

To improve the formability of thin sheet, industrial applications of dynamic load/ultrasonic vibration (UV) were utilized in many metallic sheet forming processes. The application of 20 and 28 kHz oscillation increased the limit drawing ratio (LDR) from 2.68 to 3.01 for cold rolled steel for deep drawing. Greater accuracy and deeper cups can be formed by stopping the oscillation after the maximum punch load rather than applying the oscillation throughout the deep drawing process [16]. In the rigid bulging of microchannels, a kind of dynamic load with a frequency of 0.5Hz could improve the forming depth of 7% and decrease material thinning from 10.7% of static loading to 7.8% [17,18]. In UV-assisted microdeep drawing, the punch force decreased as the oscillation amplitude increased, and LDR increased from 1.67 to 1.83, from 1.75 to 1.92, and from 1.83 to 2 for a thickness of 50, 75, and 100 μm, respectively [19]. Further, wrinkling and cracking could be avoided through ultrasonic vibration to decrease the coefficient of friction between the sheet material and the die [20]. Using UV to form molten plastic as a flexible punch, a trapezoidal cross-section microchannel of 697.2 μm width and 248.4 μm depth was successfully replicated up to 98% on a thin T2 copper sheet of an initial thickness of 50 μm [21,22]. The UV-assisted microforming processes have shown that it was helpful in improving the quality of microparts. However, the mechanism of deformation behavior should be investigated in detail, especially for the thin sheet when UV is applied.

One of reasons dynamic load/ultrasonic can improve formability is the acoustic softening effect, as shown in many research studies. High frequency vibration was found to significantly reduce the apparent static shear stress necessary for the plastic deformation of metals [23]. Further, the amount of reduction was directly proportional to the acoustic energy input to the specimen; this yielding of metals due to ultrasonic irradiation has the possibility to be more efficient than other methods, e.g., heating [24]. To realize the mechanism of the softening effect of ultrasonic vibration, a significant reduction in subgrain formation was found and analyzed from the viewpoint of acoustic energy, which can be attributed to its ability to enhance dislocation dipole annihilation to cause dislocations to travel longer distances [25]. A modeling framework for acoustic plasticity was proposed to model the acoustic softening considering the acoustic energy intensity based on the crystal plasticity theory [26]. An opposite result by a dislocation dynamics simulation indicated that the acoustic effect was associated with extensive enhancement of subgrain formation [27]. However, the effects of UV on the properties of materials have not been investigated considering dimensions of microparts in UV-assisted microforming processes. The mechanism of stress reduction was investigated, and the evolution of the microstructure was observed through experiments and simulation. However, research was seldom carried out on the small specimens. From the viewpoint of acoustic energy, the density is changed during the miniaturization of specimens. Additionally, the density of microdefects in metallic specimens, e.g., dislocations and grain boundaries, becomes smaller in small specimens, which is considered to absorb the acoustic energy. Investigations should be done to analyze the effect of grain size and specimen size on deformation behavior in UV-assisted deformation.

In the investigation, UV-assisted uniaxial tensile tests were carried out utilizing GB 5052 thin sheets with different thicknesses from 50 to 100 µm. Further, the effects of UV parameters, e.g., vibration amplitude and duration, were also studied. Properties of materials, e.g., yield stress, tensile strength, elongation, and hardening exponent, were obtained from the UV-assisted experiments. The coupling effects of UV and specimen dimension were analyzed from the viewpoint of acoustic energy in activating dislocations.

2. Materials and Methods

2.1. Experimental Set-Up

In this investigation, a kind of ultrasonic-vibration-assisted uniaxial tension device was utilized as shown in Figure 1 [28], which was developed by our group based on a testing machine (UTM6104, Shenzhen Suns Technology Stock Co., Ltd., Shenzhen, China) with a 10 kN load cell. A vibrator was designed, including an ultrasonic transducer and an ultrasonic horn. A piezoelectric ceramic transducer (PZT-8) with a sandwich structure was selected as the ultrasonic transducer, and it could effectively transform the electric energy to mechanical longitudinal vibration. To clamp the thin sheet specimen, a groove structure was designed at the top end of the ultrasonic horn. The width and depth of the groove were selected as 0.2 and 10 mm, respectively, based on finite element analysis using the ABAQUS commercial software (v6.12, Dassault Systems SIMULIA, France). One side of the thin sheet specimen is clamped by the vibrator using the groove structure at the top end of the ultrasonic horn, which is fixed at the bottom parts of the testing machine. Another side of the specimen is clamped by a tool installed on a moving beam of the testing machine. In the uniaxial tension device, the moving beam can move upward, then a force is applied on the specimen to fix the bottom part of the specimen. During the tensile test, the vibrator can be turned on, and UV is applied on the specimen by the ultrasonic horn.

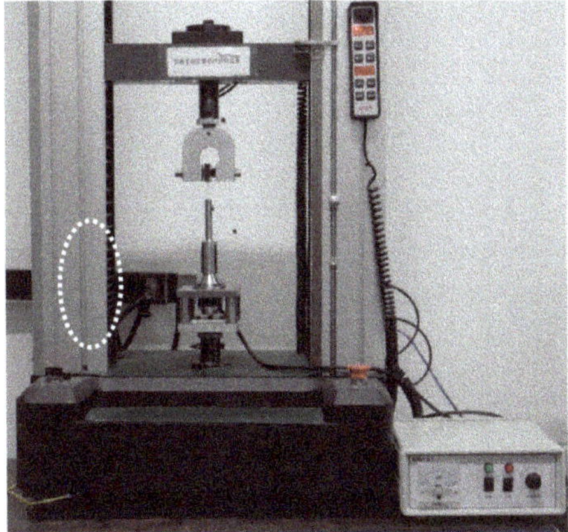

Figure 1. Ultrasonic-vibration-assisted uniaxial tension device.

Measured by the laser vibrometer Polytec OFV-5000, the longitudinal vibration frequency of the manufactured vibrator was 21.1 kHz, and its vibration amplitude could be changed from 0.48 to 15 µm depending on the input current, which could meet the requirement of uniaxial tension in these experiments.

2.2. Experimental Material

Commercial rolled aluminum alloy GB 5052 sheet was selected in this investigation for its excellent properties, e.g., higher strength alloys, excellent corrosion resistance, and high fatigue strength. The thicknesses *t* of selected rolled aluminum alloy sheets are 50, 60, 80, and 100 μm, and the dimensions of their specimens were designed according to the similar theory as shown in Figure 2. To study the effect of the grain size and reduce the effect of strain hardening induced by the rolling process, heat treatment experiments for a thickness of 50 μm were carried out using a tubular vacuum heat treatment furnace (T1200) at temperatures of 200, 300, 400, 500, and 600 °C for 1 h of the holding time with 10 °C/min of the heat rate, respectively. After heat treatment, specimens were electrochemically polished under a voltage of 5 V and current of 0.5 A for 8 s using a solution of 10 mL of $HClO_4$ and 60 mL of C_2H_5OH. The obtained microstructures under different temperatures are shown in Figure 3, and the grain size as shown in Table 1 was manually measured using the intercept method according to ASTM E112 [29]. For other thicknesses of sheets, only the temperature of 400 °C was selected to eliminate the effect of strain hardening. The heat treatment process was only used to eliminate the hardening effect during the manufacturing of the thin sheet with the rolling process. The difference of grain size for different thicknesses was ignored for the same heat treatment process. Thus, grain sizes of thin sheets with different thicknesses were not measured in this investigation.

Table 1. Grain size of a GB 5052 thin sheet after annealing treatment (50 μm in thickness).

Temperature/°C	200	300	400	500	600
Grain Size/μm	4.3	13.68	18.7	25.55	47.65

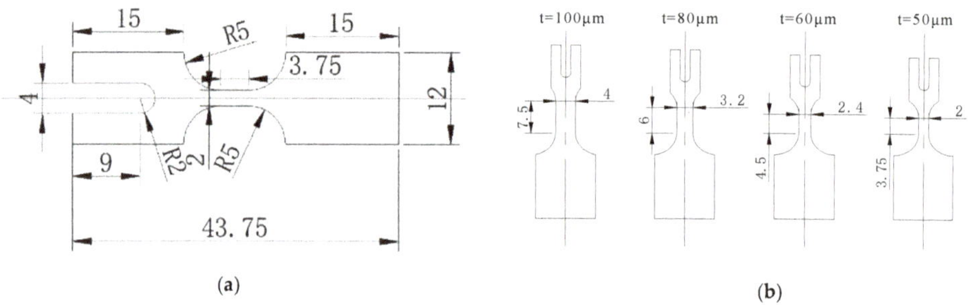

(a)   (b)

Figure 2. GB 5052 thin sheet specimens and their dimensions. (**a**) Dimensions of specimen (50 μm in thickness); (**b**) Dimensions of specimens with different thicknesses.

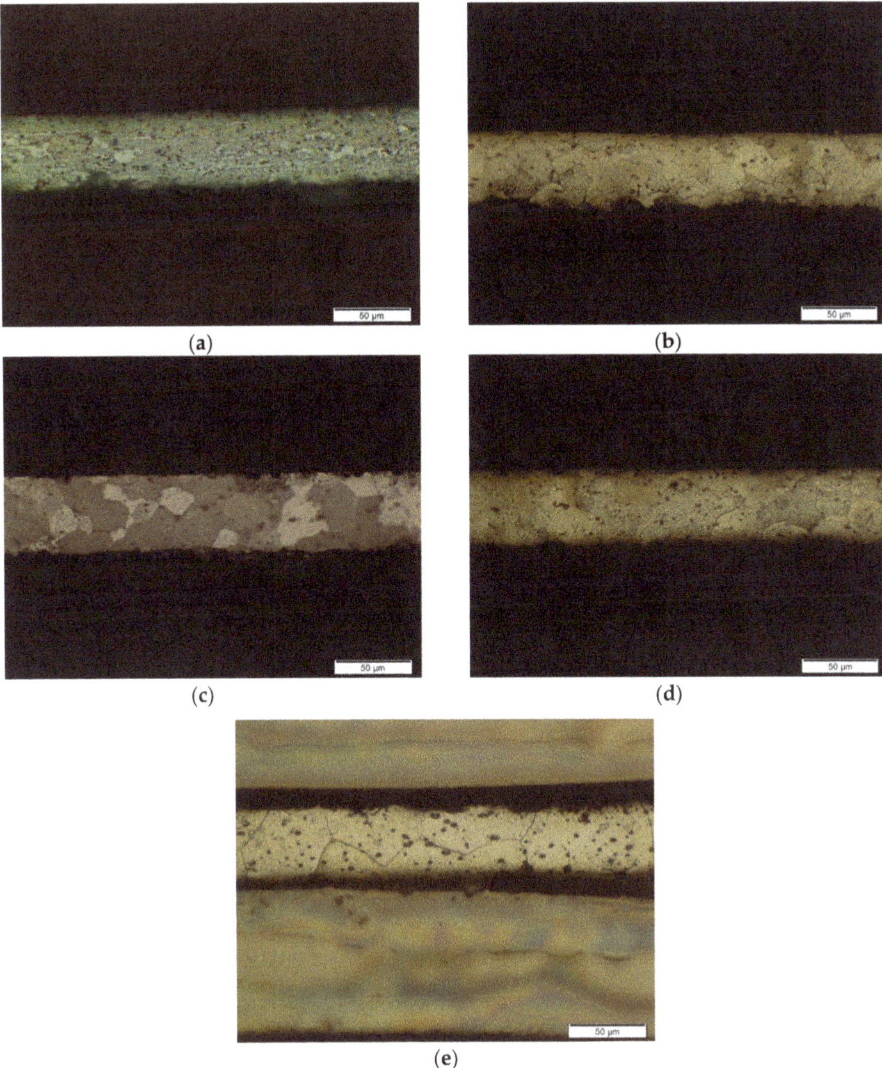

Figure 3. Microstructure of a GB 5052 thin sheet after heat treatment at different temperatures (50 μm in thickness). (**a**) 200 °C; (**b**) 300 °C; (**c**) 400 °C; (**d**) 500 °C; (**e**) 600 °C.

3. Results and Discussion

3.1. Effect of UV on Properties of Specimen with Different Thicknesses

During the miniaturization, size effects have an obvious influence on the properties of the specimen. To realize this kind of effects, GB 5052 specimens with different thicknesses were utilized after heat treatment on the temperature of 400 °C for 1 h, and uniaxial tensile tests were carried out with a punch speed of 0.3 mm/min and vibration amplitude of 0.48 μm. The experimental results are shown in Figure 4, demonstrating that the flow stress decreases with the decrease of thickness both without and with UV. This means that the size effect on flow stress occurs, which has been studied in many papers [2].

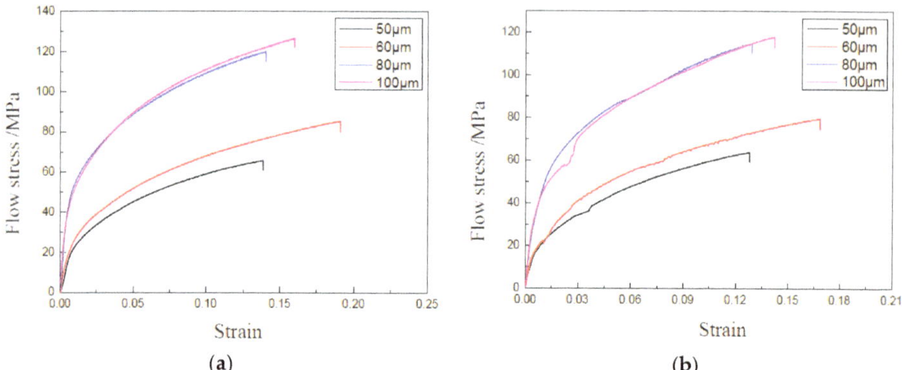

Figure 4. Curves of stress–strain of specimens with different thicknesses. (**a**) Without UV; (**b**) With UV.

To study the effect of UV on the properties of materials in detail, the yield stress, tensile strength, elongation, and hardening exponent were achieved as shown in Figure 5 based on curves shown in Figure 4. It clearly shows that the thickness of the specimen has an obvious effect on the parameters of the thin sheet. The yield stress and tensile strength decrease with the decrease of thickness, and the elongation and hardening exponent increase with the decrease of thickness. Only the elongation for a thickness of 50 μm does not follow this rule, which becomes lower.

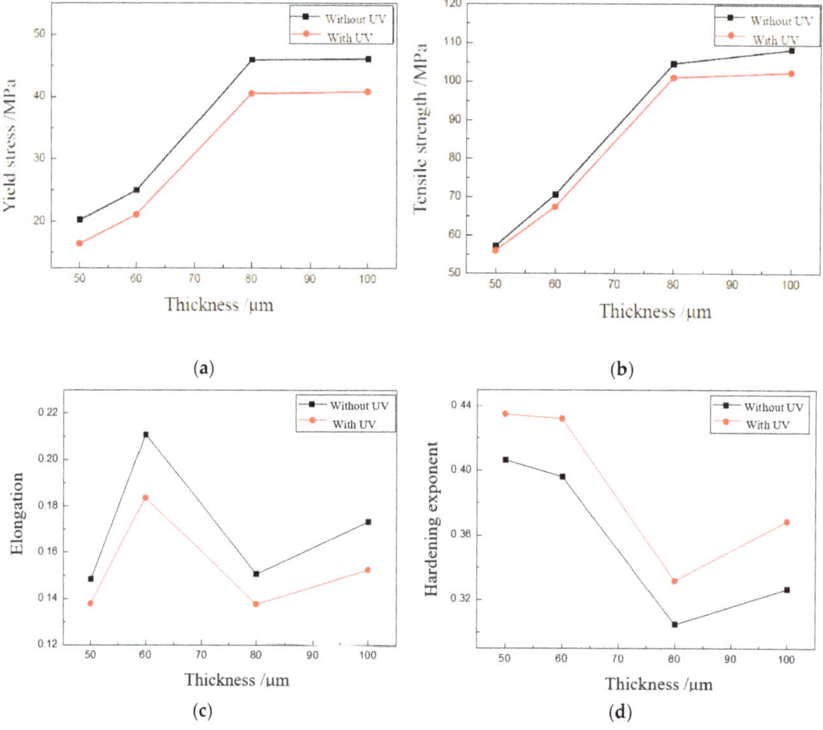

Figure 5. Effect of UV on properties of specimens with different thicknesses. (**a**) Effect of UV on yield stress; (**b**) Effect of UV on tensile strength; (**c**) Effect of UV on elongation; (**d**) Effect of UV on hardening exponent.

When UV was applied, the yield stress, tensile strength, and elongation decreased, and the hardening exponent increased, which was similar with the results shown above. For the acoustic softening effect, the flow stress can be reduced, since dislocations are activated by the energy and become easy to slip, even those on the 'hard' orientation. As a result, the yield stress and tensile strength become smaller in UV-assisted tests. When radial shrinkage occurs during tensile tests, UV can increase the extension of the crack since the acoustic energy density becomes bigger for the small cross-section area induced by the shrinkage. Then, the elongation is decreased by ultrasonic vibration. However, the proportion of uniform deformation is generally increased by UV, which is helpful in improving the critical limit. We plan to study these phenomena in future investigations.

An increase of the hardening exponent was observed in the tests by UV. The reason may be that UV activates more dislocations and promotes the interaction of dislocations. As a result, more subgrains are formed, which leads to the increase of the hardening exponent [27]

The effect of UV is almost the same for specimens with different thicknesses. This may be attributed to the fact that the acoustic energy is too small. The coupling effect of UV and dimension will be studied in depth in future investigations.

3.2. Effect of UV on Properties of Specimens with Different Grain Sizes

Uniaxial tensile tests were carried out with a punch speed of 0.3 mm/min and vibration amplitude of 0.48 μm using GB 5052 thin sheet specimens of 50 μm in thickness. The obtained curves of stress–strain for GB 5052 sheets with different grain sizes are shown in Figure 6 without and with UV, respectively. To analyze the effect of UV on the mechanical properties of GB 5052 thin sheet, yield stress, tensile strength, elongation, and harden exponent were obtained from the curves of stress–strain.

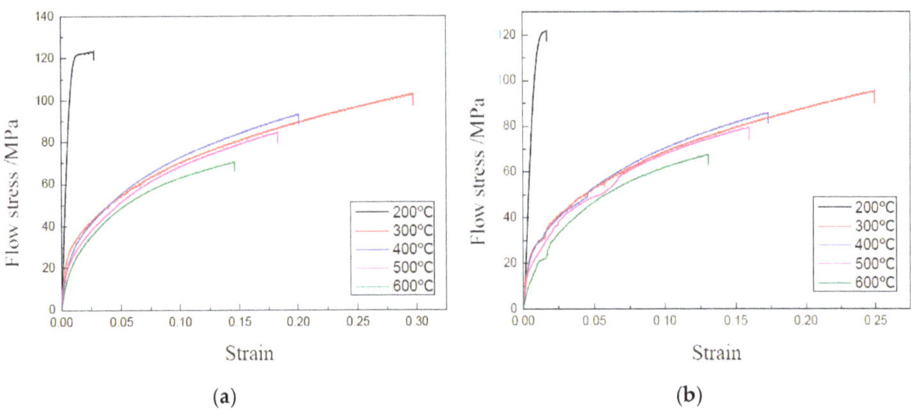

Figure 6. Curves of stress–strain of GB 5052 sheet. (**a**) Without UV; (**b**) With UV.

For the existing acoustic softening effect, the flow stress is reduced by UV. The yield stress and tensile strength are shown in Figures 7 and 8, respectively. We notice that the data in Figure 8 are different from those in Figure 5c for the same condition. The reason may be that the tests were carried out at a different time and using different batches of material.

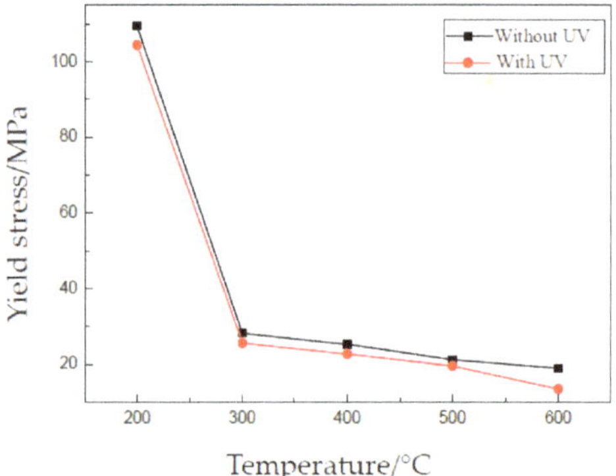

Figure 7. Effect of UV on yield stress.

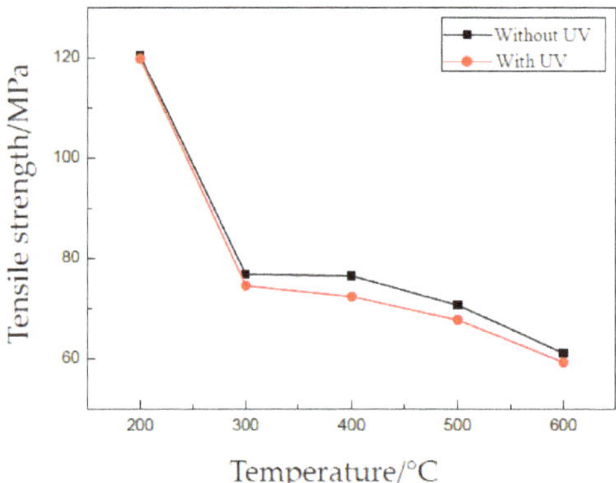

Figure 8. Effect of UV on tensile strength.

A softening effect can be observed from the two figures, and the reduction of flow stress is very small. The reason is that the softening effect depends on the level of acoustic intensity applied on specimens. A little acoustic energy used in these experiments, which is proportional to the vibration amplitude, accordingly leads to a small reduction of flow stress.

For different grain sizes, the effects of UV are almost similar. One reason is the small vibration amplitude. Another one, also important, is that the density of defects in thin sheets becomes very similar after heat treatment. Generally, defects in metallic materials can absorb acoustic energy and activate dislocations to move, which leads to a reduction of shear stress during plastic deformation. Thus, the effects of UV on yield stress and tensile strength are very small in this investigation.

The elongation of thin sheets indicates the limited plastic deformation ability of metallic materials, which is very important in thin sheet forming. The effect of UV on the elongation of thin sheet with different grain sizes is shown in Figure 9. It is found that the elongation of the thin sheet is decreased, and the reduction slightly increases with the increases of grain size. During the uniaxial tensile deformation, radial shrinkage will occur before the fracture of the specimen. The appearance of radial shrinkage means that the area of the cross-section becomes smaller. With the same vibration amplitude, as well as the same input acoustic energy, the acoustic energy density will be increased for the smaller area, which will improve the activation of dislocations and local plastic deformation. As a result, the elongation with UV becomes smaller.

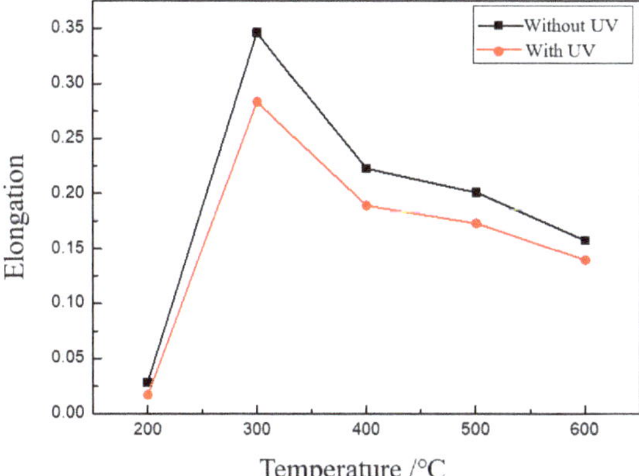

Figure 9. Effects of UV on elongation.

The hardening exponent is another important parameter which indicates the ability of uniform plastic deformation of thin sheet. The effect of UV on the hardening exponent is shown in Figure 10. The experimental results showed that UV increases the hardening exponent, especially for specimens with a bigger grain size, which means that the uniform deformation ability is increased by ultrasonic vibration. The reason is that UV as a kind of energy can activate more dislocations to move, even dislocations in an unfavorable position, which is validated by Electron Backscattered Diffraction (EBSD) analysis [25]. The more dislocations are activated, the more uniform plastic deformation can be realized.

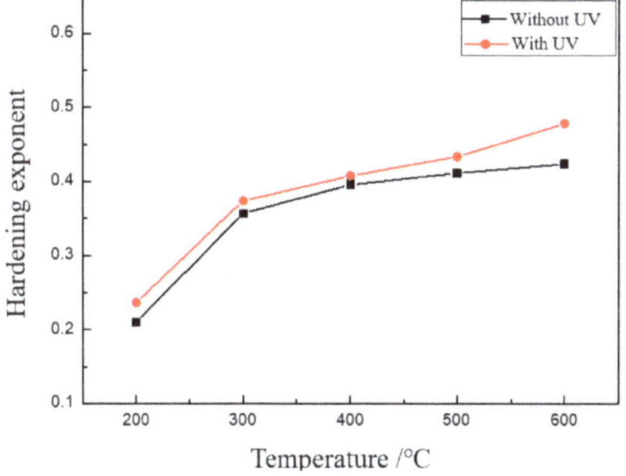

Figure 10. Effects of UV on the hardening exponent.

3.3. Effect of UV Duration on Properties of Specimen

The different duration of UV means a different acoustic energy applied on the specimen, and the effect of UV duration was analyzed in this investigation. With GB 5052 thin sheets of 50 µm in thickness treated on the temperature of 400 °C, UV-assisted uniaxial tensile experiments were carried out with a punch speed of 0.3 mm/min using a vibration amplitude of 5 µm under different durations of 20, 30, and 40 s, respectively. Experimental results showed that the flow stress can be decreased immediately during tensile deformation when UV is applied on the specimen as shown in Figure 11. This means that the acoustic softening effect occurs. The reduction of flow stress is almost the same for the same vibration amplitude. When UV is turned off, the flow stress goes back to that without UV. However, the flow stress becomes smaller after UV, which can be treated as a kind of residual effect because the acoustic energy forces dislocations to move in a preferred direction, and materials' properties undergo permanent changes [24].

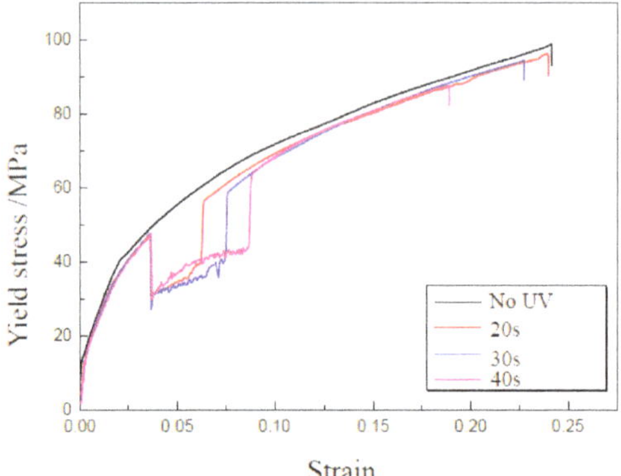

Figure 11. Effect of UV duration on flow stress.

In Figure 11, the effect of UV duration on tensile strength and elongation can be obtained, as shown in Figures 12 and 13, respectively. The tensile strength can be obviously decreased from 77.6 MPa for 0 s to 72.5 MPa for 40 s; the reduction percentage is about 6.6%. Further, the elongation was reduced from 0.73 for 0 s to 0.21 for 40 s. Although UV is turned off, its effect on the tensile strength and elongation is very clear, which is called a kind of residual effect. This may be attributed to the permanent changes of the microstructure in metallic materials after UV-assisted deformation [24]. For a longer duration of UV, more acoustic energy was applied on the specimen, and its effect on the properties increased accordingly.

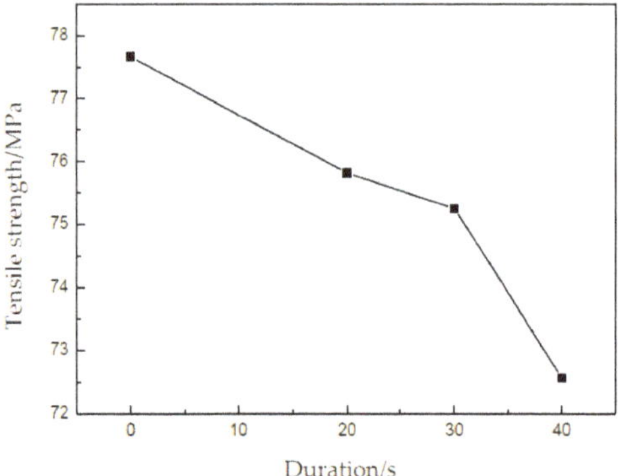

Figure 12. Tensile strength under different durations.

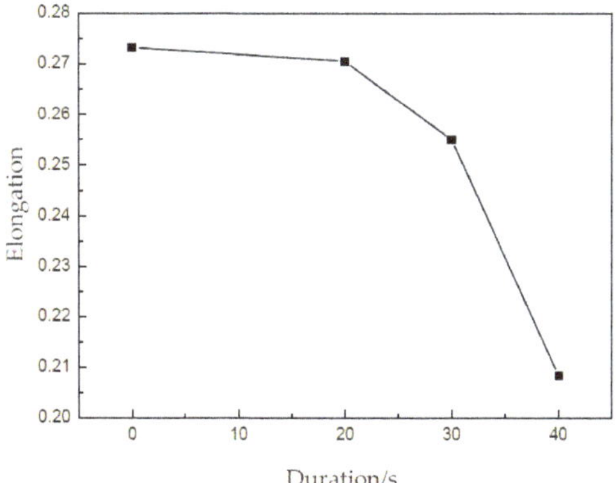

Figure 13. Elongation under different durations.

3.4. Effect of Vibration Amplitude on Properties of Specimen

As we know, acoustic energy is proportional to vibration amplitude, which has an obvious effect on the properties of materials. In this investigation, UV-assisted uniaxial tensile experiments were carried out with a punch speed of 0.3 mm/min using a duration of 30 s under different vibration amplitudes of 5, 5.6, and 6.2 μm, respectively. Further, the GB 5052 thin sheet of 50 μm in thickness treated on the temperature of 400 °C was selected, and the experimental results are shown in Figure 14. It was found that the flow stress decreases immediately. The reduction of flow stress was 16.43, 21.67, and 24.82 MPa, respectively. A higher level of acoustic intensity may provide more energy to motivate more dislocations to slide from their pinned points and also lead to a multiplication of dislocations [24]. As a result, the reduction of flow stress increases with the increase of vibration amplitude.

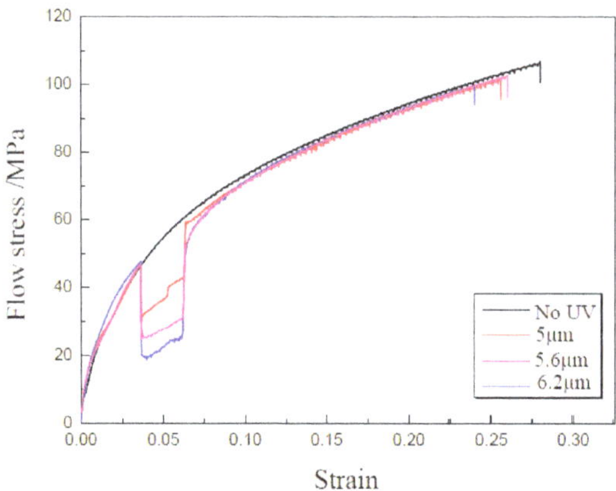

Figure 14. Curves of strain-stress under different vibration amplitudes.

4. Conclusions

In this investigation, UV-assisted uniaxial tensile tests were carried out utilizing GB 5052 thin sheets. The effects of UV on the properties of materials were analyzed in detail for specimens with different grain sizes and original thicknesses, respectively. An acoustic softening and residual effect was found in the investigation. The following conclusions can be obtained.

1. Flow stress, e.g., yield stress and tensile strength, is decreased for the acoustic effect. Additionally, the reduction of flow stress increases with the increase of vibration amplitude. The reason is that acoustic energy, which is proportional to amplitude, activates dislocations and gives the energy for dislocation slipping.

2. Since lots of dislocations, even 'hard' orientation, are activated by acoustic energy, the interactive effects of dislocations are increased. Then, the hardening exponent becomes bigger, which can improve the uniform deformation ability of thin sheets.

3. With the increased duration of UV, tensile strength and elongation are decreased. Even when UV is turned off, its effect is still observed, which is called a residual effect. The reason may be that the microstructure of the specimen is permanently changed by UV.

4. The effect of UV on the properties of specimens with different grain sizes and thicknesses is similar. The reason may be that the acoustic energy is small. The coupling effect of UV and specimen dimension will be studied in future investigations.

Author Contributions: Conceptualization, C.W.; Data curation, C.Z., X.W. and C.W.; Formal analysis, C.W., C.Z. and L.C.; Funding acquisition, C.W., X.W., W.Z. and H.H. (Haibo Han); Investigation, C.Z., W.Z., L.C. and X.W.; Methodology, C.W., X.W. and R.H.; Resources, C.W., W.Z. and H.H. (Haibo Han); Supervision, C.W.; Writing—original draft, C.W. and C.Z.; Writing—review & editing, C.W., C.Z., W.Z., and H.H. (Haidong He). All authors have read and agreed to the published version of the manuscript.

Funding: The research was funded by the Natural Science Foundation of Jiangsu Province (BK20192007), National Science Foundation of China (No. 51875128, 51805497, 51605301, 51635005), and Six Talent Peaks in Jiangsu Province (GDZB-069).

Conflicts of Interest: The authors declare no conflict of interest.

References

1. Qin, Y. Micro-forming and miniature manufacturing systems — development needs and perspectives. *J. Mater. Process. Technol.* **2006**, *177*, 8–18. [CrossRef]

2. Fu, M.; Wang, J.; Korsunsky, A. A review of geometrical and microstructural size effects in micro-scale deformation processing of metallic alloy components. *Int. J. Mach. Tools Manuf.* **2016**, *109*, 94–125. [CrossRef]

3. Wang, S.; Zhuang, W.; Cao, J.; Lin, J. An investigation of springback scatter in forming ultra-thin metal-sheet channel parts using crystal plasticity FE analysis. *Int. J. Adv. Manuf. Technol.* **2010**, *47*, 845–852. [CrossRef]

4. Máthis, K.; Köver, M.; Stráská, J.; Trojanová, Z.; Džugan, J.; Halmešová, K. Micro-Tensile Behavior of Mg-Al-Zn Alloy Processed by Equal Channel Angular Pressing (ECAP). *Materials* **2018**, *11*, 1644. [CrossRef]

5. Peng, L.; Yi, P.; Lai, X. Design and manufacturing of stainless steel bipolar plates for proton exchange membrane fuel cells. *Int. J. Hydrogen Energy* **2014**, *39*, 21127–21153. [CrossRef]

6. Xue, S.; Wang, C.; Chen, P.; Xu, Z.; Cheng, L.; Guo, B.; Shan, D. Investigation of Electrically-Assisted Rolling Process of Corrugated Surface Microstructure with T2 Copper Foil. *Materials* **2019**, *12*, 4144. [CrossRef]

7. Chan, W.; Fu, M. Experimental and simulation based study on micro-scaled sheet metal deformation behavior in microembossing process. *Mater. Sci. Eng. A* **2012**, *556*, 60–67. [CrossRef]

8. Ren, Z.; Zhang, D.; Wang, Z. Stacks with TiN/titanium as the bipolar plate for PEMFCs. *Energy* **2012**, *48*, 577–581. [CrossRef]

9. Hu, Q.; Zhang, D.; Fu, H.; Huang, K. Investigation of stamping process of metallic bipolar plates in PEM fuel cell—Numerical simulation and experiments. *Int. J. Hydrogen Energy* **2014**, *39*, 13770–13776. [CrossRef]

10. Mahabunphachai, S.; Cora, Ö.N.; Koç, M. Effect of manufacturing processes on formability and surface topography of proton exchange membrane fuel cell metallic bipolar plates. *J. Power Sources* **2010**, *195*, 5269–5277. [CrossRef]

11. Mahabunphachai, S.; Koç, M. Fabrication of micro-channel arrays on thin metallic sheet using internal fluid pressure: Investigations on size effects and development of design guidelines. *J. Power Sources* **2008**, *175*, 363–371. [CrossRef]

12. Xu, Z.; Peng, L.; Yi, P.; Lai, X. Modeling of microchannel hydroforming process with thin metallic sheets. *J. Eng. Mater. Technol.* **2012**, *134*, 021017. [CrossRef]

13. Liu, Y.; Hua, L. Fabrication of metallic bipolar plate for proton exchange membrane fuel cells by rubber pad forming. *J. Power Sources* **2010**, *195*, 3529–3535. [CrossRef]

14. Jin, C.K.; Jeong, M.G.; Gil Kang, C. Fabrication of titanium bipolar plates by rubber forming and performance of single cell using TiN-coated titanium bipolar plates. *Int. J. Hydrogen Energy* **2014**, *39*, 21480–21488. [CrossRef]

15. Elyasi, M.; Ghadikolaee, H.T.; Hosseinzadeh, M. Fabrication of metallic bipolar plates in PEM fuel cell using semi-stamp rubber forming process. *Int. J. Adv. Manuf. Technol.* **2017**, *92*, 765–776. [CrossRef]

16. Jimma, T.; Kasuga, Y.; Iwaki, N.; Miyazawa, O.; Mori, E.; Ito, K.; Hatano, H. An application of ultrasonic vibration to the deep drawing process. *J. Mater. Process. Technol.* **1998**, *80*, 406–412. [CrossRef]

17. Huang, Y.M.; Wu, Y.S.; Huang, J.Y. The influence of ultrasonic vibration-assisted micro-deep drawing process. *Int. J. Adv. Manuf. Technol.* **2014**, *71*, 1455–1461. [CrossRef]

18. Ashida, Y.; Aoyama, H. Press forming using ultrasonic vibration. *J. Mater. Process. Technol.* **2007**, *187*, 118–122. [CrossRef]

19. Luo, F.; Li, K.-H.; Zhong, J.-M.; Gong, F.; Wu, X.-Y.; Ruan, S.-C. An ultrasonic microforming process for thin sheet metals and its replication abilities. *J. Mater. Process. Technol.* **2015**, *216*, 10–18. [CrossRef]

20. Luo, F.; Wang, B.; Li, Z.W.; Wu, X.Y.; Gong, F.; Peng, T.J.; Li, K.H. Time factors and optimal process parameters for ultrasonic microchannel formation in thin sheet metals. *Int. J. Adv. Manuf. Technol.* **2017**, *89*, 255–263. [CrossRef]

21. Koo, J.-Y.; Jeon, Y.-P.; Kang, C.-G. Effect of stamping load variation on deformation behaviour of stainless steel thin plate with microchannel. *Proc. Inst. Mech. Eng. Part B: J. Eng. Manuf.* **2013**, *227*, 1121–1128. [CrossRef]

22. Park, W.T.; Jin, C.K.; Gil Kang, C. Improving channel depth of stainless steel bipolar plate in fuel cell using process parameters of stamping. *Int. J. Adv. Manuf. Technol.* **2016**, *87*, 1677–1684. [CrossRef]

23. Blaha, F.; Langenecker, B. Elongation of zinc monocrystals under ultrasonic action. *Die Naturwissenschaften* **1955**, *42*, 556. [CrossRef]

24. Langenecker, B. Effects of Ultrasound on Deformation Characteristics of Metals. *IEEE Trans. Sonics Ultrason.* **1966**, *13*, 1–8. [CrossRef]

25. Dutta, R.; Petrov, R.; Delhez, R.; Hermans, M.; Richardson, I.; Böttger, A. The effect of tensile deformation by in situ ultrasonic treatment on the microstructure of low-carbon steel. *Acta Mater.* **2013**, *61*, 1592–1602. [CrossRef]

26. Yao, Z.; Kim, G.Y.; Wang, Z.; Faidley, L.; Zou, Q.; Mei, D.; Chen, Z. Acoustic softening and residual hardening in aluminium: Modeling and experiments. *Int. J. Plast.* **2012**, *39*, 75–87. [CrossRef]

27. Siu, K.; Ngan, A.; Jones, I. New insight on acoustoplasticity – Ultrasonic irradiation enhances subgrain formation during deformation. *Int. J. Plast.* **2011**, *27*, 788–800. [CrossRef]

28. Wang, C.; Liu, Y.; Shan, D.; Guo, B.; Han, H. Investigations on mechanical properties of copper foil under ultrasonic vibration considering size effect. *Procedia Eng.* **2017**, *207*, 1057–1062. [CrossRef]

29. *ASTM E112-13. Standard Test Methods for Determining Average Grain Size*; ASTM International: West Conshohocken, PA, USA, 2013.

Article

Ultrasonic Vibration Facilitates the Micro-Formability of a Zr-Based Metallic Glass

Guangchao Han [1,2], Zhuo Peng [1], Linhong Xu [1] and Ning Li [3,4,*]

[1] School of Mechanical Engineering and Electronic Information, China University of Geosiences, Wuhan 430074, China; hgc009@cug.edu.cn (G.H.); 13237140699@163.com (Z.P.); xulinhong@cug.edu.cn (L.X.)
[2] Shanxi Key Laboratory of Non-Traditional Machining, Xi'an Technological University, Xi'an 710032, China
[3] State Key Laboratory of Advanced Welding and Joining, Harbin Institute of Technology, Harbin 150001, China
[4] State Key Laboratory of Materials Processing and Die & Mold Technology, Huazhong University of Science and Technology, Wuhan 430074, China
* Correspondence: hslining@mail.hust.edu.cn; Tel.: +86-278-755-9606; Fax: +86-278-755-4405

Received: 31 October 2018; Accepted: 12 December 2018; Published: 17 December 2018

Abstract: Thermoplastic microforming not only breaks through the bottleneck in the manufacture of metallic glasses, but also offers alluring prospects in microengineering applications. The microformability of metallic glasses decreases with a reduction in the mold size owing to the interfacial size effect, which seriously hinders their large-scale applications. Here, ultrasonic vibration was introduced as an effective method to improve the microformability of metallic glasses, owing to its capabilities of improving the material flow and reducing the interfacial friction. The results reveal that the microformability of supercooled $Zr_{35}Ti_{30}Cu_{8.25}Be_{26.75}$ metallic glasses is conspicuously enhanced by comparison with those under quasi-static loading. The more intriguing finding is that the microformability of the Zr-based metallic glasses can be further improved by tuning the amplitude of the ultrasonic vibration. The physical origin of the above scenario is understood, in depth, on the basis of ultrasonic vibration-assisted material flow, as demonstrated by the finite element method.

Keywords: metallic glasses; thermoplastic microforming; ultrasonic vibration; formability

1. Introduction

Thermoplastic microforming (TPMF) has broken through the bottleneck in the manufacture of metallic glasses (MGs), providing an alternative way to fabricate MG parts/components with high precision and excellent mechanical properties, which offers alluring prospects of MGs in microengineering applications [1–5], while the microformability of MGs is seriously hindered by the interfacial effect and high viscosity in the supercooled liquid region (SCLR, a temperature window between glass transition temperature T_g and crystallization temperature T_x). The low viscosity and the spatiotemporally homogeneous flow are also regarded as critical parameters that significantly affect the thermoplastic formability of MGs in the supercooled liquid state [6–9].

To improve the microformability of supercooled liquid MGs, processing parameters such as temperature and strain rate are usually considered as crucial factors, because these parameters determine the flow characteristics of the MGs. In our previous research [10], we revealed an inherent relationship between the thermoplastic formability and the flow characteristics, namely, Newtonian flow facilitates the forming capability, while thermoplastic forming in a non-Newtonian flow regime tends to be difficult. However, it is noted that the MGs would have a very short forming time window during the high-temperature forming process, which induces the risk of possible crystallization during plastic forming. To reduce the viscosity of MGs and avoid the possible crystallization, rapid temperature increases [11,12] and fast cooling [2] have been introduced to avoid the interaction with time–temperature–transformation (TTT) curve. On the other hand, in order to reduce the interfacial effect, the hot rolling method [13] and the thermoplastic blowing method [14] have been tried to reduce the physical contact area between MGs and the mold, while these methods are hard to be applied in the microforming of MGs. The method of adding lubricant has also been found to be very limited for reducing the interfacial friction. Therefore, a new method is urgent necessary to reduce the interfacial friction and lower the viscosity of MGs in the supercooled liquid region.

In recent years, vibration loading has been introduced to improve the microforming capacity of supercooled liquid MGs. It was found that low-frequency vibration ($f \leq 10$ Hz) could effectively improve the plastic strain of MGs, owing to the vibration-assisted viscosity reduction and homogeneous flow [6]. Lateral extrusion experiments of Zr55 ($Zr_{55}Cu_{30}Al_{10}Ni_5$) MGs with low-frequency vibrations also revealed an enhanced microformability under vibration loading [15]. Ultrasonic vibration with high loading frequency of above 20 Hz has also been widely applied in metal microforming processes because of its superiority in enhancing formability via a reduction of the forming force and a decrease in friction at the interface [16–18], originating from the ultrasonic stress-softening effect, stress superposition effect, and periodic separation with ultrasonic vibration. Studies on ultrasonic vibration have been performed for shear formability of MGs [19]. This technique also exhibits potential applications in micro/nanoscale forming of MGs. By increasing loading frequency to about 20 KHz, Ma et al. [20,21] used high frequency ultrasonic beating method to fabricate micro- to macroscale structures, avoiding crystallization and oxidation of MGs, while the systemic investigation of ultrasonic vibration on microformability of metallic glasses is still lacking, and the underlying physical origin for ultrasonic-assisted thermoplastic forming of metallic glasses still remains unanswered.

In this work, ultrasonic vibrational loading was introduced as an innovative method to improve the thermoplastic microformability of $Zr_{35}Ti_{30}Cu_{8.25}Be_{26.75}$ MGs in a supercooled liquid state. It will be shown that the microformability of the supercooled liquid Zr-based MGs improves with increasing ultrasonic vibration amplitude, exhibiting an ultrasonic vibration-enhanced thermoplastic microformability. The physical mechanism of the phenomenon is rationalized in terms of the evolution of free volume concentration (c_f) and flowing unit volume, which is further understood with assistance of finite-element-method (FEM) simulation. The present results provide an effective method to enhance the thermoplastic microformability of MGs.

2. Experimental Setup and Procedure

2.1. Materials

A $Zr_{35}Ti_{30}Cu_{8.25}Be_{26.75}$ (Zr35) MG system was selected for this research because of its excellent oxidation resistance, wide supercooled liquid region, and good glass-forming ability. Alloy cylinders with dimensions of ϕ 3 mm × 150 mm were fabricated by arc melting a mixture of pure Zr, Ti, Be, and Cu metals (purity > 99.5%) under a Ti-gettered argon atmosphere, followed by jet casting into a copper mold. The glassy structure of the as-cast alloy was verified by X-ray diffraction (XRD, Philips X'Pert Pro, Amsterdam, The Netherlands). The thermal response was determined by differential scanning calorimetry (DSC, TAQ2000, TA Instruments, New Castle, DE, USA) at a heating rate of 20 K·min^{-1}, showing a glass transition temperature (T_g) of 303.5 °C with a wide supercooled liquid region of 145 °C. The isothermal crystallization experiments revealed that the incubation time is more than 300 min at 370 °C [2,6,10,22]. The Zr35 samples with dimensions of ϕ 3 mm × 3 mm (strength of 1560 MPa) were sectioned from the bars, and the single factor experiments were completed for this research, such as ultrasonic power output, temperature, and forming velocity.

2.2. Ultrasonic Microextrusion System

To meet the requirements of ultrasonic vibration and equipment installation, we designed and fabricated a 20 kHz ultrasonic vibration system by ourselves. The ultrasonic vibration system included a TJS-3000 ultrasonic generator, two YP5020-4D ultrasonic transducers, two ultrasonic step horns (which were all produced by Hangzhou Success Ultrasonic Equipment Co., Ltd., Hangzhou, China), and a special porous sonotrode. In contrast to standard ultrasonic devices with a vertical arrangement, we designed an ultrasonic vibration system with a horizontally symmetrical arrangement, wherein a cylindrical ultrasonic horn and a microforming indenter were connected in sequence to the surface center of the porous sonotrode, as shown in Figure 1. The porous sonotrode could convert the horizontal input vibrations of the ultrasonic transducers into vertical output vibrations [23], and the indenter could have vertical ultrasonic resonance vibrations with the sonotrode. Additionally, the ultrasonic vibration system could be easily mounted with a press machine through two flange support seats without the need for specially designed mounting structures. The ultrasonic vibration modes in the vertical direction of both the unloaded ultrasonic vibration system and the forming tool (cylindrical ultrasonic horn and indenter)-loaded system were simulated by ANSYS software (version 13.0), and the results are shown in Figure 2. The results show that the frequency of the ultrasonic system is reduced from 19,661 Hz (shown in Figure 2a) to 19,573 Hz (shown in Figure 2b) with the increase in loading (frequency tracking range of the TJS-3000 ultrasonic generator (Hangzhou Success Ultrasonic Equipment Co., Ltd., Hangzhou, China) is 20 ± 0.5 kHz), and the maximum ultrasonic amplitude can be obtained at the end of the indenter (shown in Figure 2b with blue color), which means that the ultrasonic system is insensitive to load fluctuations and is fit for vertical ultrasonic micro-extrusion processes.

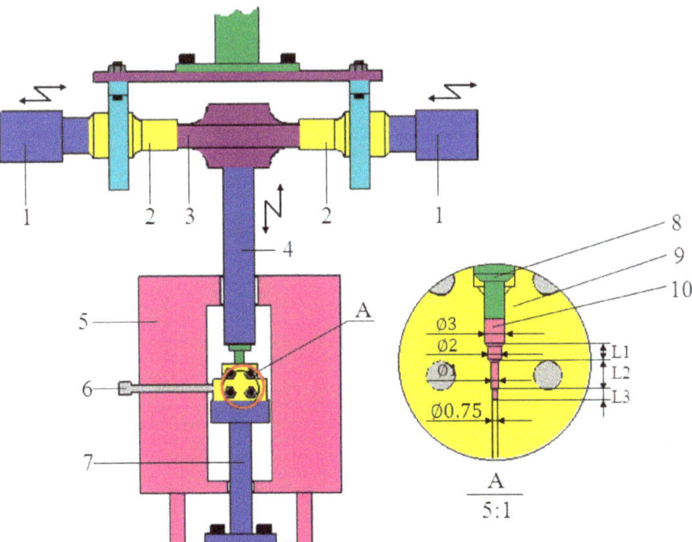

Figure 1. A schematic of the ultrasonic microextrusion system.

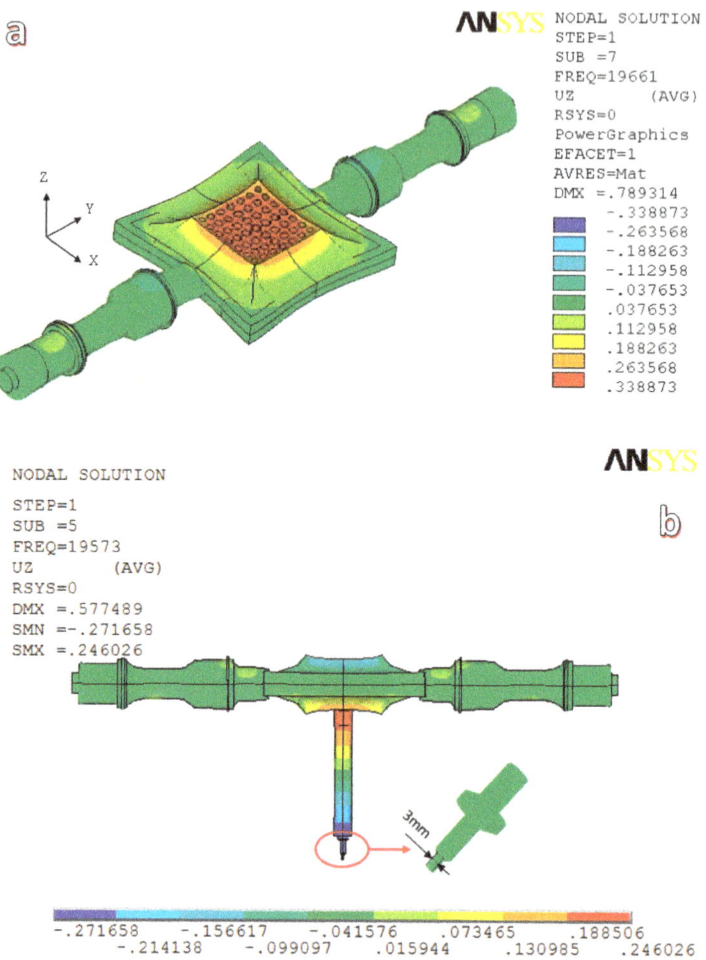

Figure 2. Models of the ultrasonic microextrusion systems in different situations: (**a**) model of the unloaded ultrasonic system; (**b**) model of the ultrasonic system loaded with a forming tool.

In addition to the ultrasonic vibration system, the microextrusion device was composed of an M-4050 universal test machine produced by Shenzhen REGER (Shenzhen, China) and a KSY-6D-16 electric heating device produced by Wuhan Yahua Electric Furnace Co., Ltd. (Wuhan, China) (the temperature range was 50–1000 °C, and the temperature control precision was ±1 °C). The ultrasonic amplitude of the indenter was adjusted by the power output percentage of the ultrasonic generator, and the power output percentages in the experiments were set to 30%, 40%, 50%, and 60%. The corresponding ultrasonic amplitudes at the end of the indenter in the vertical direction were detected by a V100 laser vibration meter from RION Co., Ltd. (Tokyo, Japan) and were 12, 16, 20, and 24 μm, respectively. The mold and the indenter were made from H13 hot-work die steel (Guangzhou Hengwei Electromechanical Equipment Co., Ltd., Guangzhou, China). The two parts of the mold were clamped together by bolts, and the microextrusion mold cavity was composed of 3 circular subsection slots, L_1, L_2, and L_3, whose diameters were 2, 1, and 0.75 mm, and lengths 2, 4, and 23.5 mm, respectively. The structure diagram of the mold and the indenter are also shown in Figure 1. The experimental temperature was set at 370, 380, and 390 °C in supercooled liquid region, while the microextrusion rate was set at 0.36 mm/min. The experimental microextrusion rate was set at 0.12, 0.24, 0.36 mm/min, while the temperature was set at 380 °C. During the ultrasonic microextrusion experiments, the mold was first heated to the setting temperature, and then the Zr35 sample was put into the mold, keeping the temperature for about 10 min, and the indenter contacted the sample with a pre-load of about 200 N. Then, the ultrasonic generator turned on, and the indenter started producing vertical ultrasonic vibration in addition to depressing the sample.

3. Results

3.1. Effect of Ultrasonic Power Output

Figure 3 illustrates the microextrusion results under different ultrasonic power outputs (which corresponds to various ultrasonic vibration amplitudes for the indenter) and the filling lengths of the 3 different circular slots and the entire extrusion length are shown in Table 1. From Figure 3a, the extrusion length (L) increases with increasing ultrasonic power output, such as L is only 3.43 mm for the supercooled liquid MGs formed under static loading, while the value of L increases to 15.42 mm under ultrasonic loading with power output of 60%. The detailed data are summarized in Table 1, from which the Zr35 alloy can be pushed into the L_3 circular slot (0.75 mm diameter) when the ultrasonic power output is greater than 30%, which indicates that a sufficiently large ultrasonic vibration amplitude is necessary for enhancing microfilling capacities of the Zr35 MG.

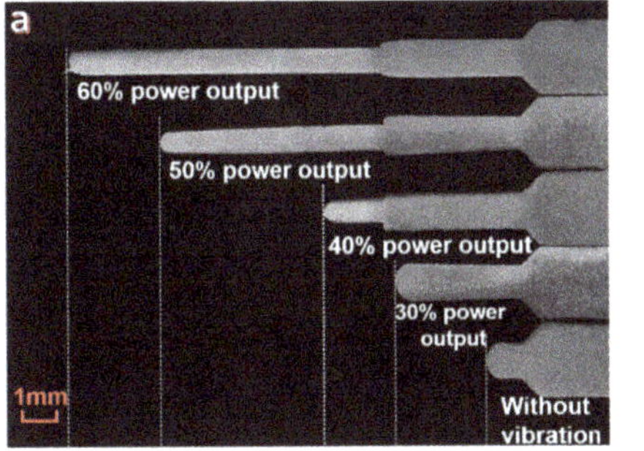

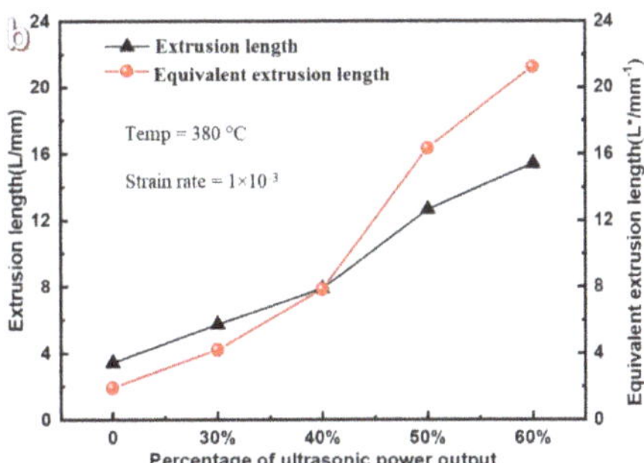

Figure 3. Microphotos of extruded Zr35 alloy (**a**) and recalculated microextrusion lengths of samples at various ultrasonic power outputs (**b**).

Table 1. Microextrusion length of Zr35 alloy at various ultrasonic output powers.

Ultrasonic Output Power	First Section Slot Length (L_1) (mm)	Second Section Slot Length (L_2) (mm)	Third Section Slot Length (L_3) (mm)	Whole Extrusion Length ($L = L_1 + L_2 + L_3$) (mm)
0	2	1.43	0	3.43
30%	2	3.72	0	5.72
40%	2	4	1.89	7.89
50%	2	4	6.66	12.66
60%	2	4	9.42	15.42

According to the previous research [24], the microforming ability of MGs on mold filling in the SCLR can be described as,

$$P = 32\eta \times v \frac{L}{d^2} \tag{1}$$

where P is the flow stress of the MGs, η is the apparent viscosity of a fluid, v is the fluid velocity, d is the diameter of the cylinder groove, and L is the fluid filling depth.

Accordingly, a similar approach is introduced to describe the microfilling capacity of MG at various sections with different sectional areas. L_i^* is defined as the equivalent filling length of the different subsections, assuming that the whole equivalent microextrusion length is defined as L^*, then,

$$L^* = \sum_{i=1}^{3} L_i^* = \frac{L_1}{d_1^2} + \frac{L_2}{d_2^2} + \frac{L_3}{d_3^2} \tag{2}$$

where L_1, L_2, and L_3 are the microfilling lengths of the different subsection slots shown in Figure 1, and d_1, d_2, and d_3 are the corresponding diameters of these circular slots.

The extrusion lengths and the corresponding equivalent extrusion lengths are calculated, and summarized and described in Figure 3b. With increasing ultrasonic power output, it is clear that the equivalent extrusion lengths increase faster than the extrusion lengths. When the ultrasonic power output increased from 0% to 60%, the equivalent microextrusion lengths increased from 1.93 to 21.25 mm, which is 11 times longer than that of static loading, while the corresponding microextrusion length is just 4.5 times greater than that under static loading. Therefore, the equivalent extrusion length more objectively reflects the variation trends of the microfilling capacities of MGs at the micro- and nanoscales.

The true stress–strain curves of Zr35 MG extruded with different ultrasonic power outputs are depicted in Figure 4. The results show that the flow stress gradually reduces with increasing ultrasonic power output. Compared with the static loading process, the true stress is maximally decreased by 12.49%–60.17% when the ultrasonic power output increases from 30% to 60%.

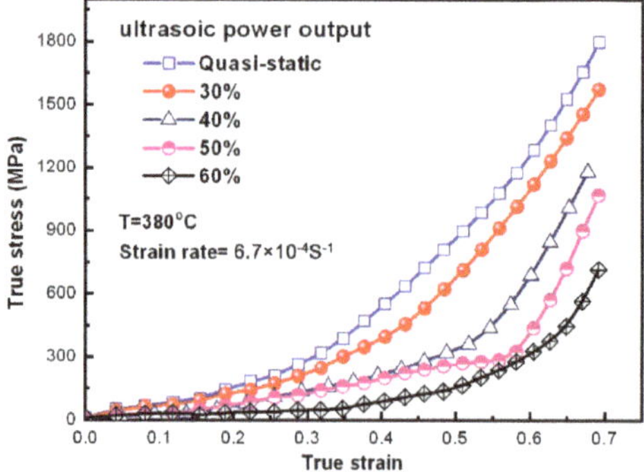

Figure 4. True stress–strain curves for different ultrasonic power outputs.

3.2. Ultrasonic Loading under Various Temperatures

The ultrasonic microextrusion results with different supercooled liquid temperatures are shown in Figure 5, where the microextrusion rate is 0.36 mm/min and the ultrasonic power output is 40%. Figure 5a shows the microextrusion length (L) of Zr35 alloy at different supercooled liquid temperatures. Overall, the extrusion length increases with increasing supercooled liquid temperatures, wherein L with ultrasonic vibration at 370 or 380 °C is larger than that with static loading at 380 or 390 °C, which are marked with red and blue circles, respectively (the extrusion lengths for different temperatures are shown in Table 2). This phenomenon indicates that the ultrasonic vibration can improve the microfilling capacity of Zr35 MG similarly with temperature raising in the SCLR.

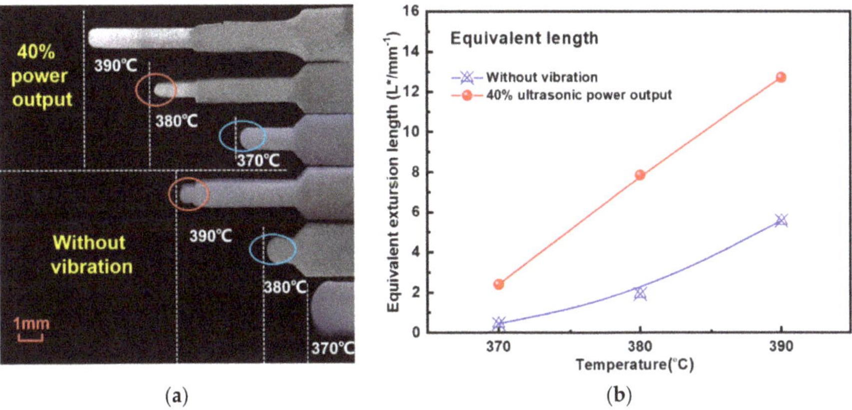

(a)   (b)

Figure 5. Ultrasonic microfilling length of Zr35 at different supercooled liquid temperatures: (a) microextrusion length (*L*); (b) equivalent extrusion length.

Table 2. Extrusion length of Zr35 at different supercooled liquid temperatures.

Extrusion Length and Temperature	Without Ultrasonic Vibration			With 40% Ultrasonic Output Power		
Supercooled liquid temperature (°C)	370	380	390	370	380	390
First section length (L_1/mm)	1.81	2.00	2.00	2.00	2.00	2.00
Second section length (L_2/mm)	0	1.43	4.00	1.90	4.00	4.00
Third section length (L_3/mm)	0	0	0.60	0	1.89	4.61
Whole extrusion length ($L = L_1 + L_2 + L_3$/mm)	1.81	3.43	6.60	3.90	7.89	10.61

The equivalent extrusion lengths at different supercooled liquid temperatures are summarized in Figure 5b. The results reveal that the equivalent microextrusion lengths at 40% ultrasonic power output increase by 2, 5.83, and 7.13 respectively, compared with those of static loading under supercooled liquid temperatures of 370, 380, and 390 °C.

The true stress–strain curves of Zr35 extruded at different supercooled liquid temperatures are shown in Figure 6. The results illustrate that the flow stress of Zr35 gradually reduces with increasing supercooled liquid temperatures. Compared with the static loading process, the true stress at 40% ultrasonic power output is maximally decreased by 23.67%, 34.36%, and 34.92%, respectively, when the supercooled liquid temperatures are 370, 380, and 390 °C.

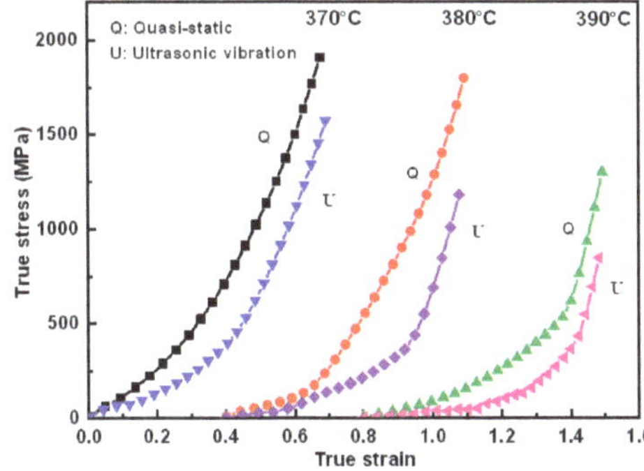

Figure 6. True stress–strain curves at different supercooled liquid temperatures.

3.3. Effect of Microextrusion Rate

Figure 7 illustrates the ultrasonic microextrusion lengths with various microextrusion rates, which is also summarized in Table 3, wherein the temperature is 380 °C and the ultrasonic power output is 40%. It is clear that the extrusion length reduces with the increasing of microextrusion speed. On the other hand, L is only 5.02 mm for the extrusion rate of 0.12 mm/min under static loading, while the value of L increases to 10.68 mm when the ultrasonic loading with power output of 40%, exhibiting an ultrasonic vibration dependence.

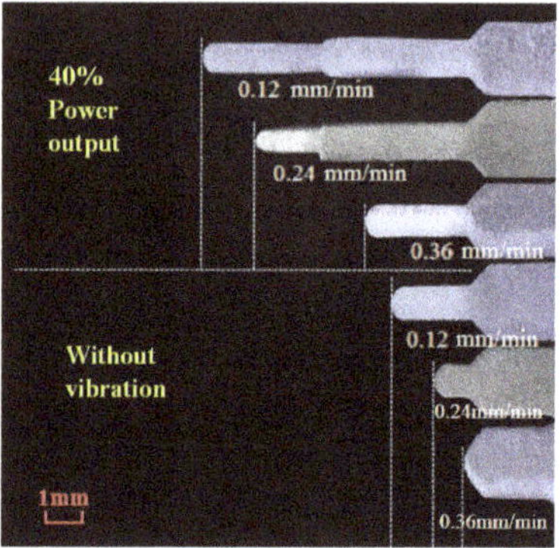

Figure 7. Ultrasonic extrusion length of Zr35 MG at various microextrusion rates.

Table 3. Extrusion length of Zr35 at different microextrusion rates.

Extrusion Rate and Length	Without Ultrasonic Vibration			With 40% Ultrasonic Output Power		
Microextrusion rate (mm/min)	0.12	0.24	0.36	0.12	0.24	0.36
First section length (L_1/mm)	2.00	2.00	2.00	2.00	2.00	2.00
Second section length (L_2/mm)	3.02	1.93	0.78	4.00	4.00	3.93
Third section length (L_3/mm)	0	0	0	4.68	2.39	0
Whole extrusion length ($L = L_1 + L_2 + L_3$/mm)	5.02	3.93	2.78	10.68	8.39	5.93

Finally, the glassy structure of the microformed Zr35 alloys were characterized with X-ray diffraction, and one of the representative results of 380 °C and 60% ultrasonic power output is illustrated in Figure 8, wherein only a wide dispersion peak appears, demonstrating that the Zr35 alloy remains as amorphous structure after ultrasonic uniaxial compression.

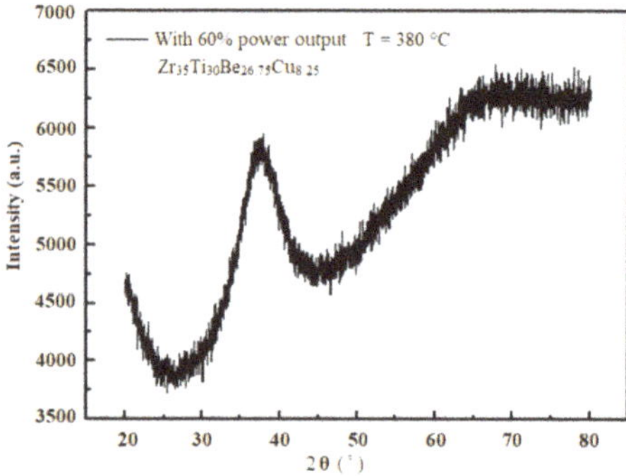

Figure 8. XRD results of ultrasonic uniaxial compressed Zr35 sample.

4. Discussion

The above phenomena can be analyzed in terms of ultrasonic energy transmission. Actually, it is very difficult to directly measure the ultrasonic vibration characteristics of MGs during the ultrasonic microextrusion process. Therefore, the ultrasonic vibration energy of the alloy obtained from the ultrasonic microextrusion process could be estimated according to the ultrasound energy density between the ultrasonic vibrating indenter and the Zr35 sample.

According to Yao [25], the coefficient of the sound power transmission from the H13 indenter to the Zr35 sample can be expressed as in Equation (3).

$$\alpha_t = \frac{4\rho_{H13}c_{H13}\rho_{Zr35}c_{Zr35}}{\left(\rho_{H13}c_{H13} + \rho_{Zr35}c_{Zr35}\right)^2} \tag{3}$$

where C_{H13} and C_{Zr35} are the longitudinal vibration wave velocities of H13 die steel and MG samples, respectively. According to the material properties of H13 steel [26] and Zr35 [27], the value of C_{Zr35} is 2861.5 m/s and the value of C_{H13} is 6072 m/s. Therefore, the value of α_t is 0.739.

The sound energy density gained by the Zr35 sample can be calculated by Equation (4). The sound energy densities of the Zr35 samples at different ultrasonic power outputs are shown in Table 4.

$$E_n = \frac{1}{2}\xi^2_{H13}\omega^2\rho_{H13}\alpha_t \tag{4}$$

E_n is the sound energy density of Zr35, ξ_{H13} is the vibration amplitude of the indenter, and ω is the excitation angular frequency ($\omega = 2\pi f = 125{,}600$ rad/s).

Table 4. The ultrasonic energy densities and amplitudes at different ultrasonic power outputs.

Ultrasonic Power Output	Ultrasonic Amplitudes ξ_{H13} (μm)	Energy Density E_n (kJ/m^3)
0	0	0
30%	12	7.41
40%	16	13.173
50%	20	20.583
60%	24	29.64

Based on the ultrasonic softening and thermal activation theories [25], the reduction of the material flow stress caused by the ultrasonic vibrations can be attributed to the ultrasonic volume effect, which is comprised of the acoustic softening and stress superposition effects. The total flow stress reduction by the ultrasonic volume effect can be expressed as in Equation (5).

$$\Delta\sigma_s = -K(\alpha_t\xi)^n - 2\beta\hat{\tau}(E_n/\hat{\tau})^m \tag{5}$$

The negative sign indicates stress reduction, β and m are experimental constants of the material ($m > 0$), $\hat{\tau}$ is the shear threshold of the material, which is equivalent to the shear strength of the material at absolute zero, α_t is the sound power transmission coefficient, n is the hardening index and the value is 1, and K is the material strength coefficient, which is the slope of the curve of stress reduction and ultrasonic amplitude caused by the stress superposition.

The parameters of β, m, $\hat{\tau}$, K, and n are intrinsic parameters of the material, so the shear stress drop, $\triangle\sigma_s$, is only related and proportional to the ultrasonic energy density, E_n, and ultrasound amplitude, ξ. With an increase in the indenter's ultrasonic vibration amplitude, the ultrasonic energy density, E_n, absorbed by the Zr35 sample, should increase. Moreover, the value of the shear stress drop, $\triangle\sigma_s$, also increases, and the flow stress of Zr35 appears to decrease, which suggests that the ultrasonic volume effect on MGs is more obvious. Thus, the microforming capacity of Zr35 is greatly enhanced, and the equivalent whole extrusion length, L^*, becomes longer.

The phenomenon of true stress–strain various with ultrasonic power output in Figure 4 can also be illustrated with the Equation (5). With increasing of power output, the ultrasonic vibration amplitude of the indenter is increased, while the stress superposition effect generated by the high frequency intermittent impact between the indenter and Zr35 sample is also strengthened, which possibly increases the decrease of the true stress of MGs. It is worth noting that previous research revealed that as for the metallic glasses, the vibration induced softening responsible for the reduction of stress [6]. On the other hand, part of the absorbed ultrasonic energy converted into heat [28], which can improve the material flow of Zr35 alloy in the SCLR. Furthermore, the ultrasonic vibration of the indenter can generate the ultrasonic softening effect in Zr35 [29], which will reduce the deformation resistance and improve the microfilling capacity of Zr35. Therefore, improvement in the microfilling capacity of MGs should be the combined effect of the ultrasonic softening effect and increasing temperature, and should be more obvious for MGs at lower temperatures. This is the reason why the ultrasonic microextrusion process can have a similar effect on improving the microforming capacity of MGs in the SCLR as increasing temperature. The increase in the microfilling capacity of Zr35 by ultrasonic vibration reaches a maximum value at 390 °C.

Furthermore, the phenomenon of true stress–strain with respect to supercooled liquid temperature in Figure 6 can be explained with free volume theory and viscosity variation of MGs. The higher temperature can lead to the increment of free volume concentration in the Zr35 and the decrease of viscosity, which facilitates the thermoplastic forming of Zr35 alloy. When the ultrasonic vibration is superimposed on the Zr35 samples, atomic diffusion increases and, therefore, there is a more conspicuous softening phenomenon.

5. FEM Analysis

In order to analyze the ultrasonic flow deformation mechanism of MGs in the SCLR, ABAQUS software was used to simulate the effects of the ultrasonic vibration amplitude on the flow deformation behavior of Zr35 during ultrasonic TPMF in the SCLR. A constitutive relation based on the free volume model was selected [30] and expressed as:

$$\ln \tau_s = (m + n)\left[\frac{W}{KT} + (\ln \gamma - \ln \gamma_0)\right] + \ln\left[s\left(\frac{1}{4}\right)^m\right] \tag{6}$$

where $\tau_s = 1/2\sigma$ is the shear stress, $m = n = -1.31 \times 10^{-3}T + 1.00478$ and $s = 21.6 + 866\{1 + \exp[(T - 674)/13.5]$ are the temperature-dependent parameters, and $\gamma_0 = 5.07 \times 10^{33} \text{ s}^{-1}$ is the reference strain rate of Vitreloy-1 BMG (i.e., $Zr_{41.2}Ti_{13.8}Cu_{12.5}$–$Ni_{10}Be_{22.5}$ BMG). The Zr-based MGs studied in the present work have a similar composition to the Vitreloy-1 BMG, and the above parameters can be used in the calculation.

To eliminate the influence of friction on the flow deformation behavior of the MGs, the uniaxial tensile process of the MGs was simulated to study the ultrasonic-assisted flow deformation mechanism of the MGs. The sample for the FEM simulations was set to be a cylindrical Zr35 sample with a diameter of 3 mm and a height of 10 mm. An axisymmetric two-dimensional simulation model was used for quasi-static analysis because of the symmetrical structural characteristics of the sample. The upper end of the sample was stretched, and the lower end was fixed. The tensile stroke was set at 4 mm.

In the ultrasonic uniaxial tensile simulation experiments of Zr35, the supercooled liquid temperature was set to 370 °C, the ultrasonic frequency was set to 20 kHz, and the ultrasonic amplitudes were set to 0, 12, and 24 μm, in turn. A 20 mm/s drawing speed was applied to the upper end of the sample, and a sinusoidal displacement was superimposed onto the lower end of the sample for simulating the ultrasonic vibration. When the vibration amplitude was 12 μm, the superimposed displacement could be calculated as follows:

$$S = 0.012 \sin 125600t \tag{7}$$

The distribution of simulated equivalent stress of uniaxial tensile with ultrasonic vibration amplitudes of 0, 12, and 24 μm are shown in Figure 9a–c. The maximum equivalent stress at the different vibration amplitudes is 159.7, 138.2, and 114.0 MPa, respectively, which is a reduction of 13.5% and 28.6% with increasing ultrasonic amplitude. The distribution area of the maximum equivalent stress also decreased, which is shown in the red area of Figure 8. Additionally, the value of the equivalent strain is shown in Figure 10a–c, which was enhanced by the increasing ultrasonic vibration amplitude. The maximum equivalent strain increased by 16.74% and 33.38% when the ultrasonic vibration amplitude was increased to 12 and 24 μm, respectively. The maximum strain area exists in the upper and middle parts of the sample.

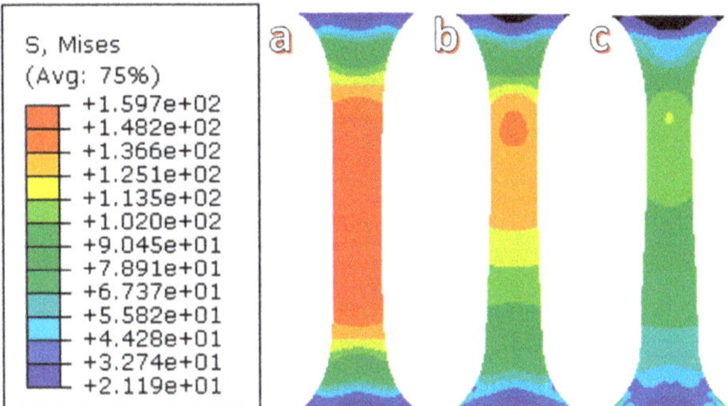

Figure 9. The distribution of the simulated equivalent stress the uniaxial tensile of Zr35 at different ultrasonic vibration amplitudes: (**a**) 0 μm; (**b**) 12 μm; and (**c**) 24 μm.

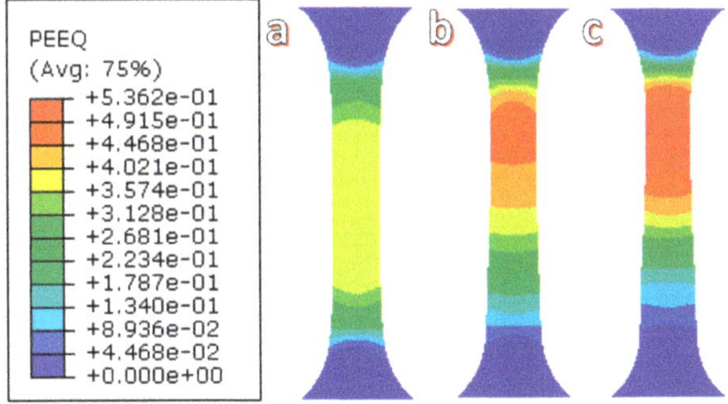

Figure 10. The distribution of the simulated equivalent strain of the uniaxial tensile of Zr35 at different ultrasonic vibration amplitudes: (**a**) 0 μm; (**b**) 12 μm; and (**c**) 24 μm.

The reason for this phenomenon is mainly attributed to the ultrasonic softening effect of the material caused by the assisted ultrasonic vibration. In the process of ultrasonic plastic deformation, the stress field generated by the internal deformation of the material is superimposed over the periodic stress field of the ultrasonic vibration. With an increase in the ultrasonic amplitude, the effect of ultrasonic softening becomes more obvious, the plastic forming of the material becomes easier, and more free volume is generated in the plastic deformation process of the MGs. The higher free-volume concentration in the MGs means a lower viscosity for the MGs, together with smaller flow units in the sample under assisted ultrasonic vibration. The more homogeneous spatiotemporal distribution of flow units facilitates the thermoplastic microformability of the MGs, based on the above experimental results and theoretical analysis.

The profiles of the corresponding velocity distributions obtained from the above FEM simulation results are shown in Figure 11. The results show that the velocity is distributed heterogeneously along the uniaxial tensile direction due to the unidirectional stress. To further distinguish differences in the material flow during the ultrasonic uniaxial tensile process at different vibration amplitudes (i.e., 0, 15, and 30 μm), the velocities (*V*) at positions A–C (with an amplitude of 0 μm), A1–C1 (with an amplitude of 15 μm) and A2–C2 (with an amplitude of 30 μm) are calculated and compared, as shown in Figure 11a. As delineated in Figure 11b,c, the flowing velocity has a value ranging from 19.85 to 1869.99 mm/s at positions A1–C1, and a value ranging from 421.71 to 3874.21 mm/s at positions A2–C2, which are two orders of magnitude greater than the velocities with vibration in the A–C areas (0.21–19.85 mm/s). This indicates that a larger ultrasonic vibration amplitude corresponds to a greater change in the flow rate of the amorphous alloy.

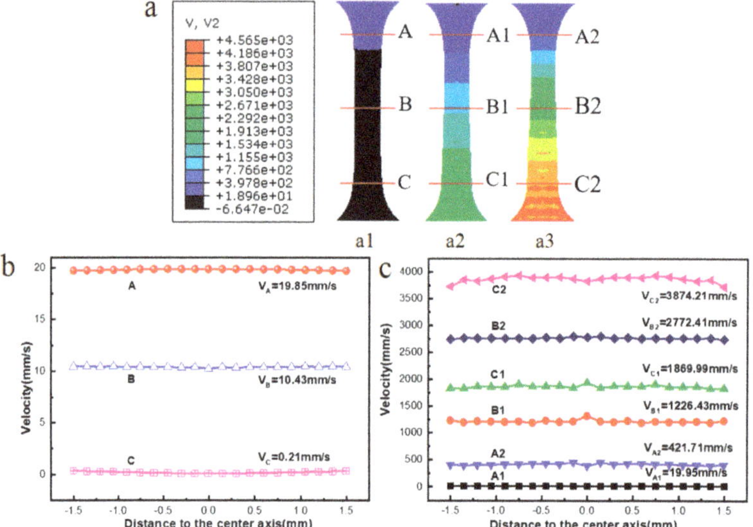

Figure 11. The simulated velocity distribution under three ultrasonic amplitudes by assuming a temperature of 370 °C: (**a1**) 0 μm; (**a2**) 15 μm; and (**a3**) 30 μm. (**b,c**) present the corresponding speed distribution at various positions.

Kim proposed a free-volume constitutive model to characterize the internal atom changes in amorphous alloys by studying the deformation behavior of amorphous alloys [30]. The relationship between free volume and strain rate is expressed in Equation (8).

$$v_f^* = \left[\frac{1}{v_{fe}} - (c \ln \gamma + l) \right]^{-1} \tag{8}$$

Here, v_f^* is the steady state free volume, $v_{fe} = (T - T_0)/DT_0$ is the balanced free volume, and c and l are temperature-related parameters.

From the above Equation (8), it can be concluded that an increase in the strain rate will lead to an increase in the steady-state free volume and a decrease in the viscosity of the amorphous alloy, and the occurrence of plastic deformation will also be easier.

In general, higher ultrasonic vibration amplitudes can achieve higher flow velocities and easier material flow of Zr35 MGs in the SCLR. Thus, the microplastic forming capacity of Zr35 can be gradually improved by increasing the vibration amplitude, which is in agreement with the analysis of the equivalent stress and strain above.

6. Conclusions

In this work, the ultrasonic vibration was introduced to the microextrusion processing of MGs. The ultrasonic microformability of Zr35 in the SCLR was studied with experiments and FEM. The conclusions are as follows.

- Ultrasonic vibration is an effective method to improve the thermoplastic microformability of MGs by superimposing the assisted ultrasonic vibration onto the microextrusion tool.
- With increasing ultrasonic power output, the equivalent microfilling length of Zr35 can be increased by 11-fold, and the true stress can be decreased by 60.17%.
- Sufficient ultrasonic power output (>30%), which means a sufficiently large ultrasonic vibration amplitude of the microextrusion tool, is the essential condition to obtain better microformability of Zr35, which can be successfully packed into a circular slot with a 0.75 mm diameter.
- Ultrasonic vibration of the microextrusion tool can produce a similar effect as temperature increase on improving the microformability of Zr35 in the SCLR.
- Larger ultrasonic vibration amplitudes of the tool can generate more free volume in the plastic microforming process of Zr35 MGs, which can then obtain a higher flow velocity and easier material flow.

Author Contributions: Data Curation, G.H. and L.X.; Formal Analysis, Z.P.; Supervision, N.L.; Writing—Original Draft Preparation, G.H.; Writing–Review & Editing, G.H. and N.L.

Funding: This research was funded by the Science and Technology Support Program of Hubei Province, China (No. 2015BAA019), the Open Research Fund Program of Shanxi Key Laboratory of Non-Traditional Machining (No. 2017SXTZKFJG01), and the Open Research Subject of Key Laboratory of Manufacturing and automation (No. szjj2017-006), and supported by the fund of the State Key Laboratory of Solidification Processing in NWPU (No. SKLSP201611).

Conflicts of Interest: The authors declare no conflicts of interest.

References

1. Khan, M.M.; Nemati, A.; Rahman, Z.U.; Shah, U.H.; Asgar, H.; Haider, W. Recent advancements in bulk metallic glasses and their applications: A review. *Crit. Rev. Solid State Mater. Sci.* **2018**, *43*, 233–268. [CrossRef]
2. Li, N.; Chen, W.; Liu, L. Thermoplastic micro-forming of bulk metallic glasses: A review. *JOM* **2016**, *68*, 1246–1261. [CrossRef]
3. Qiao, J.C.; Yao, Y.; Pelletier, J.M.; Keer, L.M. Understanding of micro-alloying on plasticity in $Cu_{46}Zr_{47-x}Al_7Dy_x$ ($0 \leq x \leq 8$) bulk metallic glasses under compression: Based on mechanical relaxations and theoretical analysis. *Int. J. Plasticity* **2016**, *82*, 62–75. [CrossRef]
4. Wang, X.; Dai, W.; Zhang, M.; Gong, P.; Li, N. Thermoplastic micro-formability of TiZrHfNiCuBe high entropy metallic glass. *J. Mater. Sci. Technol.* **2018**, *34*, 2006–2013. [CrossRef]
5. Li, N.; Zhang, J.; Xing, W.; Ouyang, D.; Liu, L. 3D printing of Fe-based bulk metallic glass composites with combined high strength and fracture toughness. *Mater. Des.* **2018**, *143*, 285–296. [CrossRef]
6. Li, N.; Xu, X.; Zheng, Z.; Liu, L. Enhanced formability of a Zr-based bulk metallic glass in a supercooled liquid state by vibrational loading. *Acta Mater.* **2014**, *65*, 400–411. [CrossRef]
7. Lee, K.S.; Kim, S.; Lim, K.R.; Hong, S.H.; Kim, K.B.; Na, Y.S. Crystallization, high temperature deformation behavior and solid-to-solid formability of a Ti-based bulk metallic glass within supercooled liquid region. *J. Alloy. Compd.* **2016**, *663*, 270–278. [CrossRef]
8. Li, C.; Zhu, F.; Zhang, X.; Ding, J.; Yin, J.; Wang, Z.; Zhao, Y.; Kou, S. The rheological behavior and thermoplastic deformation of Zr-based bulk metallic glasses. *J. Non-Cryst. Solids* **2018**, *492*, 140–145. [CrossRef]
9. Hong, S.H.; Kim, J.T.; Mun, S.C.; Kim, Y.S.; Park, H.J.; Na, Y.S.; Lim, K.R.; Park, J.M.; Kim, K.B. Influence of spherical particles and interfacial stress distribution on viscous flow behavior of Ti–Cu–Ni–Zr–Sn bulk metallic glass composites. *Intermetallics* **2017**, *91*, 90–94. [CrossRef]
10. Li, N.; Chen, Y.; Jiang, M.; Li, D.; He, J.; Wu, Y.; Liu, L. A thermoplastic forming map of a Zr-based bulk metallic glass. *Acta Mater.* **2013**, *61*, 1921–1931. [CrossRef]

11. Johnson, W.L.; Kaltenboeck, G.; Demetriou, M.D.; Schramm, J.P.; Liu, X.; Samwer, K.; Kim, C.P.; Hofmann, D.C. Beating crystallization in glass-forming metals by millisecond heating and processing. *Science* **2011**, *332*, 828–833. [CrossRef] [PubMed]

12. Kaltenboeck, G.; Demetriou, M.D.; Roberts, S.; Johnson, W.L. Shaping metallic glasses by electromagnetic pulsing. *Nature Commun.* **2016**, *7*, 10576. [CrossRef] [PubMed]

13. Martinez, R.; Kumar, G.; Schroers, J. Hot rolling of bulk metallic glass in its supercooled liquid region. *Scr. Mater.* **2008**, *59*, 187–190. [CrossRef]

14. Schroers, J.; Hodges, T.M.; Kumar, G.; Raman, H.; Barnes, A.J.; Pham, Q.; Waniuk, T.A. Thermoplastic blow molding of metals. *Mater. Today* **2011**, *14*, 14–19. [CrossRef]

15. Li, J.; Zheng, Z.; Wu, X.; Li, J. A study on micro-forming ability of Zr55 bulk metallic glass under low frequency vibrating field. *J. Plasticity Eng.* **2015**, *22*, 118–124. (In Chinese)

16. Huang, Y.; Wu, Y.; Huang, J. The influence of ultrasonic vibration-assisted micro-deep drawing process. *Int. J. Adv. Manuf. Technol.* **2014**, *71*, 1455–1461. [CrossRef]

17. Bai, Y.; Yang, M. Investigation on mechanism of metal foil surface finishing with vibration-assisted micro-forging. *J. Mater. Process. Technol.* **2013**, *213*, 330–336. [CrossRef]

18. Han, G.; Li, K.; Peng, Z.; Jin, J.; Sun, M.; Wang, X. A new porous block sonotrode for ultrasonic assisted micro plastic forming. *Int. J. Adv. Manuf. Technol.* **2017**, *89*, 2193–2202. [CrossRef]

19. Michalski, M.; Lechner, M.; Gruber, M.; Merklein, M. Influence of ultrasonic vibration on the shear formability of metallic materials. *CIRP Annals* **2018**, *67*, 277–280. [CrossRef]

20. Liang, X.; Ma, J.; Wu, X.Y.; Xu, B.; Gong, F.; Lei, J.G.; Peng, T.J.; Cheng, R. Micro injection of metallic glasses parts under ultrasonic vibration. *J. Mater. Sci. Technol.* **2017**, *33*, 703–707. [CrossRef]

21. Luo, F.; Sun, F.; Li, K.; Gong, F.; Liang, X.; Wu, X.; Ma, J. Ultrasonic assisted micro-shear punching of amorphous alloy. *Mater. Res. Lett.* **2018**, *6*, 545–551. [CrossRef]

22. Duan, G.; Wiest, A.; Lind, M.L.; Li, J.; Rhim, W.-K.; Johnson, W.L. Bulk metallic glass with benchmark thermoplastic processability. *Adv. Mater.* **2007**, *19*, 4272–4275. [CrossRef]

23. Han, G.; Li, K.; Wang, X. A Double Transducer Driving Ultrasonic Vibration Platform. Chinese Patent ZL 201520071898.1, 8 July 2015.

24. Chiu, H.M.; Kumar, G.; Blawzdziewicz, J.; Schroers, J. Thermoplastic extrusion of bulk metallic glass. *Scr. Mater.* **2009**, *61*, 28–31. [CrossRef]

25. Yao, Z.; Kim, G.-Y.; Faidley, L.; Zou, Q.; Mei, D.; Chen, Z. Effects of superimposed high-frequency vibration on deformation of aluminum in micro/meso-scale upsetting. *J. Mater. Process. Technol.* **2012**, *212*, 640–646. [CrossRef]

26. Wang, C.; Liu, Y.; Guo, B.; Shan, D.; Zhang, B. Acoustic softening and stress superposition in ultrasonic vibration assisted uniaxial tension of copper foil: Experiments and modeling. *Mater. Des.* **2016**, *112*, 246–253. [CrossRef]

27. He, J.; Li, N.; Tang, N.; Wang, X.; Zhang, C.; Liu, L. The precision replication of a microchannel mould by hot-embossing a Zr-based bulk metallic glass. *Intermetallics* **2012**, *21*, 50–55. [CrossRef]

28. Lu, J.; Wu, X.; Wu, Z.; Liu, Z.; Guo, D.; Lou, Y.; Ruan, S. Microstructure and mechanical properties of ultrafine-grained Al-6061 prepared using intermittent ultrasonic-assisted equal-channel angular pressing. *J. Mater. Eng. Perform.* **2017**, *26*, 5107–5117. [CrossRef]

29. Liu, Y.; Suslov, S.; Han, Q.; Hua, L.; Xu, C. Comparison between ultrasonic vibration-assisted upsetting and conventional upsetting. *Metall. Mater. Trans. A* **2013**, *44*, 3232–3244. [CrossRef]

30. Xu, X.; Li, N.; Liu, L. Intrinsic or extrinsic size-dependent deformation behavior of a Zr-based bulk metallic glass in supercooled liquid state? *J. Alloy. Compd.* **2014**, *589*, 524–530. [CrossRef]

 *materials*

Article

A Numerical Study of Slip System Evolution in Ultra-Thin Stainless Steel Foil

Zhongkai Ren [1], Wanwan Fan [1], Jie Hou [2] and Tao Wang [1,*]

[1]   College of Mechanical and Vehicle Engineering, Taiyuan University of Technology, Taiyuan 030024, China; zhongkai_0808@126.com (Z.R.); tyutfww@163.com (W.F.)
[2]   Analysis and Test Center, Taiyuan University of Science and Technology, Taiyuan 030024, China; houjie@yeah.net
*   Correspondence: tyutwt@163.com; Tel.: +86-187-3517-0186

Received: 13 May 2019; Accepted: 3 June 2019; Published: 5 June 2019

Abstract: In order to quantitatively describe the effect of the initial grain orientation on the inhomogeneous deformation of 304 austenitic stainless steel foil during tension, a three-dimensional uniaxial tension model was established, based on the crystal plasticity finite element method (CPFEM) and Voronoi polyhedron theory. A three-dimensional representative volume element (RVE) was used to simulate the slip deformation of 304 stainless steel foil with five typical grain orientations under the same engineering strain. The simulation results show that the number and characteristics of active slip systems and the deformation degree of the grain are different due to the different initial grain orientations. The slip systems preferentially initiate at grain boundaries and cause slip system activity at the interior and free surface of the grain. The Brass, S, and Copper oriented 304 stainless steel foil exhibits a high strain hardening index, which is beneficial to strengthening. However, the Cube and Goss oriented 304 stainless steel foil has a low deformation resistance and is prone to plastic deformation.

Keywords: ultra-thin foil; slip system evolution; tensile process; crystal plasticity; numerical simulation; grain orientation

---

1. Introduction

Precision stainless steel foil with thickness ranging from 0.01 to 0.1 mm is a high-end steel product, which has been widely used in micro-manufacturing, micro-electromechanical, and other sophisticated micro-products [1]. With the increasing miniaturization of components, higher requirements are put forward for high-quality precision stainless steel, such as good ductility, high dimensional accuracy, and low surface roughness [2,3]. When the foil thickness is reduced to microscale, the anisotropy of the material increases, and the size effects become significant [4,5]. In this case, the grain morphology, size, orientation, and grain boundary have important impacts on the plastic deformation behavior [6,7]. Therefore, the traditional macroscopic plastic deformation theory appears to be no longer applicable to the plastic deformation analysis of the stainless steel foil in microscale.

The crystal plasticity finite element method (CPFEM) is widely used to homogenize the discrete dislocations and analyze the macroscopic effects of dislocation slip, based on the continuum theory, which reflects the microstructural characteristics of materials [8–11]. The mechanisms involved in the macro and micro-deformation of non-ferrous metal materials during tensile deformation have been studied by many scholars using CPFEM. Si et al. [12] established a polycrystalline model based on CPFEM and Voronoi polyhedron theory, and studied the effect of grain size on inhomogeneous deformation and grain rotation during the deformation process of polycrystalline aluminum. The results showed that the plastic deformation occurred preferentially at grain boundaries, and the multi-system slip caused local hardening near grain boundaries. Hama and Takuda [13,14] used CPFEM to analyze the hardening and unloading behavior of magnesium alloy strips during tensile deformation. They found that the pyramidal slip system dominated in the loading process, and only the base slip systems were activated during unloading. Ritz and Dawson [15] studied the effect of grain morphology on tensile deformation of aluminum alloys using CPFEM, and pointed out that grain morphology had little effect on the anisotropy of elastic deformation but significantly affected the micro-stress distribution in the grain interior and at the grain boundary. Pi et al. [16] analyzed the effects of polycrystalline model type and tensile strain rate on deformation, necking, and texture evolution of polycrystalline aluminum during the tensile process by CPFEM. Their results showed that the stress increased with the increase in strain rate, and the calculated results by CPFEM were closer to the experimental results.

In order to balance the relationship between the total central processing unit (CPU) calculation time and the model size, the representative volume element (RVE) model with the smallest number of grains was adopted by many scholars to reflect the mechanical behavior of macroscopic samples. Nakamachi et al. [17] proposed a crystalline homogenization algorithm based on multi-scale asymptotic series expansion, established the RVE model for a micro polycrystal structure that could satisfy the periodicity condition of crystal orientation distribution, and verified the accuracy of the RVE model through analytical and statistical examinations. Jigh et al. [18] applied the RVE model to predict the mechanical behavior of metal foams with real microstructure, and studied the effects of different boundary conditions on mechanical properties. Zheng et al. [19] established the RVE model based on CPFEM to predict the influence of structural parameters and relative density of multi-stage honeycomb on its equivalent elastic parameters. The accuracy of the RVE model was verified by comparing with the elastic parameters calculated by the relevant theoretical analytical formula. Kamiński et al. [20] used the iterative and generalized stochastic perturbation technique as a finite element method to determine the effective elasticity tensor for rubber reinforced with the carbon black particles, and applied this to the homogenization problem of the RVE model in such a composite. This study showed some significant differences between numerical and analytical homogenization methods in the context of geometrical uncertainty in the RVE of the composite.

Based on the above literature reviews, it is clear that grain size and morphology as well as strain rate do have an important effect on the plastic deformation behavior of materials. At present, research on the effect of grain orientation on the tensile deformation of 304 austenitic stainless steels, based on the three-dimensional CPFEM model, is still limited. In the present paper, a three-dimensional RVE uniaxial tension model is established based on the CPFEM and the Voronoi polyhedron theory. The effects of initial grain orientation on the inhomogeneous deformation and evolution of slip systems of 304 stainless steel foil during uniaxial tension are analyzed in detail.

2. Crystal Plasticity Model

The main physical mechanism of plastic deformation of the face-centered cubic (FCC) metal at room temperature is dislocation slip [21]. According to the visco-plasticity criterion proposed by Schmid's law [22], the relationship between the shear strain rate $\dot{\gamma}^{\alpha}$ and the resolved shear stress τ^{α} of each slip system is established in the form of the exponential function as:

$$\dot{\gamma}^{\alpha} = \dot{\gamma_0}^{\alpha} f^{\alpha}\left(\frac{\tau^{\alpha}}{\tau_c^{\alpha}}\right), \tag{1}$$

where the resolved shear stress τ^{α} is the driving force of the slip motion. The slip system α starts when the resolved shear stress reaches the critical value τ_c^{α}.

The material rate sensitivity formula, established by Hutchinson [23], is as follows:

$$f^{\alpha} = \frac{\tau^{\alpha}}{\tau_c^{\alpha}} \times \left|\frac{\tau^{\alpha}}{\tau_c^{\alpha}}\right|^{n-1}, \tag{2}$$

where $\dot{\gamma_0}^{\alpha}$ is the reference shear strain rate of slip system α, n is the rate sensitive coefficient, n = 0 and n = ∞ correspond to viscoelasticity and rate-independent materials, respectively, τ^{α} is the resolved shear stress of the slip system α, τ_c^{α} represents the dislocation slip resistance or critical resolved shear stress of the slip system α, which is the accumulation of the shear strain of the corresponding slip system on the hardening of the slip system α. τ_c^α is expressed by:

$$\dot{\tau}_c^\alpha = \sum_{\beta=1}^{N} h_{\alpha\beta} \left| \dot{\gamma}^\beta \right|, \tag{3}$$

where $\dot{\gamma}^\beta$ is the shear strain rate of the slip system β and $h_{\alpha\beta}$ is the latent hardening coefficient.

In the present paper, the hardening model proposed by Bassani and Wu [24,25] was used to describe the interactions between different slip systems with different slip coefficients, which can accurately reflect the slipping process of all three hardening stages of FCC metal. The self-hardening moduli $h_{\alpha\alpha}$ can be expressed as follows:

$$h_{\alpha\alpha} = \left[(h_0 - h_s)\mathrm{sech}^2\left(\frac{(h_0 - h_s)\gamma^\alpha}{\tau_s - \tau_0} + h_s \right) \right] \times G\left(\gamma^\beta; \beta \neq \alpha\right) \tag{4}$$

$$G\left(\gamma^\beta; \beta \neq \alpha\right) = 1 + \sum_{\beta=1\beta\neq\alpha}^{N} f_{\alpha\beta} \tanh\left(\frac{\gamma^\beta}{\gamma_0} \right), \tag{5}$$

where h_0 is the initial hardening modulus; h_s is the hardening modulus in the easy slip stage; τ_0 is the initial critical resolved shear stress; τ_s is the critical shear stress saturation value; γ_0 is the reference shear strain; and $f_{\alpha\beta}$ is the interaction coefficient between the α and β slip system. The factor depends on the geometric relationship of the two slip systems, which is represented by five constants $a_i(i = 1 \sim 5)$.

The latent hardening moduli $h_{\alpha\beta}$ is given by:

$$h_{\alpha\beta} = q h_{\alpha\alpha}, \tag{6}$$

where q is the latent hardening parameter.

The above constitutive model was compiled into the user-defined material subroutine (UMAT) of ABAQUS finite element software by Fortran language for subsequent finite element simulation [26], and the combination of crystal plasticity constitutive theory and finite element method was realized. A numerical integration method using implicit differential equations was adopted to solve the rate-dependent crystal plasticity equation. The UMAT used the ABAQUS internal interface to communicate with the main solver. The specific process is shown in Figure 1.

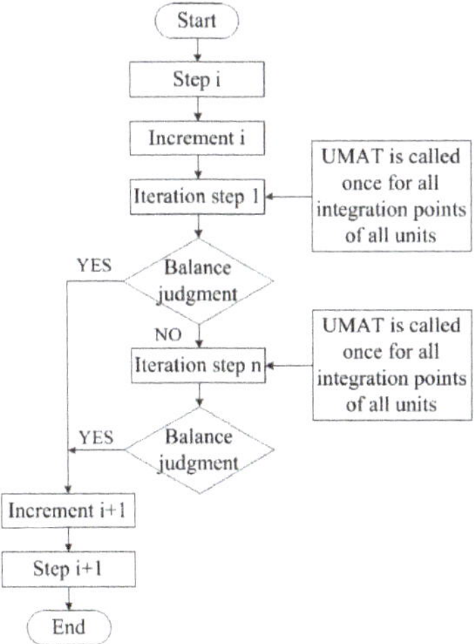

Figure 1. The internal operation flow chart of ABAQUS finite element software.

3. Finite Element Model for Stainless Steel Foil

X-ray diffraction (XRD) phase analysis of the ultra-thin 304 stainless steel foil (0.05 mm) showed that the main phase constituent was FCC austenite. The elastic constants of single crystal austenite are $C_{11} = 209$ GPa, $C_{12} = 133$ GPa, $C_{44} = 121$ GPa [27]. Franciosi et al. [28] obtained the interaction coefficients between stainless steel slip systems by hardening experiments, which could be chosen as $a_1 = 0.625$, $a_2 = 0.045$, $a_3 = 0.045$, $a_4 = 0.137$, $a_5 = 0.122$. The remaining unknown parameters could be determined by comparing the simulation results of RVE uniaxial tension with the measured stress–strain curves.

The average grain size and grain orientation information needed for simulation were obtained by the electron backscatter diffraction (EBSD) tests of the ultra-thin 304 stainless steel foil. EBSD diagrams of 304 stainless steel foil samples measured from the rolling direction (RD) and transverse direction (TD) are shown in Figure 2. The average grain size measured was 15 μm. The maximum density pole of the {111} pole figure was only 1.97 m.r.d (multiple random density), which indicates that the annealed 304 foil had little texture and the grain orientations were almost random. Therefore, the random grain orientations were applied to the polycrystalline model based on the Voronoi polyhedron theory.

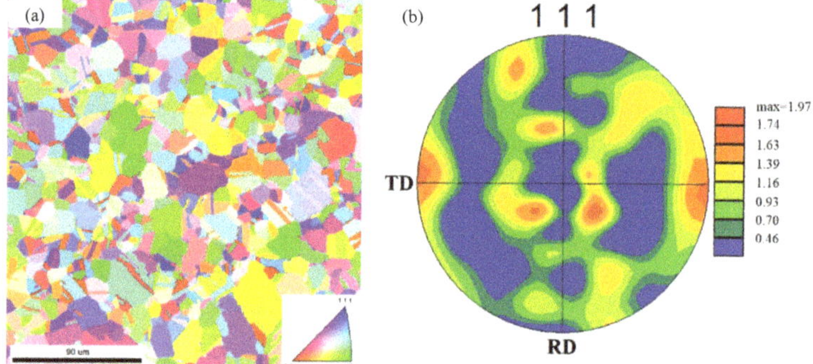

Figure 2. Electron backscatter diffraction (EBSD) mapping of the transverse direction (TD) measured from the rolling direction (RD) of 304 stainless steel foil: (**a**) grain orientation map; (**b**) {111} pole figure.

The RVE model was established based on ABAQUS Standard. The effects of the number of grains and the number of elements in the grains on the tensile deformation accuracy of RVE model were analyzed, as shown in Figure 3. The RVE model contained 512 grains with a grain size of about 15 μm, and each grain contained 125 hexahedral elements (C3D8I), with average element dimensions of $3 \times 3 \times 3$ μm³ (Figure 4b), which could guarantee the calculation accuracy. Different colors in the model represent different grain orientations, and the aggregates of the same orientation are one grain.

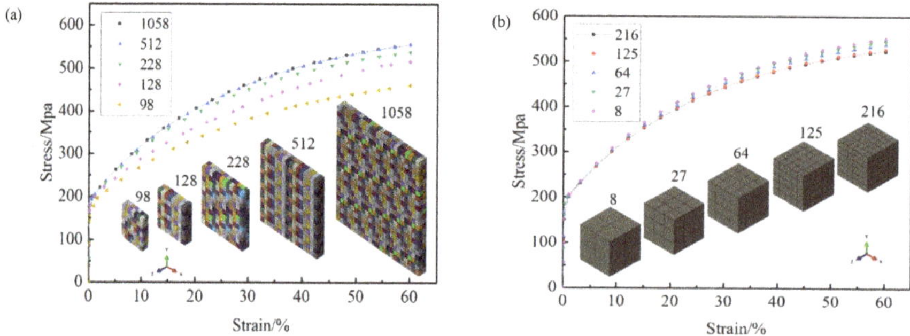

Figure 3. Effect of grain number (**a**) and unit number (**b**) on the representative volume element (RVE) model.

The boundary conditions of the RVE uniaxial tension model are shown in Figure 4a:

(1) The back surface center node was set at X = 0, Y = 0, Z = 0 to avoid the rigid motion of the RVE model as a whole.

(2) The back surface was set at X = 0. There was no constraint in the Y and Z directions, so that plastic deformation could occur in the Y and Z directions.

(3) The front surface was set in a uniform tensile displacement along in the X-direction.

All calculations were run on a workstation (2.6 GHz processor) with 20 multiple processors in parallel. The maximum number of increments was 100,000, to ensure convergence of the RVE model. It took around 5 h to complete one tension test of the RVE with 60% engineering strain.

According to the international standard ASTM-E345-16 [29], 0.05 mm thick 304 stainless steel foil tensile specimens were prepared and stretched using an INSTRON-5969 universal tensile testing machine (Instron corporation, Norwood, MA, USA). In the experiment, the stretching rate was 0.03 mm/min. The RVE model established above was subjected to uniaxial tensile simulation to an engineering strain of 60%. The obtained stress–strain curves were compared with the curves measured by tensile tests. The reasonable crystal plastic parameters were determined by optimization (see Table 1), and the comparison results shown in Figure 5 were obtained.

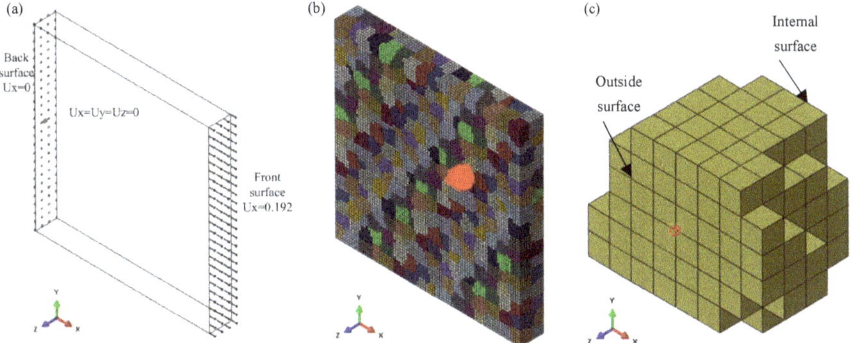

Figure 4. 304 stainless steel RVE tensile model: (**a**) boundary conditions in the simulation; (**b**) RVE model; (**c**) G411.

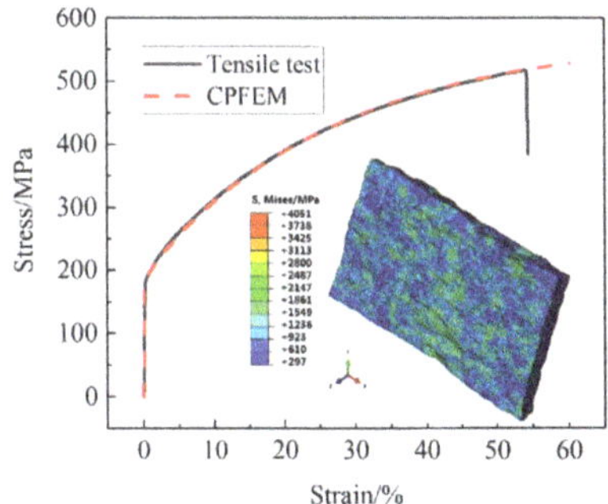

Figure 5. Comparison of the stress–strain curves between RVE simulation and experimental results.

Table 1. Constitutive model parameters of ultra-thin 304 stainless steel foil.

n	$\dot{\gamma}$	h_0	τ_1	τ_0	h_s	γ_0	q
60	0.0001	245	30	74	105	0.001	1.0

To study the effect of grain orientation on the motion of slip systems, five typical grain orientations (see Table 2) were assigned to the 411 grain (G411, Figure 4c) in the RVE model, and the other grains were given a random orientation, as shown in Figure 6. Then, the stress–strain curves of uniaxial tension to 60% engineering strain were obtained, as shown in Figure 7. It is clear that the five stress–strain curves coincide completely, indicating that a change of single grain orientation does not affect the mechanical properties of the stainless steel ultra-thin foil. The accuracy of the RVE model established in this paper is verified.

The 304 stainless steel had an FCC crystal structure. The dislocation slip motion occurred on the 12 slip systems {111} <110>, as shown in Table 3. The [100] direction of the crystal was consistent with the X-axis, with [010] and [001] corresponding to the Y-axis and Z-axis, respectively. According to the theory of crystal plasticity, the initiation of the slip system is always controlled by the shear strain rate on the slip system. The negative value of shear strain rate indicates that the slip should be along the opposite direction to the slip system [30].

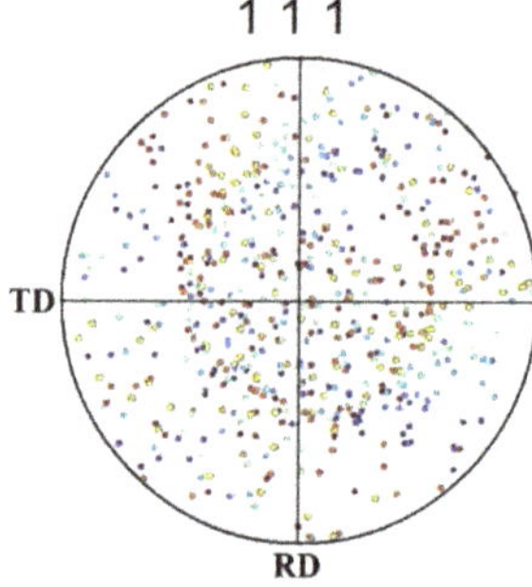

Figure 6. {111} pole figure before RVE deformation with random orientation.

Table 2. Five typical grain orientations of G411.

Number	Grain Orientation	Euler Angles	Miller Indices	Angle to X-Direction
No.1	Brass	$(35°, 45°, 90°)$	$(1\,0\,1)\,[\bar{1}\,\bar{2}\,1]$	110.56°
No.2	Copper	$(90°, 35°, 45°)$	$(1\,1\,2)\,[\bar{1}\,\bar{1}\,1]$	125.26°
No.3	S	$(61°, 34°, 64°)$	$(2\,1\,3)\,[\bar{1}\,\bar{2}\,1]$	116.09°
No.4	Cube	$(0°, 0°, 0°)$	$(0\,0\,1)\,[1\,0\,0]$	0°
No.5	Goss	$(0°, 45°, 90°)$	$(1\,0\,1)\,[0\,\bar{1}\,0]$	90°

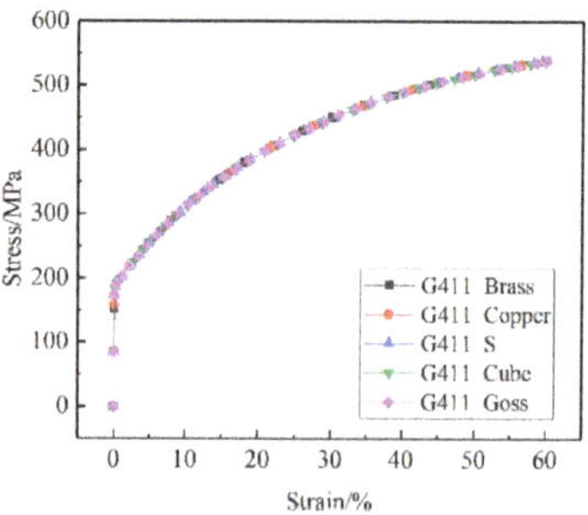

Figure 7. Stress–strain curves of G411 in the RVE model under five different grain orientations.

Table 3. Slip systems of face-centered cubic FCC metal.

Slip Plane	Slip Direction	Slip System	Slip Plane	Slip Direction	Slip System
$(1\,1\,1)$	$[0\,\bar{1}\,1]$	a1	$(1\,\bar{1}\,1)$	$[0\,1\,1]$	c1
	$[1\,0\,\bar{1}]$	a2		$[1\,1\,0]$	c2
	$[\bar{1}\,1\,0]$	a3		$[1\,0\,\bar{1}]$	c3
$(\bar{1}\,1\,1)$	$[1\,0\,1]$	b1	$(1\,1\,\bar{1})$	$[0\,1\,1]$	d1
	$[1\,1\,0]$	b2		$[1\,0\,1]$	d2
	$[0\,\bar{1}\,1]$	b3		$[\bar{1}\,1\,0]$	d3

4. Results and Discussion

4.1. Microscopic Stress and Strain Distribution of G411

Figure 8 shows the stress distribution of G411 under five typical grain orientations for the RVE models with 60% tensile engineering strain. It can be seen that the Brass, Copper, and S oriented grains had an obvious rotation phenomenon. The stress at the internal grain boundary was highly concentrated and the maximum stress value was high, even up to 2507 MPa for the Copper-oriented grain. The grain morphologies of the Cube and Goss oriented grains were similar after tensile deformation, which was mainly elongated along the tensile direction. The stress distribution was relatively uniform and the maximum stress value was small. The maximum stress value of the Cube oriented grain was 2004 MPa. It can be concluded that the different initial grain orientations led to the different degrees of the grain deformation under the same deformation conditions, and the microscopic stress distribution and the maximum stress value differed greatly.

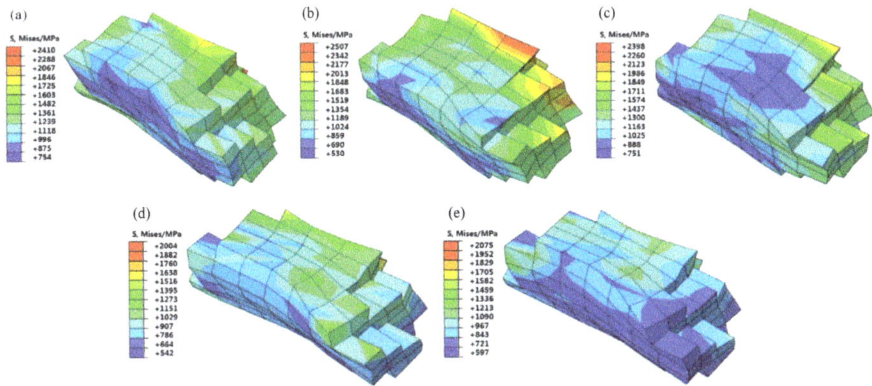

Figure 8. Grain morphology and Mises stress distribution after the G411 deformation: (**a**) Brass; (**b**) Copper; (**c**) S; (**d**) Cube; (**e**) Goss.

Figure 9 shows the accumulative slip distribution of G411 under five typical grain orientations for the RVE models with 60% tensile engineering strain. As can be seen, the cumulative shear strain distributions of five typical oriented grains were not uniform, due to the restraint of adjacent grains, and there were obvious strain gradient characteristics. The strain was mainly concentrated in the boundary region with the internal grain contact. The Cube and Goss oriented grains had large deformations along the X-axis tensile direction. The cumulative shear strain distribution appeared to be more uniform and the cumulative shear strain value was small. The maximum cumulative shear strain value of the Cube oriented grain was 1.999. It can be seen that the initial grain orientation directly affected the cumulative shear strain distribution of the grains.

Considering the deformation coordination of polycrystalline, the grain deformation can be affected by the constraints of adjacent grains. Due to the different initial grain orientations, the order of grain deformation varies with the starting sequence of slip systems in different parts of the grain, so the deformation degree of the grain is different [31]. Compared with the resistance overcome by the slip of the grain internal dislocation, the slip at the grain boundary needs to overcome the resistance caused by the coordination of intergranular deformation, so the stress of polycrystalline plastic deformation is concentrated at the grain boundary [32].

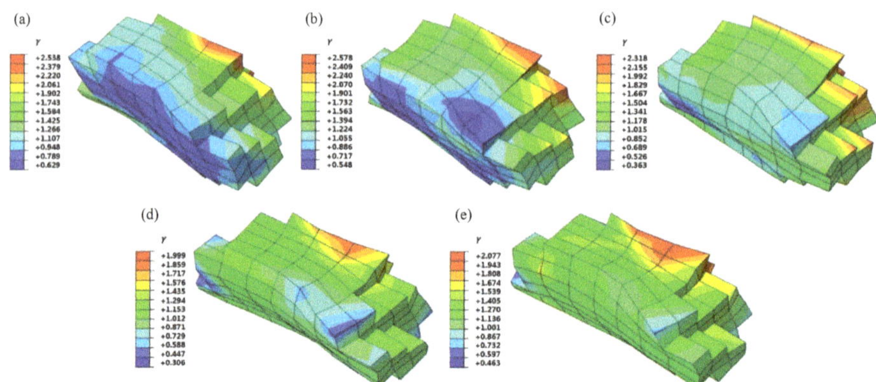

Figure 9. Cumulative shear strain distribution after the G411 deformation: (**a**) Brass; (**b**) Copper; (**c**) S; (**d**) Cube; (**e**) Goss.

4.2. Slip System Evolution of G411 with the Typical Grain Orientations

In order to investigate the relationship between grain orientation and shear strain rate of the slip systems during tensile deformation, a node on the outside surface of G411, shown in Figure 4c, was selected for analysis, and the results are shown in Figure 10. When the G411 grain orientation was Brass, Copper, or S-type, the number and motion state of the activated slip systems at the node were similar, because of the approximate angles between the three given orientations and the tensile direction. The slip systems a2 and c3 were activated in the above three orientations. Cube and Goss oriented grains had the largest number of activated the slip systems—eight. For the Cube oriented grain, the shear strain rates of slip systems a2, c2, c3, and d2 were large and slipped along their positive directions. The shear strain rates of slip systems a3, b1, b2, and d3 were small and slipped along their opposite directions. For the Goss oriented grain, the slip systems a1, b3, c1, and c2 slid along their opposite directions and the shear strain rates were large, while the slip systems a3, b2, d1, and d3 slid along their positive directions and the shear strain rates were small. It can be concluded that the number and motion state of the activated slip systems at the same position varied greatly because of the different initial grain orientations. The Cube and Goss oriented grains had the largest number of activated the slip systems in the uniaxial tensile stress state, which was beneficial to the occurrence of dislocation slip movement.

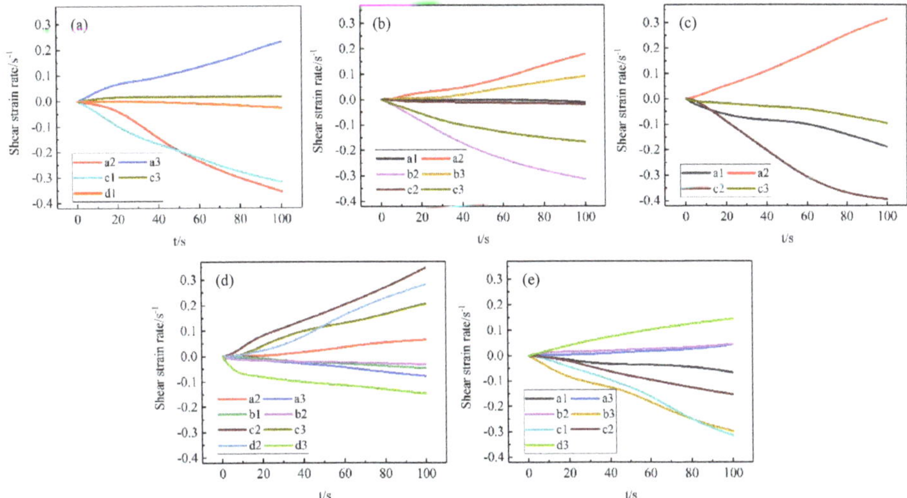

Figure 10. Shear strain rates of a node on the G411: (**a**) Brass; (**b**) Copper; (**c**) S; (**d**) Cube; (**e**) Goss.

The starting process of the 12 slip systems of FCC metal during uniaxial tension deformation is shown in Figure 11. When the grain orientation was Cube type, the local coordinate system of the crystal coincided with the global coordinate system of the sample and the angle between grain orientation and tensile force direction was zero. It can be seen that at this time the slip systems of a1, b3, c1 and d1 were perpendicular to the direction of the force. Thus, shear strain did not occur and the shear strain rate was zero. For the Goss grain orientation, the crystal coordinate system did not coincide with the sample coordinate system and the angle between grain orientation and tensile force direction was 90°. At this time, the slip systems a2, b2, c3, d2 were perpendicular to the tensile direction and the shear strain rate was zero. As shown in Figure 10d,e, the other eight slip systems were activated except for the above four slip systems perpendicular to the tensile stress. The shear strain rates of each slip system are different due to the influence of grain morphology and adjacent grains. In summary, the movement of the slip systems in the process of tensile deformation was analyzed by simulation and theory, and the accuracy of the crystal plasticity finite element model was verified.

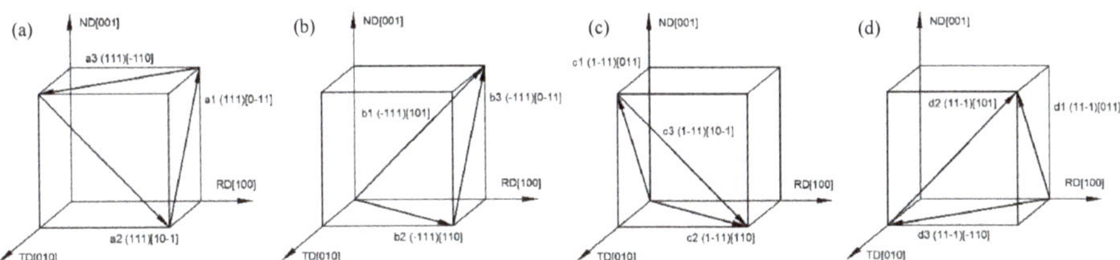

Figure 11. Illustration of {111} planes for initial tensile process: (**a**) (111) slip plane; (**b**) (-111) slip plane; (**c**) (1-11) slip plane; (**d**) (11-1) slip plane.

Figure 12 shows the evolution of the shear strain rate of the slip systems a2 and c3 during uniaxial tensile deformation of the Brass oriented G411. It can be observed that the shear strain rate distribution of the slip systems was inhomogeneous. The slip appea red in the opposite direction inside the single crystal grain, which is conducive to the formation of sub-crystallines. The shear strain rate distribution of different slip systems was also various. The a2 and c3 slip systems were activated first at the grain boundaries. With the rotation of the slip surface and direction of the tensile deformation, the internal and free surface slip systems started. The slip strain rate distribution of the slip system a2 varied greatly from −0.321 to 1.956. The outside surface nodes mainly moved along the opposite direction of the slip system a2. The difference of the shear strain rate of the slip system c3 in the whole grain was only 1.662, and the distribution was relatively uniform. It can be seen that the slip systems were activated at the grain boundary first and then caused the movement of the internal and free surface slip systems, and finally formed a slip band in the whole grain. At the same time, the above conclusions can be drawn from the analysis of the shear strain rate evolution process of the other four typical grain orientation slip systems.

Affected by grain orientation, morphology, and polycrystalline deformation compatibility, the number of the activated slip systems and the magnitude of shear strain at different locations are quite different. The motion state of the activated slip systems during plastic deformation is also different. These differences lead to inhomogeneous deformation of the grain and contribute to the stress–strain concentration zone [33].

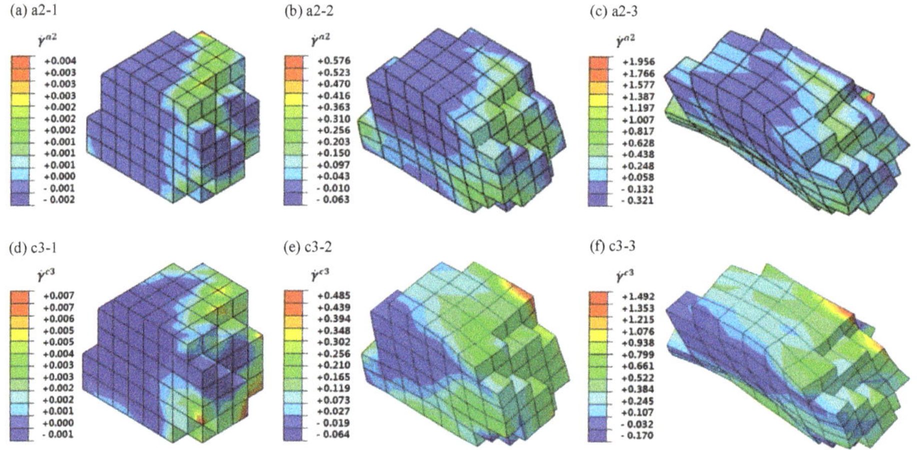

Figure 12. Shear strain rate along active slip system a2 (a2-1, a2-2, a2-3) and c3 (c3-1, c3-2, c3-3) for the G411 of Brass with the tensile time 0.5 s (a2-1, c3-1), 30 s (a2-2, c3-2), 100 s (a2-3, c3-3).

4.3. Stress and Strain Curves of Typical Texture Orientation Dominant Polycrystalline

An RVE model with a typical texture orientation was established. The number of grains with each dominant orientation accounted for 50% of the total grains. The simulated stress–strain curves are shown in Figure 13. As can be seen, the yield strength of the RVE model with the dominant texture orientation was approximately 193 MPa, but its deformation resistance varied greatly. In the plastic deformation stage, the stress–strain curves of Brass, S, and Copper oriented dominant models were higher than those of random grain oriented models. The stress–strain curves of Cube oriented and Goss oriented RVE models coincide basically and are significantly lower than those of random oriented RVE models. Under the 60% strain condition, the stress values of the Brass, S, and Copper oriented dominant models are higher than those of random grain oriented model. The maximum stress value of the Copper-type orientation-dominated model is 618 MPa. The stress values of Cube-type and Goss-type orientation-dominated models are approximately 495 MPa, which is lower than that of the random grain orientation model (538 MPa). The results show that Brass, S, and Copper oriented stainless steel materials have large energy storage, large plastic deformation resistance, and a high strain hardening index, which are beneficial to strengthening. Cube-oriented and Goss-oriented stainless steels exhibit low strength and low deformation resistance, and are prone to plastic deformation.

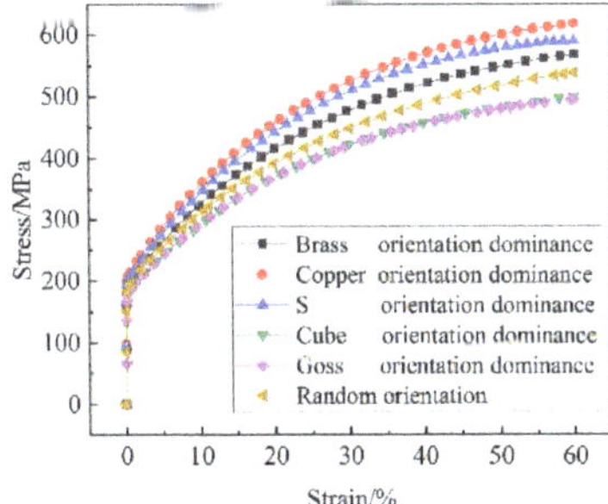

Figure 13. Comparison of stress and strain curves of RVE models with typical texture orientation.

5. Conclusions

(1) Based on the crystal plasticity theory and microscopic characterization analysis, a crystal plastic RVE model of 304 stainless steel foil was established. By comparing the tensile stress–strain curves between the RVE simulation and experiment, the calibrated crystalline plastic parameters can accurately reflect the mechanical properties of 304 stainless steel foil. Combined with the simulation and theoretical analysis of the evolution of slip systems during the tensile deformation process, the accuracy of the crystal plastic finite element model was verified.

(2) The crystal plasticity RVE model was used to simulate the effect of typical grain orientation on the inhomogeneous deformation of ultra-thin 304 stainless steel foil during tension. The results show that the numbers of start-up and movement state of the slip systems at the same location are quite different due to the various initial grain orientations, resulting in a great difference in grain deformation degrees and stress–strain distribution. The effect of grain orientation on the plastic deformation of polycrystals is mainly manifested in the mutual constraints and coordination of various grain deformation processes. During the tensile deformation process, the stress distribution and the shear stress with different slip systems are different due to the different initial grain orientation. Therefore, the initiating sequence of the slip systems is different, resulting in different deformation orders and deformation degrees. The grains in favorable orientations slip first, and the grains in unfavorable orientations are difficult to slip.

(3) The change of stress–strain curves in the tensile process of 304 stainless steel foil RVE model with typical orientation was analyzed by CPFEM. The results show that the Brass, S, and Copper oriented dominant models exhibit a high strain hardening index and are beneficial to metal strengthening. However, Cube and Goss oriented dominant models have lower strength and low deformation resistance, so 304 stainless steel foil with these orientations are prone to plastic deformation.

Author Contributions: Z.R. conceived and guided the simulation analysis; W.F. wrote the paper; J.H. provided experiment data, materials and experimental equipments; T.W. revised the paper.

Funding: This research was funded by the National Natural Science Foundation of China (Grant No. 51804215), the China Postdoctoral Science Foundation (Grant No. 2019M651074), the Shanxi Province Science and Technology Major Projects (Grant No. 20181102015, 20181101008), the Taiyuan City Science and Technology Major Project (Grant No. 170203) and the Natural Science Foundation of Shanxi Province (Grant No. 201801D221221).

Conflicts of Interest: The authors declare no conflict of interest.

References

1. Xiao, H.; Ren, Z.; Liu, X. New mechanism describing the limiting producible thickness in ultra-thin strip rolling. *Int. J. Mech. Sci.* **2017**, *133*, 788–793. [CrossRef]
2. Chen, Z.J.; Chen, X.; Zhou, T.P. Microstructure and mechanical properties of J55ERW steel pipe processed by on-line spray water cooling. *Metals* **2017**, *7*, 150. [CrossRef]

3. Daming, N.; Zhen, L.; Kaifeng, Z. Grain size effect of commercial pure titanium foils on mechanical properties, fracture behaviors and constitutive models. *J. Mater. Eng. Perform.* **2017**, *26*, 1283–1292. [CrossRef]

4. Fu, M.W.; Chan, W.L. Geometry and grain size effects on the fracture behavior of sheet metal in microscale plastic deformation. *Mater. Des.* **2011**, *32*, 4738–4746. [CrossRef]

5. Grogan, J.A.; Leen, S.B.; McHugh, P.E. Influence of statistical size effects on the plastic deformation of coronary stents. *J. Mech. Behav. Biomed. Mater.* **2013**, *20*, 61–76. [CrossRef] [PubMed]

6. Furushima, T.; Tsunezaki, H.; Manabe, K.I.; Alexsandrov, S. Ductile fracture and free surface roughening behaviors of pure copper foils for micro/meso-scale forming. *Int. J. Mach. Tools Manuf.* **2014**, *76*, 34–48. [CrossRef]

7. Chen, S.D.; Liu, X.H.; Liu, L.Z. Symmetric and asymmetric rolling pure copper foil: crystal plasticity finite element simulation and experiments. *Acta Metall. Sin. (Engl. Lett.)* **2015**, *25*, 1024–1033. [CrossRef]

8. Ye, W.; Efthymiadis, P.; Pinna, C.; Ma, A.; Shollock, B.; Dashwood, R. Experimental and modelling study of fatigue crack initiation in an Aluminium beam with a hole under 4-point bending. *Int. J. Solids Struct.* **2018**, *138*, 87–96. [CrossRef]

9. Ma, X.; Zhao, J.; Du, W.; Zhang, X.; Jiang, L.; Jiang, Z. An analysis of ridging of ferritic stainless steel 430. *Mater. Sci. Eng. A* **2017**, *685*, 358–366. [CrossRef]

10. Bishoyi, B.; Debta, M.K.; Yadav, S.K.; Sabat, R.K.; Sahoo, S.K. Simulation of texture evolution during uniaxial deformation of commercially pure Titanium. *IOP Conf. Ser. Mater. Sci. Eng.* **2018**, *338*, 012038. [CrossRef]

11. Singh, J.; Kim, M.S.; Choi, S.H. The effect of initial texture on micromechanical deformation behaviors in Mg alloys under a mini-V-bending test. *Int. J. Plast.* **2019**, *117*, 33–57. [CrossRef]

12. Si, L.Y.; Lü, C.; Tieu, A.K.; Liu, X.H. Simulation of polycrystalline aluminum tensile test with crystal plasticity finite element method. *Trans. Nonferrous Metals Soc. China* **2007**, *17*, 1412–1416. [CrossRef]

13. Hama, T.; Takuda, H. Crystal-plasticity finite-element analysis of inelastic behavior during unloading in a magnesium alloy sheet. *Int. J. Plast.* **2011**, *27*, 1072–1092. [CrossRef]

14. Hama, T.; Takuda, H. Crystal plasticity finite-element simulation of work-hardening behavior in a magnesium alloy sheet under biaxial tension. *Comp. Mater. Sci.* **2012**, *51*, 156–164. [CrossRef]

15. Ritz, H.; Dawson, P.R. Sensitivity to grain discretization of the simulated crystal stress distributions in FCC polycrystals. *Modell. Simul. in Mater. Sci. Eng.* **2009**, *17*, 015001. [CrossRef]

16. Pi, H.C.; Han, J.T.; Zhang, C.G.; Tieu, A.K.; Jiang, Z.Y. Modeling uniaxial tensile deformation of polycrystalline Al using CPFEM. *J. Univ. Sci. Technol. Beijing* **2008**, *15*, 43–47. [CrossRef]

17. Nakamachi, E.; Tam, N.N.; Morimoto, H. Multi-scale finite element analyses of sheet metals by using SEM-EBSD measured crystallographic RVE models. *Int. J. Plast.* **2007**, *23*, 450–489. [CrossRef]

18. Jigh, H.G.J.; Frsi, M.A.; Hosseini, T.H. Pre-diction of the stress–strain behavior of open-cell Aluminum foam under compressive loading and the effects of various RVE boundary conditions. *J. Mater. Eng. Perform.* **2018**, *7*, 2576–2585. [CrossRef]

19. Zhang, L.; Jing, W.G.; Wu, Z.H. Finite element simulation on mechanical properties of hierarchical honeycombs. *J. Nan-chang Hangkong Univ. (Nat. Sci.)* **2016**, *30*, 22–27.

20. Kamiński, M.; Sokołowski, D. Dual probabilistic homogenization of the rubber-based composite with random 373 carbon black particle reinforcement. *Comp. Struct.* **2016**, *140*, 783–797. [CrossRef]

21. Rehrl, C.; Völker, B.; Kleber, S.; Antretter, T.; Pippan, R. Crystal orientation changes: A comparison between a crystal plasticity finite element study and experimental results. *Acta Mater.* **2003**, *51*, 2379–2386. [CrossRef]

22. Smallman, R.E. Plasticity of crystals with special reference to metals by E. Schmid and W. Boas. *J. Appl. Crystallogr.* **2014**, *25*, 587–588.

23. Hutchinson, J.W. Bounds and self-consistent estimates for creep of polycrystalline materials. *Proc. R. Soc. London A Math. Phys. Sci.* **1976**, *348*, 101–127. [CrossRef]

24. Wu, T.Y.; Bassani, J.L.; Laird, C. Latent hardening in single crystals I. theory and experiments. *Proc. R. Soc. London A Math. Phys. Sci.* **1991**, *435*, 1–19. [CrossRef]

25. Bassani, J.L.; Wu, T.Y. Latent hardening in single crystals II. analytical characterization and predictions. *Proc. R. Soc. London A Math. Phys. Sci.* **1991**, *435*, 21–41. [CrossRef]

26. Kysar, J.W. *Addendum to a User-Material Subroutine Incorporating Single Crystal Plasticity in the ABAQUS Finite Element Program*; Mech report 178; Division of Engineering and Applied Sciences, Harvard University: Cambridge, MA, USA, 1997.

27. Simmons, G.; Wang, H. *Single Crystal Elastic Constants and Calculated Polycrystal Properties*; MIT Press: Cambridge, UK, 1971.
28. Franciosi, P.; Berveiller, M.; Zaoui, A. Latent hardening in FCC crystals. In *Strength of Metals and Alloys*; Haasen, P., Gerold, V., Kostorz, G., Eds.; Elsevier Ltd.: Amsterdam, The Netherlands, 1979; Volume 1, pp. 23–28.
29. *Methods for Tension Testing of Metallic Foil*; ASTM E345-08; American Society of Testing and Materials: West Conshohocken, PA, USA, 2016.
30. Pan, W.K.; Wang, Z.Q.; Zhang, Y.W. Discrete slip model of crystal and finite element analysis. *Acta mech. Sin.* **1997**, *29*, 278–286.
31. Zhang, C.; Zhang, L.W.; Shen, W.F.; Xia, Y.N.; Yan, Y.T. 3d crystal plasticity finite element modeling of the tensile deformation of polycrystalline ferritic stainless steel. *Acta Metall. Sin. (Engl. Lett.)* **2017**, *30*, 79–88. [CrossRef]
32. Si, L.Y. Simulation of the Texture Evolution During Cold Deformation of FCC Metal with Crystal Plasticity FEM. Ph.D. Thesis, Northeastern University, Shengyang, China, June 2009.
33. Chen, S.D.; Liu, X.H.; Liu, L.Z.; Song, M. Crystal plasticity finite element simulation of slip and deformation in ultra-thin Copper strip rolling. *Acta Metall. Sin.* **2016**, *52*, 120–128.

Article

Elucidation of Shearing Mechanism of Finish-type FB and Extrusion-type FB for Thin Foil of JIS SUS304 by Numerical and EBSD Analyses

Yohei Suzuki [1,*], Tomomi Shiratori [1], Ming Yang [2] and Masao Murakawa [3]

[1] Komatsuseiki kosakusho. Co., Ltd., 942-2 Shiga, Suwa, Nagano 392-0012, Japan
[2] Tokyo Metropolitan University, 6-6, Asahigaoka, Hino, Tokyo 191-0065, Japan
[3] Nippon Institute of Technology, 4-1 Gakuenndai, Miyashiro, Minamisaitama, Saitama 345-8501, Japan
* Correspondence: y-suzuki@komatsuseiki.co.jp; Tel.: +81-266-52-6100

Received: 3 June 2019; Accepted: 1 July 2019; Published: 3 July 2019

Abstract: A numerical analysis using FE (finite element) analysis was performed to clarify the shearing mechanism in the process of extrusion-type fine blanking (FB) for a thin foil of JIS SUS304 in this study. Extrusion-type FB, in which a negative clearance between the punch and the die has been developed and investigated experimentally to improve the quality of the sheared surface in the blanking of thin foils. The resultant sheared surface for extrusion-type FB indicated an almost completely sheared surface, and the fracture portion on the sheared surface was much smaller than that in conventional FB, the so-called finish-type FB. The material flow and fracture criteria in extrusion-type FB were analyzed in comparison with those in finish-type FB. The differences in material flow and so-called critical fracture value were verified for the two processes. The principal stress near the shearing surface has mostly compressive components in extrusion-type FB due to its negative clearance, and the critical fracture value was also less than that in finish-type FB, in which the principal stress near the shearing surface has mostly tensile components. Furthermore, SEM observation with EBSD (electron back-scatter diffraction) analysis of the shearing surface was performed to verify the phenomena. Reductions in deformation-induced crystal orientation rotation and martensite transformation in extrusion-type FB were confirmed in comparison with those in finish-type FB from the analysis results.

Keywords: fine blanking; metallic microgear; finite element analysis; electron backscatter diffraction; critical fracture value

---

1. Introduction

Recently, the notable miniaturization of electronic devices and medical instruments has been observed. Accordingly, the miniaturization of their components and their high-precision capability is also required [1]. The fabrication of microparts with complex profiles using stamping press machines is attractive owing to its high productivity. Blanking is one of the typical stamping processes used to rapidly cut parts with complex profiles from sheet metal, and fine blanking (FB) [2] can be used to make a part with a fine shearing surface. However, problems due to the miniaturization of parts arise in microblanking. As parts become smaller, the sheet becomes thinner. In general, the clearance between the punch and the die is expressed as the ratio to the thickness of the sheet, so the absolute value of the clearance also decreases as the sheet thickness decreases [3]. According to findings on the FB process at the macro scale, the clearance is assumed to less than 1% of the thickness t [4]. Since a sheet of 0.1 mm thickness or less is usually used in microblanking, the clearance becomes 1 μm or less and, as a consequence, the manufacturing accuracy of the die and the positioning accuracy of the punch and die are approaching their limits [5]. Furthermore, when the thickness of the work material decreases, fracture is likely to occur owing to the effects of the surface roughness of the material and the crystal grain size [6–9], which affects the product accuracy. To solve these problems, the authors previously developed a novel FB process and demonstrated its advantages by an experimental comparison with the conventional FB process using a general-purpose precision stamping press machine [10]. The conventional FB process is finish-type FB with an extremely narrow positive tool clearance and a counterpunch. The newly developed process is extrusion-type FB with a negative tool clearance and a counterpunch pressure pad. By comparing the results, the developed process was found to be superior from the most important viewpoint of suppressing the fracture of the surface. However, the material flow and fracture mechanism have not yet been clarified, and furthermore, the most important parameters in the process are unknown.

In this study, an FE analysis is performed to clarify the material flow and fracture mechanism in extrusion-type FB in comparison with finish-type FB, particularly from the viewpoint of suppressing fracture or cracking on the shearing surface during the FB process. Furthermore, the mechanism of fracture suppression was verified by SEM observation with EBSD (electron back-scatter diffraction) analysis, which enables the observation of the material deformation state using the information obtained by the orientation analysis of the crystallinity of the shearing surface of samples.

2. FEM Simulation Model and Conditions

Figure 1 shows SEM images of microgears fabricated by finish-type FB and extrusion-type FB [10]. As described previously, the shearing surface of the microgear processed by extrusion-type FB is smoothly sheared without fracture in comparison with that processed by finish-type FB.

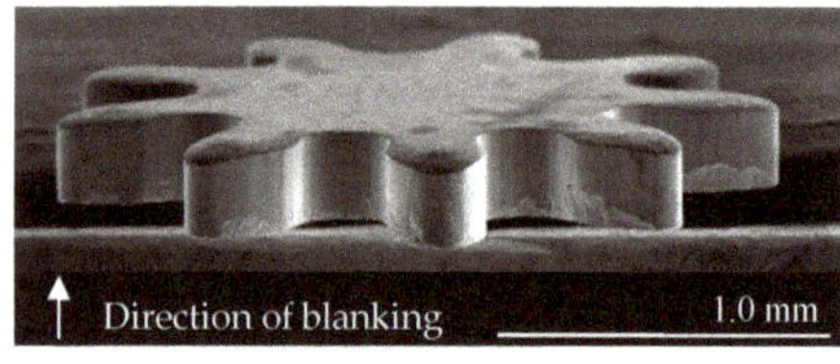

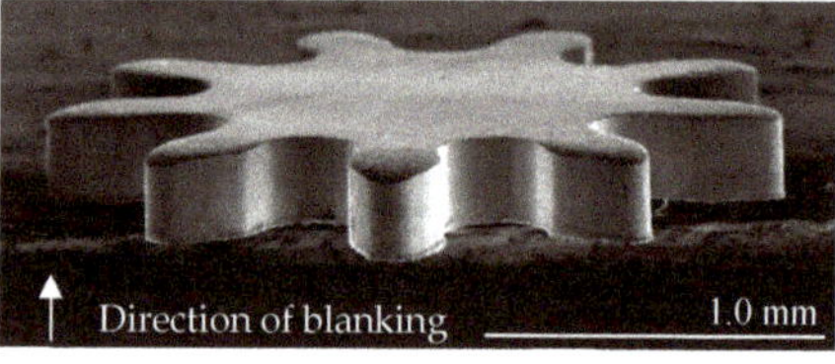

(a) finish-type FB   (b) extrusion-type FB

Figure 1. SEM images of the microgear [10]: (**a**) finish-type FB; (**b**) extrusion-type FB.

To elucidate the mechanism of fracture suppression, we developed an analysis model for both FB methods for use in FE simulation, and the model and its conditions are shown in Figure 2 and Table 1, respectively. The microgear was assumed to be axisymmetric about the gear centerline, and a two-dimensional model was applied in this simulation to reduce the calculation time. The commercially available FEM (Finite element method) code DEFORM2D (Ver.11.1) was utilized for the simulation. The punch force Fp was processed as a constant displacement type, i.e., Fp was calculated as the reaction force Fm of the workpiece. The workpiece used was a JIS SUS304 sheet with a thickness t of 0.178 mm, and a microgear workpiece having an outer diameter of φDw of 3.5 mm was used to correspond to the experiment at conditions of the press in the previous report [10]. In our present FE analysis, the outer peripheral edge of the workpiece was assumed to be constrained, as shown in Figure 2, since a coil material having sufficient width relative to the size of the microgear was used in the experiment, and there was no material flow in the width direction of the material. A die diameter φD_{d1} of 1.752 mm (tool clearance Cl of 2 μm) was used for finish-type FB, and a die diameter φD_{d2} of 1.732 mm (clearance Cl of −8 μm) was used for extrusion-type FB. The punch diameter φDp of 1.748 mm and the counterpunch diameter φDc of 1.720 mm were common to both processes. The punch, die and blank holder/stripper were assumed to be rigid bodies, and the workpiece was assumed to be elastoplastic. Approximately 15,000 four-node rectangular elements were generated on the workpiece. Since the deformation is concentrated in an extremely narrow range in the deformation area around the cutting edge of the tool, there is significant distortion in the elements [11]. The simulation result in a previous FB study, which revealed the relationship between the mesh size and the clearance *Cl*, was referred to for the determination of the mesh size [12]. Specifically, the mesh size in the analysis was set to about 1 μm, which is smaller than the clearance *Cl*. In such a case where the deformation of the elements was apparently large, a remeshing function was adopted to rebuild the meshes.

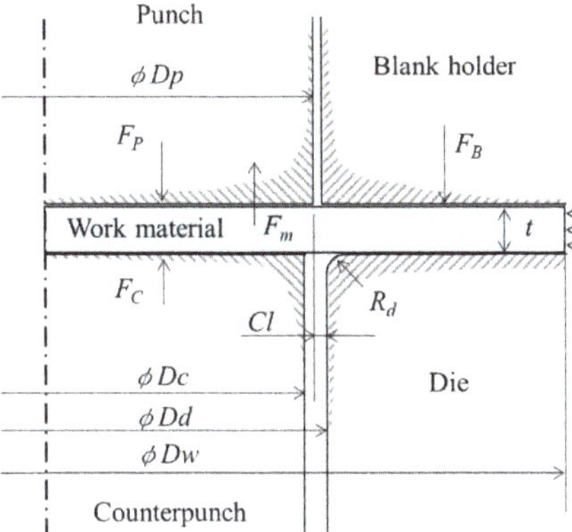

Figure 2. FEM simulation model.

Table 1. FEM simulation conditions.

Simulation Model Type	Axisymmetric Model
Object type	Workpiece: elasto-plastic (φD_w:3.5 mm) Punch /Die: rigid (φD_p:1.748 mm, φD_{d1}:1.752 mm, φD_{d2}:1.732 mm) Blank holder /Stripper: rigid Counterpunch: rigid (φD_c:1.720 mm)
Clearance (*Cl*)	Finish-type FB: 2 μm Extrusion-type FB: −8 μm
Blank holder force (F_B) Counterpunch force (F_C) Blanking force (F_P)	1000 N (50% of maximum blanking force) 400 N (20% of maximum blanking force) Non-constant value
Radii of tool cutting edges	R_p = 0.00 mm, R_d = 0.01 mm
Work material (Workpiece)	SUS304 t = 0.178 mm Young's modulus: 193 GPa Poisson's ratio: 0.3
Fracture criterion equation	Cockcroft and Latham
Friction coefficient (μ)	0.08

The material constants were derived from the data available from an actual experimental tensile test giving the flow stress-plastic strain curve shown in Figure 3. Furthermore, to verify and observe the fracture process in both FB methods, we used the damage value C (critical fracture value [13]) obtained from the ductile fracture criterion [14] of Cockcroft and Latham defined as the value of the integral corresponding to the evolution of the maximum principal stress as follows:

$$C = \int_0^{\bar{\varepsilon}} \frac{\bar{\sigma}_{max}}{\bar{\sigma}} d\bar{\varepsilon} \tag{1}$$

where C: critical fracture value, $\bar{\sigma}_{max}$: maximum principal stress, $\bar{\sigma}$: effective stress, and $\bar{\varepsilon}$: effective strain.

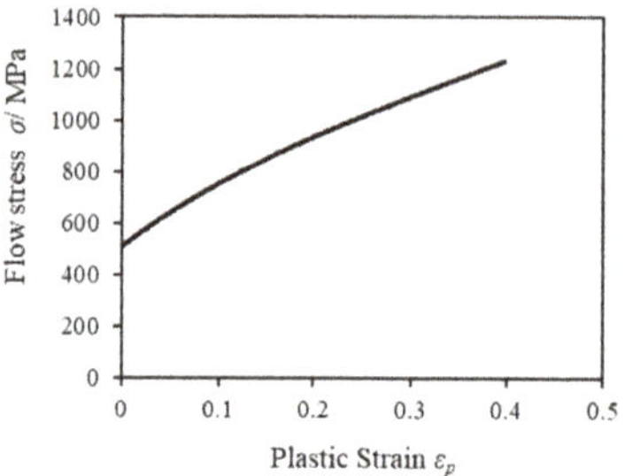

Figure 3. Flow stress-plastic strain curve.

The friction coefficient was assumed to be 0.08, referring to the value recommended by DEFORM2D for cemented carbide dies. Although there was a concern that the friction may change significantly during the blanking process, as a result of examining them in the report of Sasada et al. [15]. It was reported that the FEM analysis agrees well with the blanking result even if the friction coefficient was constant. Accordingly, the coefficient of friction was set constant in this study. The temperature was set to be constant in this analysis. Although temperature changes occur in the tool and material during blanking, the ratio of heat flow to the material could be considered negligibly small owing to the small volume ratio of the material to the die at the microscale.

3. FEM Simulation Results and Discussion

Since fracture occurred at a punch stroke of about 80% of the sheet thickness t according to our previous experimental results [10] for finish-type FB, the damage values (critical fracture values) at a punch stroke of 80% were investigated for both processes. Figure 4 shows a comparison of the damage value between a) finish-type FB and b) extrusion-type FB in the case of the punch penetration. In the case of finish-type FB, the result shows that damage accumulated in the region connecting the punch shoulder and die. In the case of extrusion-type FB, the damage value was also high in the punch shoulder and punch side, but it was lower in other regions including the shearing zone, which means that some allowance remains to prevent fracture. This was in agreement with the experimental results obtained with an actual stamping press machine in which extrusion type FB did not cause any breakage or separation of the workpiece, even at 80% punch penetration through the sheet thickness t [10].

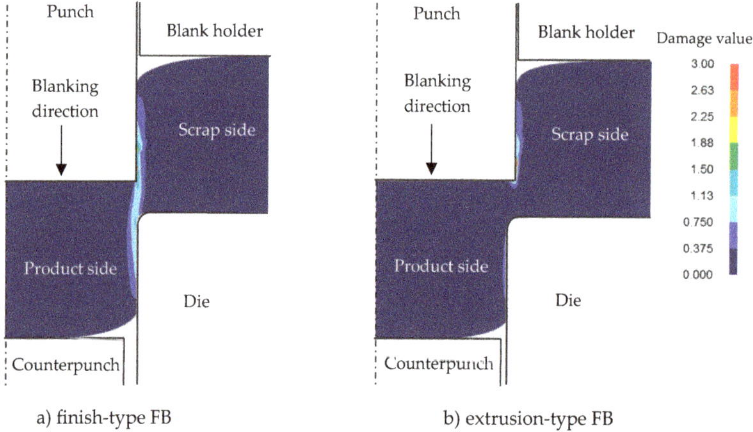

a) finish-type FB   b) extrusion-type FB

Figure 4. Comparison of damage value defined as the integration value corresponding to the evolution of the maximum principal stress $\bar{\sigma}_{max}$ for 80%t punch penetration between finish-type fine blanking (FB) and extrusion-type FB: (**a**) finish-type FB; (**b**) extrusion-type FB.

The von Mises stress and maximum principal stress distributions for both FB methods are shown in Figures 5 and 6, respectively, for the same condition of punch penetration of 80% as in Figure 4. It was found that, although the von Mises stresses for both FB methods showed similar distributions, the maximum principal stress distributions were significantly different. It can be observed that, in the case of finish-type FB, in the area between the scrap and the product (being blanked out), a tensile stress distribution appears to prevail, while in the case of extrusion type FB, a compressive stress distribution prevails. These results are consistent with the results for the damage value shown in Figure 4, in which higher maximum principal stresses corresponded to higher damage values, consistent with previous knowledge that the compression stress distribution can reduce the damage value and suppress fracture in blanking processes even if the von Mises stress is similar [16].

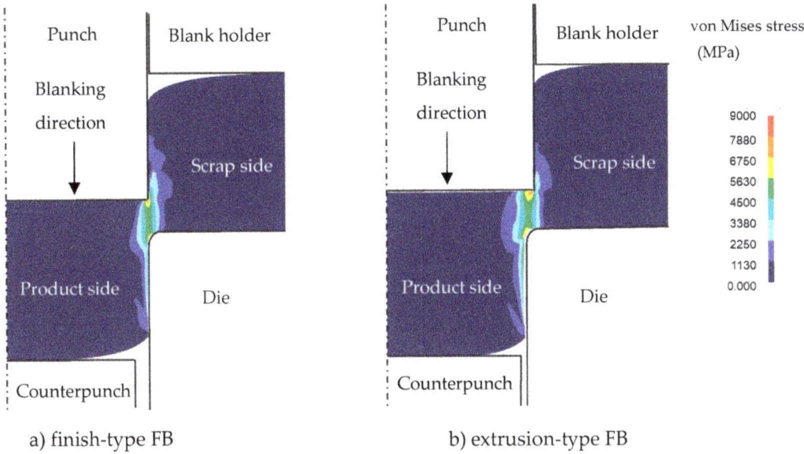

Figure 5. Comparison of von Mises stress for 80%t punch penetration between the finish-type FB and extrusion-type FB in Figure 4: (**a**) finish-type FB; (**b**) extrusion-type FB.

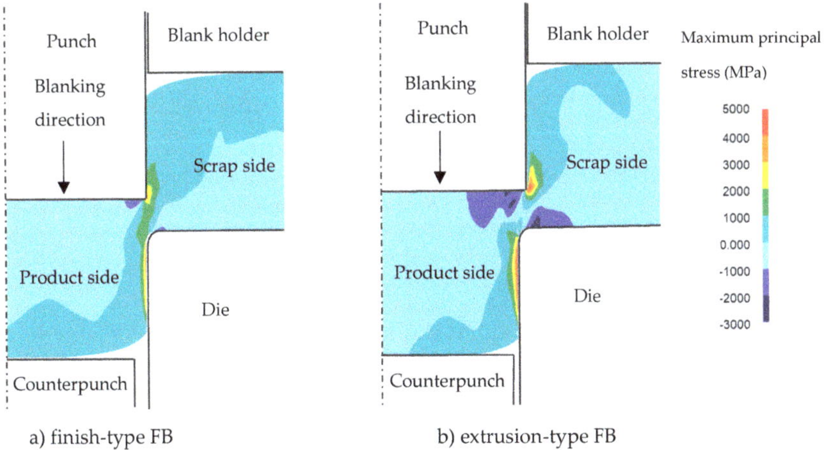

Figure 6. Comparison of maximum principal stress for 80%t punch penetration between the finish-type FB and extrusion-type FB in Figure 4: (**a**) finish-type FB; (**b**) extrusion-type FB.

From the various reports thus far, it is known that the stress distribution within a material is related to the material flow in the processing. For example, Huang et al. have shown the effect of the tool clearance (either positive or negative) on the die roll during blanking [17], and Thipprakmas et al. have clarified the mechanism of fracture during FB, in which an extremely small but positive tool clearance was used [18]. However, they have not yet clarified the fracture mechanism, in which a positive clearance and a negative clearance might cause different fracture behaviors. The material flows in both FB processes were also compared at the same punch penetration of 80% sheet thickness and Figure 7 shows the resultant material flows in the shearing zone. In the case of finish-type FB, the workpiece located immediately under the punch flows almost perpendicularly to that in the die interior portion, inducing material flow on the scrap side that was parallel to the interior material flow. Accordingly, we concluded that this material flow behavior generates a tensile stress field between the scrap side of the workpiece (which is pressed down by the blank holder) and the blanked-out workpiece, resulting in a consequently high cumulative damage value, ultimately leading to the breakage of the workpiece. In the case of extrusion type FB, a change in the direction of the workpiece flow is observed near the die corner B. In particular, immediately adjacent to point B, owing to the material flow towards the die portion B, the workpiece behaves as if it were self-crushed. As a result, we consider that a compressive stress field is generated where fracture is suppressed.

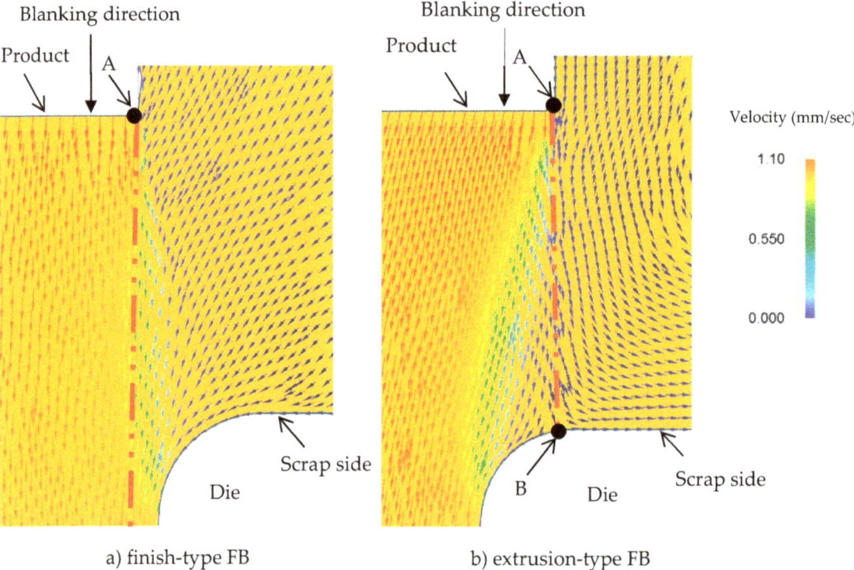

Figure 7. Comparison of material flow for 80%*t* punch penetration between finish-type FB and extrusion-type FB, in an enlarged view: (**a**) finish-type FB; (**b**) extrusion-type FB.

4. Discussion Based on EBSD Analysis

The shearing surfaces were observed by SEM with EBSD (electron back-scatter diffraction) to clarify the shearing mechanism for both FB methods. EBSD analysis is an effective methodology for evaluating the deformation state of grains in a workpiece. In particular, the KAM (kernel average misorientation) is a value that indirectly represents the degree of plastic deformation via the amount of crystal orientation and can correspond to the equivalent plastic strain in plastic working [19]. Furthermore, it has been used to clarify the fracture generation mechanism during microshear processing [20]. Since it is known that JIS SUS304 transforms from a ductile austenite phase to a brittle martensitic phase depending on the processing strain, information on the phase transformation will be also useful for understanding the fracture mechanism.

For sample preparation for EBSD analysis, first of all, the cross section of workpieces of the microgear half-blanked by the punch that penetrated about 60% of the sheet thickness for both FB methods was fabricated. Next, it was cut to the area shown in Figure 8, and the surface was mirror-finished by mechanical polishing and ion milling. EBSD analysis was performed at an acceleration voltage of 20 kV and a measurement interval of 0.15 μm. Figure 9 shows the analysis results in the form of KAM maps for the finish and extrusion-type FB methods. Figure 10 shows the analysis results in the form of phase maps. It was found that the phase transformation from the austenite phase (red area) to the martensitic phase (green area) occurs in the areas with high strain in the KAM maps. It was found that areas with a high strain of five degrees red areas and different crystal orientations are distributed in the area connecting the punch edge and the die corner. Moreover, the phase transformations occurred in areas with high strain in both methods consistent with the results of FE analysis. However, the areas with high strain and phase transformation were concentrated in the shearing zone in the case of finish-type FB, while they shifted to the scrap side (left side) for extrusion-type FB. In these areas, compression stress is dominant and does not contribute to fracture, as described in the last section. By assuming that cracks occur along the vertical shearing line (white lines in Figures 9 and 10), a difference in strain concentration on the white line (length of about 80 μm) can be seen for the two methods. Figure 11 shows the KAM values along the line from point a to point b in Figure 9 for the two methods. In finish-type FB, the KAM values are mostly five degrees from point a to point b. On the other hand, in extrusion-type FB, the KAM values along Y axis are approximately two degrees around point a and increase toward point b. These findings indicate that strain has already accumulated in the area where shearing is going to proceed in finish-type FB, while there is less accumulation of strain in the area in extrusion-type FB. As a result, extrusion-type FB can prevent fracture from the viewpoint of the strain accumulation region and crystal grain phase field.

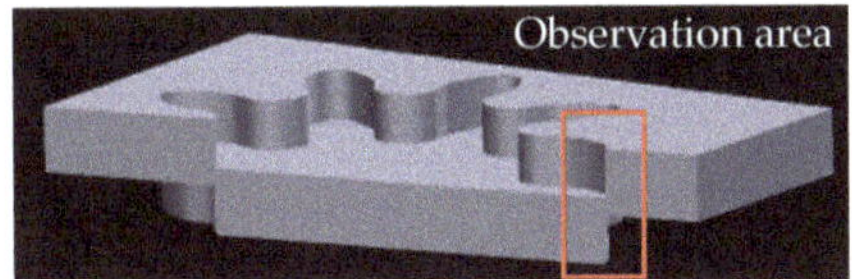

Figure 8. Observation area of microgear for electron back-scatter diffraction (EBSD) analysis.

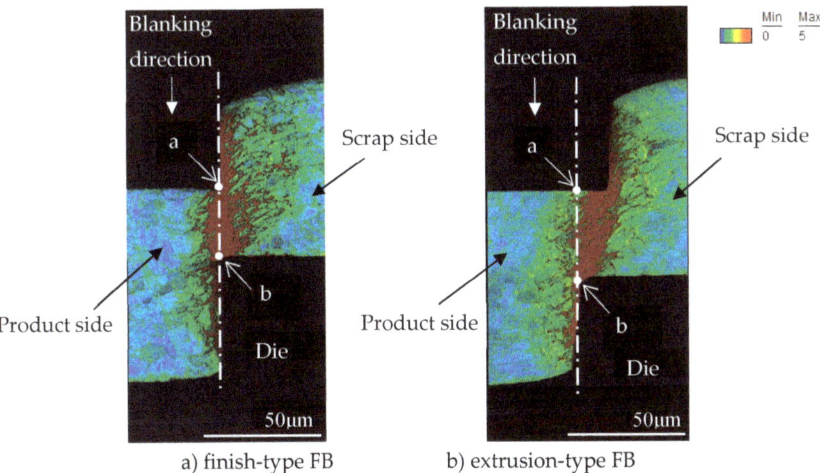

Figure 9. Comparison of kernel average misorientation (KAM) map for 60%*t* punch penetration between finish-type FB and extrusion-type FB: (**a**) finish-type FB; (**b**) extrusion-type FB.

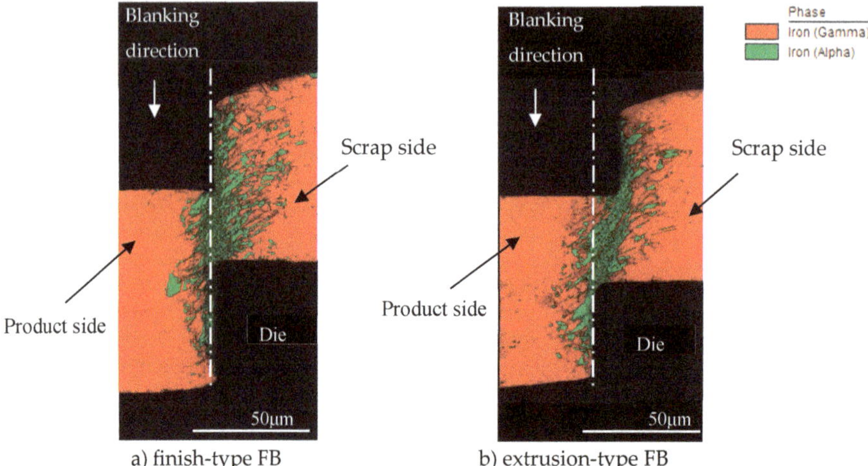

a) finish-type FB   b) extrusion-type FB

Figure 10. Comparison of phase map for 60%*t* punch penetration between finish-type FB and extrusion-type FB: (**a**) finish-type FB; (**b**) extrusion-type FB.

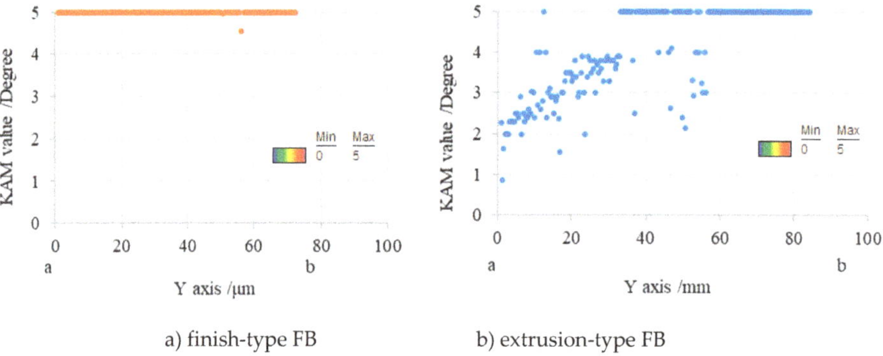

a) finish-type FB   b) extrusion-type FB

Figure 11. Comparison of KAM values between a and b of Figure 7: (**a**) finish-type FB; (**b**) extrusion-type FB.

From the discussion in Sections 3 and 4, we can conclude that the extrusion-type FB process can prevent fracture and improve the blanking quality by changing the material flow during the process. It is a high-potential process for blanking at the microscale. However, since the material is compressed in the negative clearance zone in addition to the shearing, the load on the punch in extrusion-type FB is higher than that in conventional FB. It is necessary to clarify design guidelines for the process parameters, such as the optimal negative clearance and the radius of the die corner, in further work.

5. Conclusions

An FE analysis was performed to clarify the material flow and fracture mechanism in extrusion-type FB, in which a negative clearance was applied, in comparison with that in finish-type FB, particularly from the viewpoint of suppressing the fracture and cracking on the shearing surface during the process. The critical damage value, an index showing the possibility of fracture in the workpiece during FB, was introduced as a critical fracture value. Furthermore, the SEM observation of the cross-section of blanked workpieces with EBSD analysis was performed to confirm the results of the analysis. The conclusions of this study are as follows:

(1) The perimeter of the workpiece under a punch flows almost perpendicularly towards the die interior in finish-type FB. This workpiece flow behavior generates a tensile stress field between the scrap-side workpiece (which is pressed down by the blank holder) and the blanked workpiece, with a consequent high damage accumulation value, ultimately leading to breakage of the workpiece.

(2) In contrast, in the case of extrusion-type FB, since the perimeter of the workpiece flow under the punch flows towards the end of the die radius, high compressive stress occurred in this region, which led to different material flow and reduced cumulative damage.

(3) The critical damage value for extrusion-type FB was less than that for finish-type FB owing to the difference in stress state near the shearing zone: compressive stress is dominant for extrusion-type FB, while tensile stress is dominant for finish-type FB.

(4) The difference in strain distribution between finish-type FB and extrusion-type FB was analyzed by EBSD. The relationship between the strain distribution and the occurrence of fracture was clarified.

Author Contributions: Conceptualization, Y.S., M.Y. and M.M.; Experiment, Y.S.; Writing—Original Draft, Y.S.; Writing—Review & Editing, M.Y., T.S. and M.M.

Funding: This research received no external funding.

Acknowledgments: We would like to express our sincere gratitude to the members of the R&D section of our company and also to Bastien Poitrimol, a French international student cooperating with us, for his valuable assistance in the experiment.

Conflicts of Interest: The authors declare no conflict of interest.

References

1. Yang, M. Micro integral forming technology in press die. *J. Jpn. Soc. Technol. Plast.* **2006**, *47*, 558–563. [CrossRef]
2. Nakagawa, T. *Fine Blanking*; The Nikkan Kogyo Shimbun: Tokyo, Japan, 1998; pp. 1–162.
3. Maeda, T. Design of punching die. *J. Jpn. Soc. Technol. Plast.* **1960**, *1*, 309–316.
4. Maeda, T. Experimental investigation on fine blanking 2nd Report. *J. Jpn. Soc. Technol. Plast.* **1968**, *9*, 627–636.
5. Nakano, S.; Suzuki, Y.; Aihara, T.; Shiratori, T. Visualization technology and nanometric positioning die system for micropiercing. *J. Jpn. Soc. Technol. Plast.* **2015**, *56*, 213–218. [CrossRef]
6. Xu, Z.T.; Peng, L.F.; Lai, X.M.; Fu, M.W. Geometry and grain size effects on the forming limit of sheet metals in micro-scaled plastic deformation. *Mater. Sci. Eng. A* **2014**, *611*, 345–353. [CrossRef]
7. Zhao, Y.H.; Guo, Y.Z.; Wei, Q.; Topping, T.D.; Dangelewicz, A.M.; Zhu, Y.T.; Langdon, T.G.; Lavernia, E.J. Influence of specimen dimensions and strain measurement methods on tensile stress–strain curves. *Mater. Sci. Eng. A* **2009**, *525*, 68–77. [CrossRef]
8. Zhao, Y.H.; Guo, Y.Z.; Wie, Q.; Dangelewicz, A.M.; Xu, C.; Zhu, Y.T.; Langdon, T.G.; Zhou, Y.Z.; Lavernia, E.J. Influence of specimen dimensions on the tensile behavior of ultrafine-grained Cu. *Scr. Mater.* **2008**, *59*, 627–630. [CrossRef]
9. Wang, J.L.; Fu, M.W.; Shi, S.Q. Influences of size effect and stress condition on ductile fracture behavior in micro-scaled plastic deformation. *Mater. Des.* **2017**, *131*, 69–80. [CrossRef]
10. Suzuki, Y.; Shiratori, T.; Yang, M.; Murakawa, M. Processing of metal microgear tooth profile by finish blanking and extrusion blanking and evaluation of cut-surface shape. *J. Jpn. Soc. Technol. Plast.* **2018**, *60*, 64–69. [CrossRef]
11. Yoshida, Y. Finite element method analysis of shearing process. *J. Jpn. Soc. Technol. Plast.* **2012**, *53*, 800–804. [CrossRef]
12. Tanaka, T.; Hagihara, S.; Tadano, Y.; Inada, T.; Mori, T.; Fuchiwaki, K. Application of finite element method to analysis of ductile fracture criteria for punched cutting surfaces. *Mater. Trans.* **2013**, *43*, 1697–1702. [CrossRef]
13. Thipprakmas, S.; Jin, M.; Kanaizuka, T.; Yamamoto, K.; Murakawa, M. Prediction of Fineblanking surface characteristics using the finite element method (FEM). *J. Mater. Process. Technol.* **2008**, *198*, 391–398. [CrossRef]
14. Cockcroft, M.G.; Latham, D.J. Ductility and the workability of metals. *J. Inst. Met.* **1968**, *96*, 33–39.
15. Sasada, M.; Shimura, K.; Aoki, I. Study on coefficient of friction between tool and material in Shearing. *Trans. Jpn. Soc. Mech. Eng. Ser. C* **2005**, *71*, 249–255. [CrossRef]

16. Maeda, T.; Nakagawa, T. Experimental investigation on fine blanking 1st Report. *J. Jpn. Soc. Technol. Plast.* **1968**, *9*, 618–626.
17. Huang, X.; Xiang, H.; Zhuang, X.; Zhao, Z. Improvement of die-roll quality in compound fine-blanking forming process. *Adv. Mater. Res.* **2011**, *337*, 236–241. [CrossRef]
18. Thipprakmas, S.; Jin, M.; Murakawa, M. An investigation of material flow analysis in fineblanking process. *J. Mater. Process. Technol.* **2007**, *192*, 237–242. [CrossRef]
19. Kimura, H.; Wang, Y.; Akiniwa, Y.; Tanaka, K. Misorientation analysis of plastic deformation of autenitic srainless steel by EBSD and X-ray diffraction methods. *Trans. Jpn. Soc. Mech. Eng.* **2005**, *71*, 118–124. [CrossRef]
20. Shiratori, T. Effects of Grain Size and Process Condition on Stability of Sheared Surface in Micropunching at SUS304. Ph.D. Thesis, Tokyo Metropolitan University, Tokyo, Japan, 2017.

Article

Optimum Clearance in the Microblanking of Thin Foil of Austenitic Stainless Steel JIS SUS304 Studied from Shear Cut Surface and Punch Load

Yohei Suzuki [1],*, Ming Yang [2] and Masao Murakawa [3]

[1]   Komatsuseiki Kosakusho Co., Ltd., 942-2 Shiga, Suwa, Nagano 392-0012, Japan
[2]   Tokyo Metropolitan University, 6-6 Asahigaoka, Hino, Tokyo 191-0065, Japan; yang@tmu.ac.jp
[3]   Nippon Institute of Technology, 4-1 Gakuenndai, Miyashiro, Minamisaitama, Saitama 345-8501, Japan;
     masa.murakawa@gmail.com
*   Correspondence: y-suzuki@komatsuseiki.co.jp; Tel.: +81-266-52-6100

Received: 22 December 2019; Accepted: 31 January 2020; Published: 3 February 2020

Abstract: An extrusion-type fine blanking with a negative clearance was proposed by the authors instead of standard fine blanking for creating a full-sheared surface in the micro blanking process. In this study, micro blanking experiments and finite element analyses with narrow, zero and negative clearances are carried out for the optimizing the clearance at which a shear cut surface can be finished with a full-sheared surface with the minimized punch load. Fracture criterion, hydrostatic stress and maximum punch stress for the conditions with various clearances are investigated. As a result, it was clarified that the clearance at which the cut surface does not fracture and minimization of the punch load is achieved is gained by the use of clearance -4 μm.

Keywords: punch load; cut surface quality; optimum clearance; fine blanking; blanking experimental; finite element method analysis

---

1. Introduction

In recent years, attention has been paid to micro-processing technologies suitable for mass production. However, problems, such as the lowering of material forming limits due to thinner materials [1–4] and the lowering of dimensional accuracy (e.g., the fracture that occurs on the cut surface during shearing) due to the downsizing of product dimensions, have not been solved. Various research studies have been conducted on manufacturing methods for the microfabrication of small parts, such as laser processing, etching processing, and electron beam processing [5–7]. However, these microfabrication methods are not necessarily optimal from the viewpoint of productivity and cost. The authors have selected press stamping, which is excellent in terms of both productivity and cost, and examined the feasibility of a micropart manufacturing method. More specifically, an involute tooth profile part with a microsize and a complicated shape was selected as a representative example of a microsize part, and the so-called fine blanking (FB) [8,9] for the effect of hydrostatic pressure on the ductility of the metal [10], was used. In particular, the feasibilies of the finish-type FB using the narrow clearance, and the extrusion-type FB using the negative clearance, were verified, and the relationship between the difference between the narrow and negative clearances and the shear cut surface was investigated. The results revealed that it was possible to process full-sheared surfaces by adopting negative clearance [11]. In addition, the finite element method (FEM) and material crystal analysis by Electron Back Scatter Diffraction (EBSD) have also revealed the difference in the processing mechanism between the two FB methods [12].

However, according to the above-mentioned verification results, it was predicted that the load applied to the punch tip would be increased by adopting negative clearance. This suggests that, in general, punch breakage and wear will also increase [13]. Even in conventional shearing, punch breakage and wear are two of the most important issues, and various studies on punch load in conventional shearing have been conducted so far. For example, Aoki et al. studied the tool cutting edge wear mechanism by shear experiments [13], Maeda et al. verified the progress of the tool cutting edge wear [14], Koga et al. verified the relationship between clearance and tool wear [15], and the punch wear and quality of punched products, in punching thin sheets [16–20].

FE analysis has been used to elucidate various processing mechanisms, and it has been also used for punch load analysis in blanking. For example, Nakashima et al. obtained an equivalent stress on the tool by analyzing the axisymmetric model with the tool set to an elastic body [21], and Hambi reported tool wear results in press stamping using a finite element code wear prediction model [22]. Falconnet et al. reported punch wear in the blanking of a copper thin sheet [23,24]. However, the punch load due to the difference in negative clearance, including zero clearance in the FB proposed in previous reports [11,12], has not been verified or analyzed. In this study, we aim to derive the optimal clearance at which the shear cut surface can be finished with a full-sheared surface and the punch load can be minimized by experimental verification and finite element analysis at the narrow, zero, and negative clearances.

2. Experimental Procedure

2.1. Blanking Condition

From the blanking experiment at the narrow clearance, zero clearance, and negative clearance, the conditions under which the cut surface does not fracture were verified. At the same time, the load during blanking at each clearance was measured. The material to be processed was JIS SUS304 made by TOKUSHU KINZOKU EXCEL Co., Ltd. (Tokyo, Japan), with a plate thickness t of 0.1 mm and a width of 20 mm. Table 1 shows the mechanical properties of the material. Table 2 shows the material of tools' (punch and die) composition and mechanical properties, available from Fuji Die Co., Ltd. (Tokyo, Japan) Next, Figure 1 shows a schematic diagram of the blanking process using negative clearance developed for progressive machining in actual production to explicitly explain how negative clearance punching is technically feasible, and Table 3 shows the detailed specifications of the die comprehensive blanking experiment series, particularly from the viewpoint of various tool clearances. To explain from another viewpoint, of the aforementioned comprehensive experiment series, referring first to die-set production, the punch is processed, using a grinding machine for small to micro-precision tools and high-precision production parts, and then the die is roughly processed by a wire cutting process and finally lapped to remove the deformed superficial layer on the tooth surface. In addition, to secure the positional accuracy of the tool (punch/die), JIS-SKD11, which can ensure strength and dimensional stability against heat treatment, can be used for parts such as the stripper plate, punch plate, and die plate. Jig-grinding finish was used for drilling the relevant parts where a relative position relationship is required to align the punch and the die and, additionally, shim tape was used to move the die position in 1 μm unit. For blanking experiment series, a cemented carbide punch with $Dp_1 = \varphi 1.748$ mm was combined with a cemented carbide die with an inner diameter (Dd_1), such that the clearance (CL) values between the punch and the die were 2, 0, −2, −4, and −8 μm (2%t, 0%t, −2%t, −4%t, −8%t) respectively. At a positive clearance CL, a punch stroke greater than the plate thickness of the workpiece is technically feasible, as aforementioned with reference to Figure 1, but at zero and negative clearances, if the punch stroke exceeds the plate thickness, the punch and die could come into contact with each other. Strictly speaking, since the die radius corner has actually a radius value of 0.01 mm, the punch and die do not come into contact even if the punch stroke becomes equal to the plate thickness. Therefore, in the actual process, the punch stroke was stopped once it was at 99% of the plate thickness, and the remaining 1% was made into a progressive die structure that could be removed at the next stage of the die set. The die edge was provided with a small radius portion of R and a counter punch to suppress cracking during shearing. The plate presser force was set to a maximum of about 500 N using a coil spring, and the reverse presser force was set to a maximum of about 200 N by the same method. For measuring the punch load, a load cell was provided on the upper surface of the punch, and a load displacement diagram was obtained by measuring the slide displacement of the press plate with a laser displacement meter. The press machine itself was not exclusively designed for FB purposes; a general-purpose screw servo press machine (made by DT-J515 Microfabrication Research Laboratory with pressurization capacity of 50 kN) was used, in order to have the same functions. It should be stressed here that controlling the amount of punch stroke is important to prevent interference between the punch and the die. In this experiment, the slide displacement of the press machine, measured with a laser displacement meter, and a load cell set on the upper surface of the punch, could be controlled in an amount of 1 μm unit.

Table 1. Mechanical properties of work material.

Tensile Strength (MPa)	896
0.2% Proof Stress (MPa)	583
Elongation (%)	47

Table 2. Material of the tools' composition and mechanical properties.

Composition and Mechanical Properties	Punch	Die
Composition	WC-Co	WC-Co
Hardness (HRA)	95.0	91.5
Compressive Stress (MPa)	6880	5400

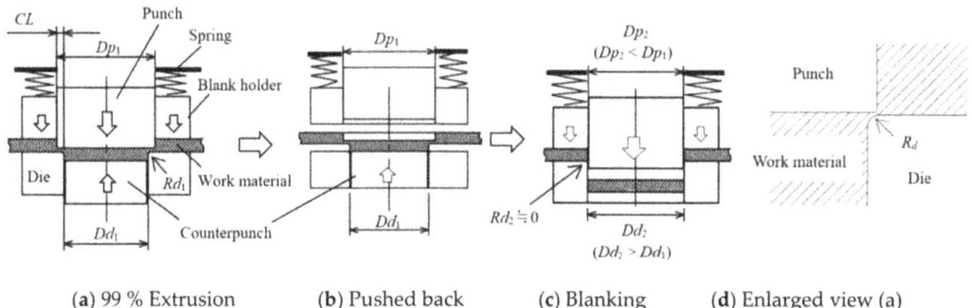

(a) 99 % Extrusion   (b) Pushed back   (c) Blanking   (d) Enlarged view (a)

Figure 1. Schematic of extrusion blanking used in the mass-production progressive die system [11].

Table 3. Blanking tool specifications.

Items	Value
Clearance CL (μm)	2, 0, −2, −4, −8 (2%t, 0%t, −2%t, −4%t, −8%t)
Punch outer diameter Dp_1 (mm)	1.748
Die inner diameter Dd_1 (mm)	1.752, 1.748, 1.744, 1.740, 1.732
Punch outer diameter Dp_2 (mm)	1.740
Die inner diameter Dd_2 (mm)	1.750
Die radius Rd_1 in 1st step (mm)	0.01
Die radius Rd_2 in 2nd step (mm)	Nearly zero
Counterpunch outer diameter Dc (mm)	1.730
Blank holder force (F_B)	500N (50% of blanking force)
Counterpunch force (F_C)	200N (20% of blanking force)

2.2. FEM Simulation Model and Conditions

Evaluation of the cut surface and punch load at each clearance, obtained by the aforementioned comprehensive blanking experiments, were considered by the following three FE analysis methods.

The first utilized the ductile fracture condition value (hereinafter referred to as damage value *C*) from the maximum tensile/compressive principal stress generated in the shearing region and evaluated and compared the cut surfaces from the blanking experiment result and the FE analysis result. The second evaluation compared hydrostatic stress, which is one of the important parameters of FB. The third evaluation compared the punch stress between samples. Figure 2 shows the employed finite element simulation model, and Table 4 shows the FEM simulation conditions for this model. The analysis model was axisymmetric, with the analysis time taken into consideration, and the commercial code DEFORM2D (Version 11.3) was used. The punch and die are assumed to be elastic bodies, and the number of elements is set to about 10,000. The blank holder/stripper and counter punch are assumed to be rigid bodies. The work material is assumed to be elastoplastic, and the number of elements is set to about 15,000. Four-node rectangular elements were generated on the tools and the work material. As is known by the skilled persons in the shearing industry, in the deformation region around the tool edge, the deformation is concentrated in a very narrow range, and a large distortion occurs in the elements. Therefore, mesh size was determined by referring to the simulation results [25] in the previous FB study, which showed the relationship between the mesh size and the clearance *CL*. Specifically, the mesh size was set to about 1 µm, which is smaller than the clearance. To prevent the interruption of analysis when an element is more deformed than the certain condition during the analysis, a remeshing function for reproducing the element available from the relevant FE code was applied. The remeshing condition was based on the possible depth of interference between the work material element and the tool boundary (specifically, this interference depth was 1 µm). Incidentally, the aforementioned FEM conditions were based on the fact that the results of the blanking experiment and the FEM analysis were in good agreement in the previous report [12]. For the punch and die, we selected cemented carbide (WC-15% Co) on the DEFORM2D software mentioned earlier. The material constant of JIS SUS304, which is the work material, was determined from the flow stress–plastic strain curve obtained from the results of the performed tensile test shown in Figure 3. The damage value *C*, obtained from Cockcroft and Latham's failure condition and expressed as Equation (1), was used to predict the fracture in shearing; in accordance with the previous report [26], this prediction was actually possible, where *C* is the damage value, σ_{max} is the maximum principal stress, σ is the equivalent stress, and ε is the equivalent strain. In addition, according to the prediction, cracks and fractures that occurred during the blanking could be expressed using the element elimination method [23]; specifically, an element was eliminated when the damage value *C* of that element reached the fracture critical value C_{cr} of 1.5 and an element with C_{cr} of 1.5 was connected to four or more elements. The set critical value C_{cr} of 1.5 is based on the fact that the fracture start was 78% of the plate thickness in the actual blanking experiment with a clearance *CL* = 2 µm The shear friction coefficient was assumed to be 0.08, referring to the value recommended by DEFORM2D for cemented carbide dies. Although there was concern that the friction may change significantly during the blanking process, as a result of examining them in the report of Sasada et al. [27], it was reported that the FEM analysis agrees well with the blanking result even if the friction coefficient was assumed to be constant.

$$C = \int_{0}^{\bar{\varepsilon}} \frac{\bar{\sigma}_{max}}{\bar{\sigma}} d\bar{\varepsilon} \tag{1}$$

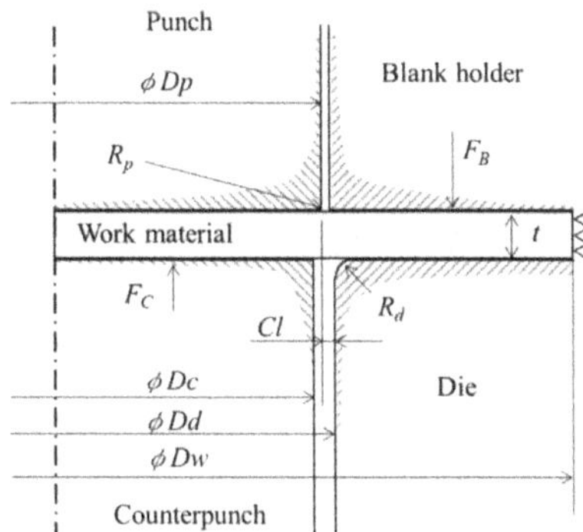

Figure 2. Finite Element Method simulation model (axisymmetric model).

Table 4. FEM simulation condition.

Simulation Model	Axisymmetric Model
Object type	Work material: elastic-plastic Punch/Die: elastic Blank holder/Stripper: rigid Counterpunch: rigid
Clearance CL (µm)	2, 0, −2, −4, −8, (2%t, 0%t, −2%t, −4%t, −8%t)
Punch outer diameter Dp (mm)	1.748
Die inner diameter Dd (mm)	1.752, 1.748, 1.744, 1.740, 1.732
Counterpunch outer diameter Dc (mm)	1.730
Work material outer diameter Dw (mm)	3.5
Tool cutting edges	$R_p = 0.002$ mm, $R_d = 0.010$ mm
Blank holder force (F_B) Counterpunch force (F_C)	500 N (50% of blanking force) 200 N (20% of blanking force)
Blanked material	JIS SUS304 $t = 0.1$ mm Young's modulus: 193 GPa Poisson's ratio: 0.3
Ductile fracture criteria	Cockcroft–Latham
Fracture critical value Ccr	1.5
Shear friction coefficient (μ)	0.08

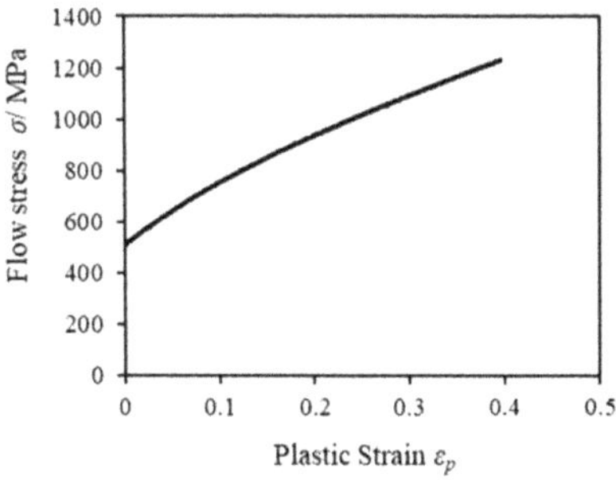

Figure 3. Flow stress–plastic strain curve.

3. Results and Discussion

3.1. Results of Blanking Experiment

First, Figure 4 shows SEM images of the cut surface of a blanked-out product at each clearance obtained in the blanking experiment. The percentage ratio of the cut surface at $CL = 2$ µm was about 16% shear droop, 62% burnished surface, and 22% fractured surface. At $CL = 0$ µm and −2 µm, the ratio of the shear droop and the fractured surface decreased, and the burnished surface ratio improved ($CL = 0$ µm: 13% shear droop, 77% burnished surface, and 10% fractured surface; $CL = -2$ µm: 12% shear droop, 78% burnished surface, and 10% fractured surface). At $CL = -4$ µm, the ratio of shear droop decreased further, and the fracture surface could not be confirmed, but, as shown in the schematic diagram of Figure 3f, shape deformation occurred along the die R (10% shear droop, 80% burnished surface, and 10% deformation). At $CL = -8$ µm and $CL = -4$ µm, the deformation values were the same (10% shear droop, 80% burnished surface, and 10% deformation). Incidentally, we assumed that these deformations in products and parts can be removed by the so-called barrel polishing, and products and parts can be used. Therefore, it was clarified that by adopting $CL = -4$ µm or −8 µm, it is possible to obtain parts with minimal shear droop and no fracture surface from the viewpoint of actual production.

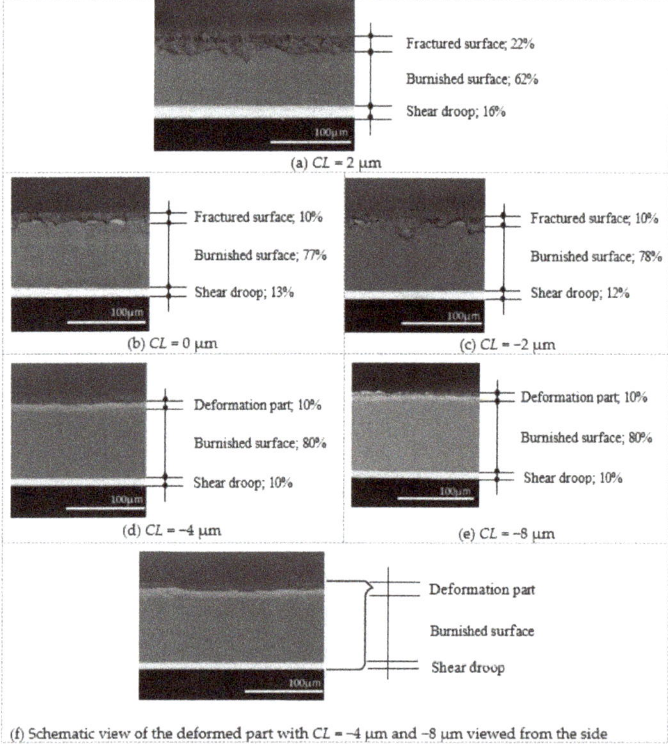

Figure 4. SEM image cut surface of various clearance blanking. (**a**) $CL = 2$ µm; 16% shear droop, 62% burnished surface, and 22% fractured surface, (**b**) $CL = 0$ µm; 13% shear droop, 77% burnished surface, and 10% fractured surface, (**c**) $CL = -2$ µm; 12% shear droop, 78% burnished surface, and 10% fractured surface, (**d**) $CL = -4$ µm; 10% shear droop, 80% burnished surface, and 10% deformation, (**e**) $CL = -8$ µm; 10% shear droop, 80% burnished surface, and 10% deformation, (**f**) Schematic view of the deformed part with $CL = -4$ µm and −8 µm viewed from the side.

Next, Figure 5 shows the blanking load stroke diagram obtained from the blanking experiment. The figure shows that the blanking load increases as clearance decreases. The shear energy obtained from the area of the load stroke diagram, which is a parameter representing the load applied to the punch, was increased by about 15% when $CL = -8$ μm compared to $CL = 2$ μm. A comparison of $CL = -4$ μm and -8 μm, at which the full-sheared product surface was obtained, showed that the blanking load and shear energy are lower at $CL = -4$ μm. Therefore, the optimum clearance obtained from the present blanking experiment is $CL = -4$ μm.

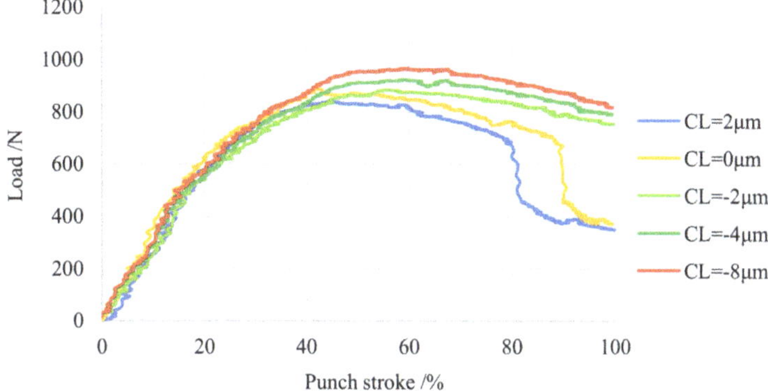

Figure 5. Load–stroke curves for various clearance.

3.2. Consideration of Cut Surface Generation in Each Clearance by FEM Analysis

Figure 6 shows the results of the cut surface at each clearance obtained by FEM analysis utilized damage value C. When the clearances were 2 μm, 0 μm, and −2 μm, the damage value C exceeded the critical fracture value C_{cr} of 1.5, indicating that the elements were erased and cracks occurred, leading to fracture. It was simulated that no fracture at the product side occurred at $CL = -4$ μm and -8 μm. Furthermore, the results of FEM analysis for each clearance are very similar to those in the actual blanking experiment; here, the damage value C is obtained from the integral value along the strain history of the maximum principal stress, as shown in the previous Equation (1). That is, the damage value C increases as the tensile stress in shearing (the maximum principal stress is positive) increases. Therefore, to prevent the damage value C from becoming excessively large, it is necessary to prevent tensile stress from acting on the shearing region as much as possible.

Therefore, we evaluated and compared the flow states of materials that are considered to affect the stress state. Figure 7 is an enlarged view of the shear deformation area at the stage where the punch has penetrated 70% of the plate thickness, which is the step immediately before the fracture starts, and the difference in material flow at each clearance is shown by the flow velocity and flow direction. The single dot and dash line in a red color, shown in the figure, is the line drawn vertically from the punch tip, and becomes the path of the punch tip as the FEM analysis progresses. When $CL = 2$ μm, the velocity, along the single dot and dash line in a red color, is almost constant and the material flows smoothly. This means that the material of scrap is compressed by the blank holder and the material of product is pressed by the punch penetration. Then, the tensile stresses are generated in narrow shear deformation area. On the other hand, when the clearance goes to zero or negative, the velocity along the single dot and dash line in red color decreases as it approaches the die. Therefore, contrary to the case of $CL = 2$ μm, it is presumed that the tensile stress along the area is low.

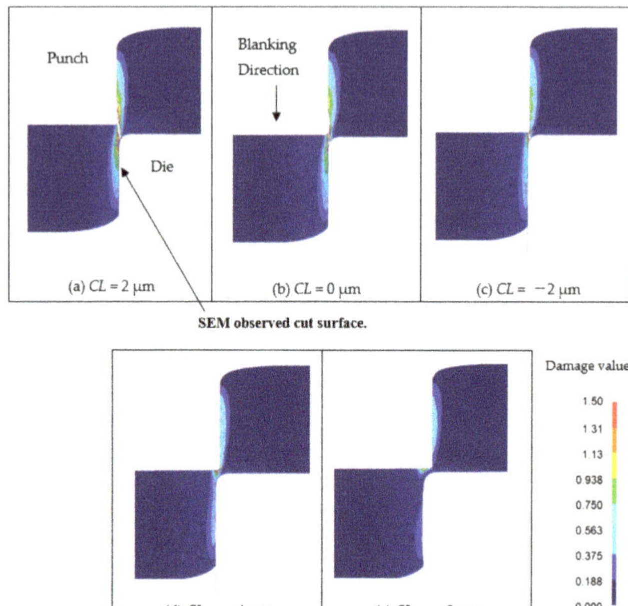

Figure 6. Comparison of damage value C of various clearance blanking. (**a**) $CL = 2$ μm, (**b**) $CL = 0$ μm, (**c**) $CL = -2$ μm, (**d**) $CL = -4$ μm, (**e**) $CL = -8$ μm.

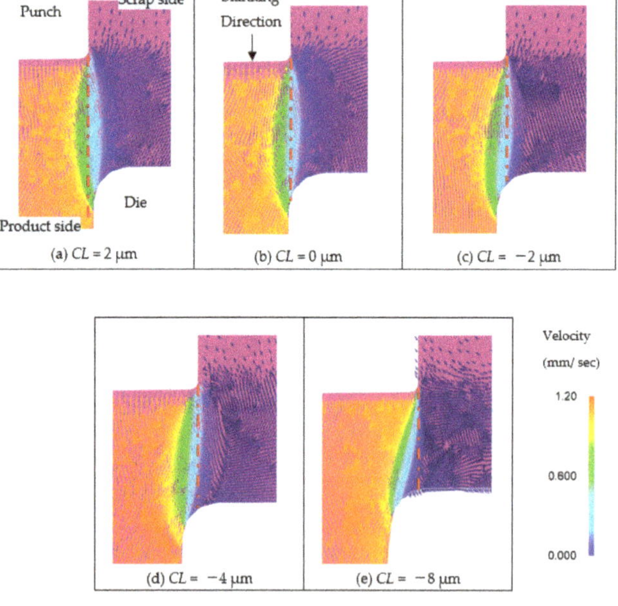

Figure 7. Comparison of material flow for 70%t punch penetration of various clearance blanking in enlarged view of shear deformation area. (**a**) $CL = 2$ μm, (**b**) $CL = 0$ μm, (**c**) $CL = -2$ μm, (**d**) $CL = -4$ μm, (**e**) $CL = -8$ μm.

Similar to the comparison of material flow, the hydrostatic stress should be also discussed. In general terms, the ductility of metals is said to improve under hydrostatic pressure [10]. However, the stress and strain results obtained by FEM analysis are calculated on the basis of the data obtained from the tensile test of the workpiece, so the metal ductility effect due to hydrostatic stress is not taken into account. However, as a guideline for die design in actual manufacturing, if the hydrostatic pressure state can be visualized, it may be useful to control the fracture surface ratio of the cut surface. Figure 8 shows the hydrostatic stress state of each clearance when the punch penetrated 70% of the plate thickness. As the clearance decreases, the hydrostatic stress decreases and the hydrostatic pressure increases. For example, hydrostatic stress value in the area connecting the punch corner and the die R corner were 0 MPa to −500 MPa at CL = 2 µm, −500 MPa to −1000 MPa at CL = 0 µm, −1000 MPa to −1500 MPa at CL = −2 µm, −1500 MPa to −2000 MPa at CL = −4 µm, and −2000 MPa to −2500 MPa at CL = −8 µm. It was shown that the clearance CL, at which the fracture does not occur in the above-mentioned blanking experiment, was −4 µm or less. The FEM analysis under the given conditions indicated that a hydrostatic stress of −1500 MPa or less is necessary. However, as mentioned above, the metal ductility effect of hydrostatic stress is not taken into account in the FEM analysis, so the relationship between the cut surface obtained in the punching experiment and the damage value in the FE analysis will be verified in the future.

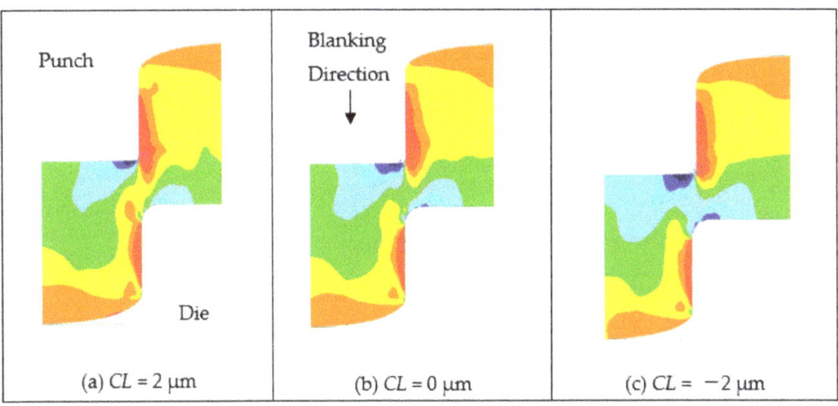

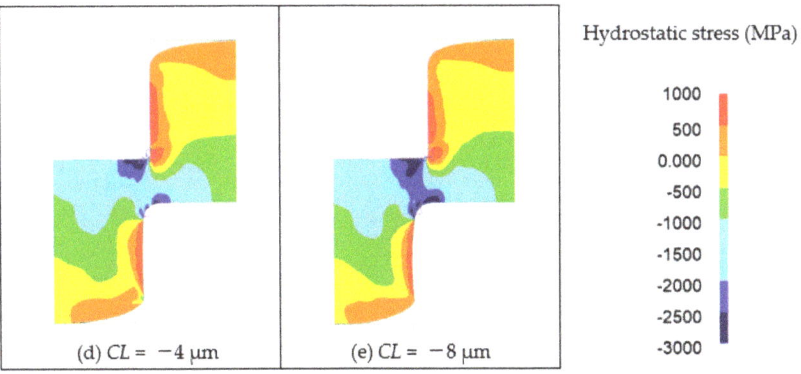

Figure 8. Comparison of hydrostatic stress for 70%t punch penetration of various clearance blanking. (**a**) CL = 2 µm, (**b**) CL = 0 µm, (**c**) CL = −2 µm, (**d**) CL = −4 µm, (**e**) CL = −8 µm.

3.3. FEM Analysis of the Load on the Punch Tip

Generally, the punch breaks and wears at its tip during blanking [14]. Therefore, the relationship between clearance and the equivalent stress applied to the punch was evaluated at Point A at the start of the tip radius portion of the punch, as shown in Figure 9. Although not shown, Point A is the point with the highest equivalent stress in the equivalent stress distribution of the punch. Figure 10 shows the evaluation and comparison results of the aforementioned knowledge [14]. The equivalent stress increased as the clearance decreased, and the maximum equivalent stress was 4200 MPa at $CL =$ 2 μm when punch stroke was about 80%. The maximum equivalent stress was highest at $CL = -8$ μm, reaching 4600 MPa when punch stroke was about 80%. Although not shown, the stress state was compression. In accordance with Table 2, which showed that fracture occurs when the compressive stress of cemented carbide reaches about 6880 MPa, in this FEM analysis, fracture compression loads of 61% and 67% were applied at $CL = 2$ μm and $CL = -8$ μm, respectively. To reduce the punch load, it is necessary to design dies and conditions such that the hydrostatic pressure stress is −1500 MPa even with a clearance $CL = -4$ μm or more. For example, the future challenge is to explore the possibility of reducing the punch load by examining the size and shape of the die R. The breakage and wear of the punch is most likely to occur when the punch returns from the workpiece [14]. Moreover, since the tensile strength is lower than the compressive strength of the cemented carbide punch [28], it is also extremely important to study the return process.

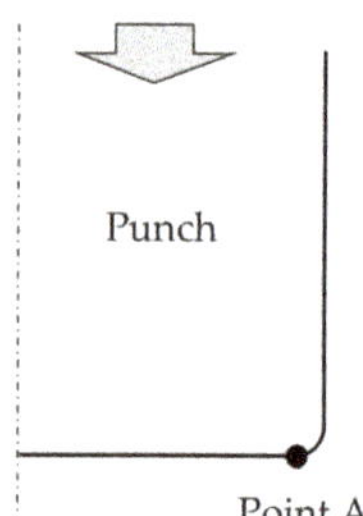

Figure 9. Stress estimation point in FEM analysis.

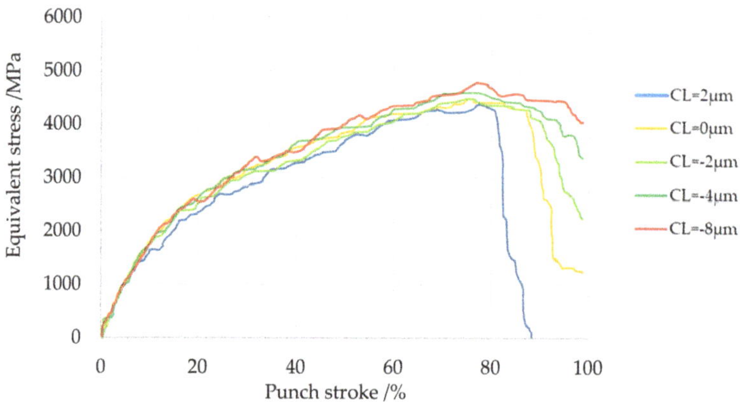

Figure 10. Equivalent stress at Point A at the tip of the punch for various clearances.

4. Conclusions

The effectiveness of negative clearance was verified from the viewpoint of the cut surface and punch load in micro component processing, where it is difficult to use so-called fine blanking. Specifically, a comprehensive blanking experiment and FEM analysis using a commercially available code were carried out, particularly from the viewpoint of the tool clearance conditions at narrow clearance, zero clearance, and negative clearance in the FB processing of the austenitic stainless steel JIS SUS304, and the following conclusions were obtained.

1. As the clearance decreases, the fractured surface of product side the cut surface decreases;
2. As the clearance decreases, the load on the punch tip increases;
3. $CL = -4$ µm is the optimum clearance to obtain a product side cut surface with no fracture and to reduce the punch load.

Author Contributions: Conceptualization, Y.S., M.Y. and M.M.; Experiment, Y.S.; Writing—Original Draft, Y.S.; Writing—Review & Editing, M.Y. and M.M. All authors have read and agreed to the published version of the manuscript.

Funding: This research received no external funding.

Acknowledgments: We would like to express our gratitude to Y. Nagasu of the Industrial Technology Center of Nagano Prefecture for his valuable guidance in the FE analysis.

Conflicts of Interest: The authors declare no conflict of interest.

References

1. Xu, Z.T.; Peng, L.F.; Lai, X.M.; Fu, M.W. Geometry and grain size effects on the forming limit of sheet metals in micro-scaled plastic deformation. *Mater. Sci. Eng. A* **2014**, *611*, 345–353. [CrossRef]
2. Zhao, Y.H.; Guo, Y.Z.; Wei, Q.; Topping, T.D.; Dangelewicz, A.M.; Zhu, Y.T.; Langdon, T.G.; Lavernia, E.J. Influence of specimen dimensions and strain measurement methods on tensile stress–strain curves. *Mater. Sci. Eng. A* **2009**, *525*, 68–77. [CrossRef]
3. Zhao, Y.H.; Guo, Y.Z.; Wie, Q.; Dangelewicz, A.M.; Xu, C.; Zhu, Y.T.; Langdon, T.G.; Zhou, Y.Z.; Lavernia, E.J. Influence of specimen dimensions on the tensile behavior of ultrafine-grained Cu. *Scr. Mater.* **2008**, *59*, 627–630. [CrossRef]
4. Wang, J.L.; Fu, M.W.; Shi, S.Q. Influences of size effect and stress condition on ductile fracture behavior in micro-scaled plastic deformation. *Mater. Des.* **2017**, *131*, 69–80. [CrossRef]
5. Sugioka, K. The state of the art and future prospect of ultrafast laser micro-processing. *J. Jpn. Soc. Precis. Eng.* **2015**, *81*, 709–713. [CrossRef]
6. Shikida, M. Overview of Etching Process for Fabricating Microstructures. *J. Surf. Finish. Soc. Jpn.* **2008**, *59*, 84–87. [CrossRef]
7. Ogiso, H.; Nakano, S. Micro Fabrication by Ion Beam. *J. Jpn. Soc. Precis. Eng.* **2004**, *70*, 1473–1476. [CrossRef]
8. Maeda, T.; Nakagawa, T. Experimental Investigations on Fine Blanking 1st Report. *J. Jpn. Soc. Technol. Plast.* **1968**, *9*, 618–626.
9. Maeda, T.; Nakagawa, T. Experimental Investigations on Fine Blanking 2nd Report. *J. Jpn. Soc. Technol. Plast.* **1968**, *9*, 627–636.
10. Bridgmen, P.W. *Studies in Large Plastic Flow and Fracture*; Harvard University Press: Cambridge, MA, USA, 1952.
11. Suzuki, Y.; Shiratori, T.; Yang, M.; Murakawa, M. Processing of metal microgear tooth profile by finish blanking and extrusion blanking and evaluation of cut-surface shape. *J. Jpn. Soc. Technol. Plast.* **2018**, *60*, 64–69. [CrossRef]
12. Suzuki, Y.; Shiratori, T.; Yang, M.; Murakawa, M. Elucidation of Shearing Mechanism of Finish-type FB and Extrusion-type FB for Thin Foil of JIS SUS304 by Numerical an EBSD Analyses. *Materials* **2019**, *12*, 2143. [CrossRef] [PubMed]
13. Aoki, I. Wear Mechanism of Blanking Tools and Factors Affecting Tool Wear. *J. Jpn. Soc. Technol. Plast.* **1986**, *27*, 140–150.
14. Maeda, T.; Matsuno, K. Wear of punching die. *J. Jpn. Soc. Technol. Plast.* **1966**, *7*, 265–273.

15. Koga, N.; Tsukakoshi, K. Effect of Clearance on Tool Wear During Blanking of High-Tensile-Strength Steel Sheets. *J. Jpn. Soc. Technol. Plast.* **2014**, *55*, 48–52.

16. Mucha, J.; Jaworski, J. The Quality Issue of the Parts Blanked from Thin Silicon Sheets. *J. Mater. Eng. Pref.* **2017**, *26*, 1865–1877. [CrossRef]

17. Mucha, J. An experimental analysis of effects of various material tool's wear on burr during generator sheets blanking. *Int. J. Adv. Manuf. Technol.* **2010**, *50*, 495–507. [CrossRef]

18. Wei, G.; Hon-Yuen, T. Effects of extended punching on wear of the WC/Co micropunch and the punched microholes. *Int. J. Adv. Manuf. Technol.* **2012**, *59*, 955–960.

19. Hernandez, J.; Franco, P.; Estrems, M.; Faura, F. Modelling and experimental analysis of the effects of tool wear on form errors in stainless steel blanking. *J. Mater. Process. Technol.* **2006**, *180*, 143–150. [CrossRef]

20. Faura, F.; Lopez, J.; Sanes, J.; Garcia, A. Tools life equation for blanking 18-8 stainless steel strips. *Rev. Metal. Madrid.* **1998**, *34*, 328–334. [CrossRef]

21. Nakajima, T.; Yoshida, Y.; Matsuno, T.; Seto, A.; Suehiro, M. Influence of tool edge shape on stress distribution at piecing punch tip. In Proceedings of the 64th Japanese Joint Conference for the Technology of Plasticity, Osaka, Japan, 1–3 November 2013; pp. 309–310.

22. Hambli, R. Blanking tool wear modeling using the finite element method. *Int. J. Mach. Tools Manuf.* **2001**, *41*, 1815–1829.

23. Falconnet, E.; Makich, H.; Chambert, J.; Monteil, G.; Picart, P. Numerical and experimental analyses of punch wear in the blanking of copper alloy thin sheet. *Wear* **2012**, *296*, 598–606. [CrossRef]

24. Falconnet, E.; Chambert, J.; Makich, H.; Monteil, G. Prediction of abrasive punch wear in copper alloy thin sheet blanking. *Wear* **2015**, *338–339*, 144–154. [CrossRef]

25. Tanaka, T.; Hagihara, S.; Tadano, Y.; Inada, T.; Mori, T.; Fuchiwaki, K. Application of finite element method to analysis of ductile fracture criteria for punched cutting surfaces. *Mater. Trans.* **2013**, *43*, 1697–1702. [CrossRef]

26. Taupin, E.; Breitling, J.; Wu, W.; Altan, T. Material fracture and burr formation in blanking results of FEM simulations and comparison with experiments. *J. Mater. Process. Technol.* **1996**, *59*, 68–78. [CrossRef]

27. Sasada, M.; Shimura, K.; Aoki, I. Study on coefficient of friction between tool and material in Shearing. *Trans. Jpn. Soc. Mech. Eng. Ser. C* **2005**, *71*, 249–255. [CrossRef]

28. Kawakami, M. Development of Nano-grained hard-metal and application examples. *Sokeizai* **2011**, *52*, 28–32.

Article

Fabrication of Micro-Punch Array by Plasma Printing for Micro-Embossing into Copper Substrates

Tomomi Shiratori [1,*], Tatsuhiko Aizawa [2], Yasuo Saito [3] and Kuniaki Dohda [4]

[1]   Komatsuseiki Kosakusho, Co. Ltd., R&D section, Suwa, Nagano 392-0012, Japan
[2]   Surface Engineering Design Laboratory, Shibaura Institute of Technology, Ohta, Tokyo 144-0045, Japan
[3]   Chuo Denshi Kogyo, Co. Ltd., Ujyou, Kumamoto 869-0512, Japan
[4]   Department of Mechanical Engineering, Northwestern University, Evanston, IL 60208, USA
*   Correspondence: shiratori@komatsuseiki.co.jp; Tel.: +81-266-52-6100

Received: 16 July 2019; Accepted: 14 August 2019; Published: 19 August 2019

Abstract: Copper substrates were wrought to have micro-grooves for packaging by micro-stamping with use of a AISI316 stainless steel micro-punch array. The micro-texture of this arrayed punch was first tailored and compiled into CAD data. A screen film was prepared to have the tailored micro-pattern in correspondence to the CAD data. A negative pattern to this screen was printed directly onto the AISI316 die substrate. This substrate was plasma nitrided at 673 K for 14.4 ks. The unprinted die surfaces were selectively nitrogen super-saturated to have sufficiently high corrosion toughness and hardness; other surfaces were masked by the prints. The two-dimensional micro-pattern on the screen was transformed into a three-dimensional nitrogen supersaturated micro-texture embedded in the AISI316 die. The printed surfaces were selectively sand-blasted to fabricate the micro-textured punch array for micro-embossing. A uniaxial compression testing machine was utilized to describe the micro-embossing behavior in copper substrates and to investigate how the micro-texture on the die was transcribed to the copper. The micro-punch array in this study consisted of three closed loop heads with a width of 75 μm and a height of 120 μm after plasma nitriding and sand-blasting. Since the nitrogen supersaturated heads had sufficient hardness against the blasting media, the printed parts of AISI316 die were removed. The micro-embossing process was described by comparison of the geometric configurations between the multi-punch array and the embossed copper plate.

Keywords: packaging; copper substrate; micro-embossing; micro-textures; plasma printing; micro-punch array; screen printing; AISI316

---

1. Introduction

A key technology in the plastic mold packaging of hollowed GaN high electron mobility transistor (HEMT) chips lies in micro-joining with sufficient interfacial integrity between the copper substrate and the plastic molds [1]. In particular, liquid crystal plastic (LCP) molds have been utilized for packaging with sufficient gross-leak proof [2–4]. The copper substrates must have micro-textures for those plastic molds to be joined with sufficiently high integrity of the interface strength between the LCP molds and copper substrate.

Authors have proposed a non-traditional method to fabricate the micro-punch with the use of low temperature plasma nitriding [5]. This plasma printing is based on the principle that a two-dimensional micro-pattern is transformed into the three-dimensional nitrogen-embedded micro-structure to be working as the micro-punch array heads [6]. The initial micro-pattern was directly printed onto the stainless steel die substrate by ink-jet printing [7], screen printing [8], and maskless lithography [9].

In the present study, a micro-punch array was designed to have multi-heads with a geometric configuration of three continuous closed loops. Tailored CAD data were first transferred to screen film for printing as a negative micro-pattern to these loops. This negative micro-pattern was screen-printed onto an austenitic stainless steel AISI316 die substrate. This die substrate was plasma nitrided at 673 K for 14.4 ks to develop the two-dimensional micro-pattern into a three-dimensional nitrogen supersaturated microstructure in the die. A micro-punch array was fabricated by mechanically removing the printed parts of the AISI316 die. Scanning electron microscopy (SEM), electron diffractive X-ray spectroscopy (EDX), and a non-contact three-dimensional measuring device were utilized to describe the evolution of the nitriding-induced microstructure by element mapping and the dimensional change, respectively. A compression testing machine was utilized to emboss the micro-punch array into the copper substrate for micro-texturing.

2. Experimental Procedure

An AISI316 die substrate with dimensions measuring 24 mm × 12 mm × 5 mm was utilized as a substrate material. Its surface was mirror-polished for plasma printing. The average roughness (*Ra*) of the AISI316 die substrate was 0.010 μm. The plasma printing procedure was composed of the screen printing, low-temperature plasma nitriding, blasting, and micro-embossing using a tensile tester. The present plasma printing process consisted of three steps, as illustrated in Figure 1. First, the negative micro-pattern of arrayed punch heads was printed as a two-dimensional mask onto a mirror-polished AISI316 substrate (Figure 1a). Figure 1b depicts the screen-printed mask pattern. Second, this printed substrate was plasma nitrided at 673 K for 14.4 ks to selectively super-saturate nitrogen onto the unprinted substrate surfaces (Figure 1c). The printed surfaces were not nitrided and maintained the same hardness as the matrix so that they could be mechanically removed with ease from the substrate to form the multi-punch array, as shown in Figure 1d.

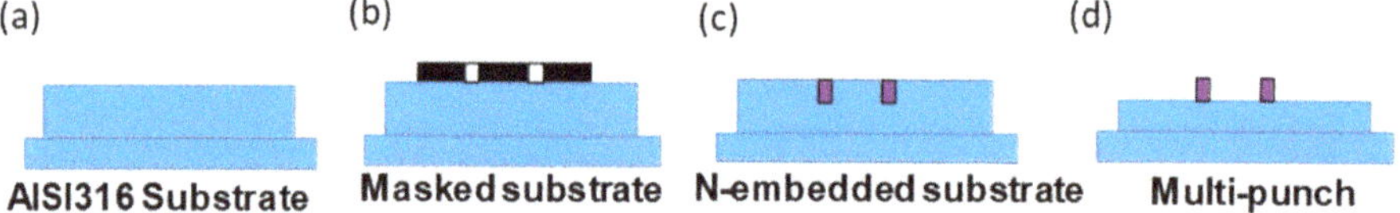

Figure 1. Plasma printing procedure in the present study. (**a**) The starting AISI316 substrate, (**b**) masked substrate by screen printing, (**c**) nitrogen-embedded substrate by plasma nitriding, and (**d**) multi-punch formed by blasting.

2.1. Screen Printing onto Die Substrate

The screen-printing system (NEWLONG, Co., Ltd., Tokyo, Japan), as shown in Figure 2, was employed to print the CAD-designed micro-pattern onto the substrate's surface. The three closed loop patterns with a width of 50 μm were directly printed onto the AISI316 die surface within the range of 19.15 mm × 8.05 mm, as depicted in Figure 3a. The details of the CAD data for three closed loops are shown in Figure 3b. These loops were aligned to have mutual distances of 150 mm between adjacent loops. In the following experiments, a screen with three closed loops was employed to print its negative pattern onto the surface of the AISI316 die. Ink for screen-printing must be optimally selected from among several candidates to have sufficient thermal resistance during plasma nitriding (at 673 K). A polymer-based ink has the risk to diminish itself during plasma nitriding at 673 K. In practical operations, specially formulated TiO$_2$ ink (Teikoku Printing Inks Mfg. Co., Ltd., Tokyo, Japan), without the use of thinning agents, was used for directly printing onto the die surface and then dried at 373 K for 1.2 ks in air.

Figure 2. Appearance of screen-printing system.

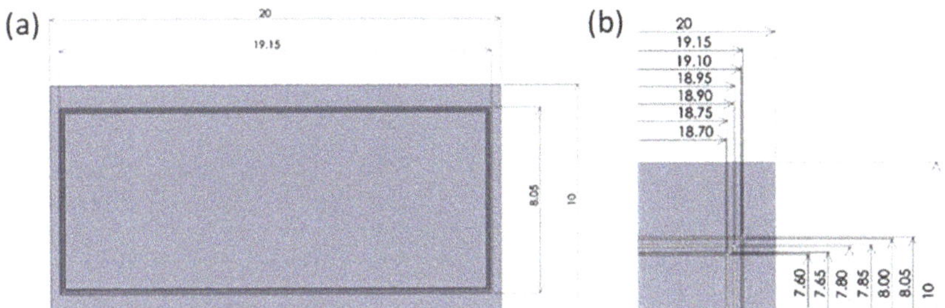

Figure 3. Geometry of a closed loop patterns on the CAD data. (**a**) General pattern of three closed loop, (**b**) details of closed loop patterns.

2.2. Low-Temperature Plasma Nitriding

A high-density RF (radio frequency)/DC (direct current) plasma nitriding system (YS-Electric Industry, Co., Ltd., Yamanashi, Japan) was utilized to selectively super-saturate nitrogen on the unprinted substrate surfaces at 673 K for 14.4 ks at 70 Pa, as shown in Figure 4. After evacuation down to 0.1 Pa, the nitrogen gas was introduced to pre-sputter the printed die surface for 1 ks under a DC-bias of −600 V. After re-evacuation, the specimen was heated to 673 K under a nitrogen atmosphere at 250 Pa. Then, the nitrogen hydrogen mixture gas was introduced with a flow rate of 160 mL/min for the nitrogen and 30 mL/min for the hydrogen, respectively. After plasma nitriding, the specimen was cooled in the chamber under a nitrogen atmosphere. The micro-printed AISI316 substrate surface was fully covered by a plasma sheath with a high nitrogen ion and NH-radical densities; enough to drive the nitrogen super-saturation at lower temperatures [5]. This selective anisotropic nitrogen-embedding process resulted in selective hardening and selective nitrogen concentrations. The printed surface remained as a matrix hardness, while the unprinted surfaces were selectively hardened to 1400 HV for the AISI316 substrates, as reported in Reference [10].

Figure 4. Appearance of the plasma nitriding machine.

2.3. Mechanical Blasting Process

Mechanical blasting equipment (Fuji Manufacturing Co., Ltd., Tokyo, Japan) was also employed to selectively remove the un-nitrided parts and the masked ink from the substrate. Owing to the hardness distribution, the nitrided areas were left as a punch head while the un-nitrided areas were completely removed by this processing. Figure 5 depicts the blasting apparatus for manual operation. The blasting rate was controlled by the shooting speed of the blasting media. The punch height was also varied by duration time. In the following blasting step, fine silica particles with an average diameter of 5 μm were utilized as the blasting medium. The shooting rate was maintained constant at 2 m/s and the shooting angle was 60 degrees. As depicted in Figure 5b, the specimen was fixed into a jig on the shooting stage for continuous shooting operation. The duration time was selected to be 300 s in the experiments. According to Reference [11], the punch height reached 120 μm by blasting the printed parts of substrate for 300 s.

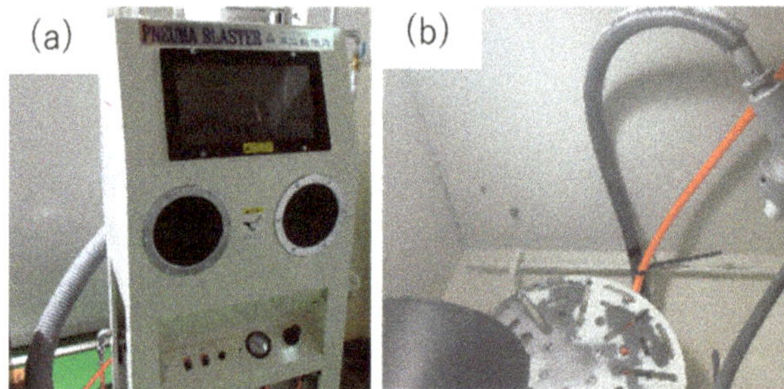

Figure 5. Mechanical blasting equipment. (**a**) Overall image of blasting equipment and (**b**) the shooting stage for blasting the substrates in manual operation.

2.4. Micro-Embossing Process

A precision universal testing machine AUTOGRAPH AGS-X 10 kN (SHIMADZU Corporation, Kyoto, Japan) was utilized for micro-embossing, as shown in Figure 6a. The plasma-printed punch array was set onto a compression test jig and embossed into the copper specimen, as shown in Figure 6b. The compression testing conditions were as follows: The compressive velocity was constant at 0.1 mm/s until the applied load reached maximum at 10 kN and the duration time was set at 10 s. An oxygen-free copper plate (20 mm × 10 mm × 1 mm) was employed as a work material for this micro-embossing. The average roughness was 0.093 μm.

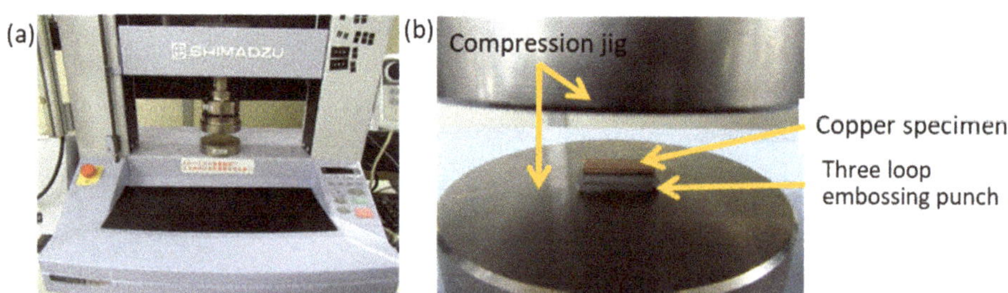

Figure 6. Micro-embossing in the uniaxial compression onto the copper plates. (**a**) Overview of the compression testing machine and (**b**) enlarged view of the micro-embossing experiment condition.

2.5. Evaluation Method for Micro-Embossed AISI316 Die Substrate

The AISI316 die substrate in each step of the plasma printing process was evaluated by a three-dimensional measurement machine (Alicona Imajing GmbH., Graz, Austria), as well as scanning electron microscopy (SEM; JSDM-IT300LV, JEOL Ltd., Tokyo, Japan). Energy dispersive X-ray spectroscopy (EDX; Pegasus, EDAX, Inc., Tokyo, Japan) was utilized for fine element mapping.

3. Experimental Results

The plasma printing procedure in Figure 1 was put into practice to shape the micro-punch array with three continuous closed loop heads for micro-embossing. A selective nitrogen embedding process using low temperature plasma nitriding was described as an essential step for plasma printing. Secondly, the micro-punch array with three continuous closed loop heads was fabricated using sand-blasting. This micro-punch was embossed into copper plates to demonstrate that plasma printing should work as an effective tool to make micro-textures in copper substrates for packaging.

3.1. Selective Nitrogen Embedding

The micro-patterned AISI316 die was plasma nitrided at 673 K for 14.4 ks at 70 Pa to demonstrate that nitrogen solutes were homogeneously embedded in the unprinted areas. Figure 7 shows a SEM image of the nitrided AISI316 die before removing the printed masks from the die surface. Three continuous closed loops were seen on the AISI316 surface and were metallic and shining while other areas were still covered by masks. The average width of the three closed loops from W_1 to W_3 was 90 μm. There were three closed loop width differences between the CAD-designed 50 μm and the demonstrated 90 μm. This phenomenon suggested that the polymer-based ink was diminished during plasma nitriding.

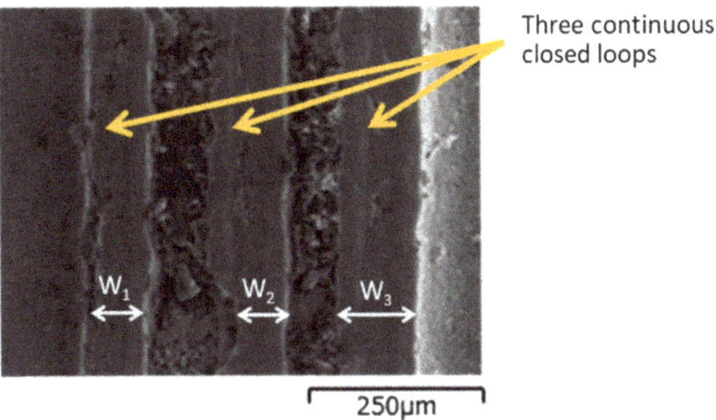

Figure 7. SEM image of the plasma nitrided AISI316 die at 673 K for 14.4 ks before removal of the masks. Three continuous closed loops were seen as three lines and the average width of lines from W_1 to W_3 is 90 μm.

SEM–EDX was utilized to describe element mapping on the nitrided AISI316. As shown in Figure 8, both titanium (Figure 8a) and oxygen (Figure 8b) are present on the AISI316 die surface, except for on the three lines. The surface other than the three lines were still covered by the TiO_2 ink. While the iron (Figure 8c), the chromium (Figure 8d), and nickel (Figure 8e) were only detected on the three lined areas. That is, the three lines were a part of the bare AISI316 matrix without TiO_2 masks. Nitrogen (Figure 8f) was uniformly detected on the whole AISI316 die surface. This proves that the masked AISI316 die was uniformly covered by the plasma sheath and homogeneously nitrided under high nitrogen ion density conditions in plasmas.

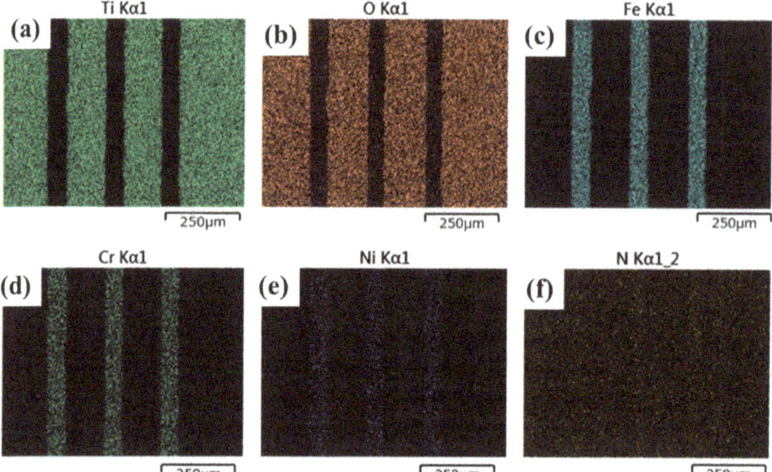

Figure 8. Element mapping analyzed by SEM–energy dispersive X-ray spectroscopy (EDX) for the nitrided AISI316 substrate. (**a**) Titanium, (**b**) oxygen, (**c**) iron, (**d**) chromium, (**e**) nickel, and (**f**) nitrogen.

3.2. Fabrication of Micro-Punch Array

Sand-blasting was utilized to mechanically remove the unnitrided parts from the AISI316 surfaces, as well as the TiO_2 masks. Figure 9 depicts the micro-punch array with three continuous closed loop heads. Figure 10 shows an enlarged SEM image and height distributions. The height distribution and surface roughness distribution of the multi-punch array was measured at line *A–A'* and line *B–B'*. From Figure 10b, the average height of three closed loop punch reached to the 117 μm. These punches have sharp edge shoulders. The average punch width reached 75 μm at the height position in Figure 10b. The surface roughness in line *B–B'* with the length of 100 μm, was measured by using the cutoff value 20 μm. After JIS B 0601, it reached *Ra* 0.035 μm.

Figure 9. Micro-textured AISI316 micro-punch array with three continuous closed loop heads on the surface.

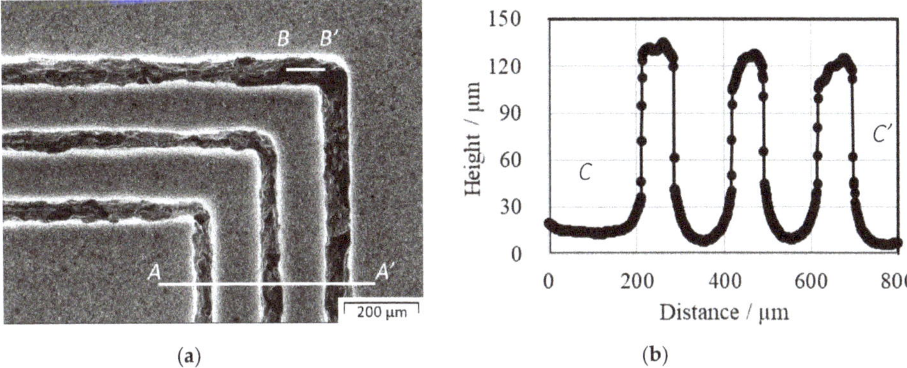

Figure 10. Enlarged SEM image and height distribution of the multi-punch array. (**a**) Enlarged SEM image of micro-textured AISI316 micro-punch array, (**b**) height distribution of the multi-punch array measured along the line *A–A'*. The average punch width at 90 µm height position is 75 µm and the average punch height is 117 µm.

The whole masked and unnitrided AISI 316 parts were removed to leave the three loop heads. Figure 11 depicts an SEM image of three loop punch heads and their nitrogen mapping. AISI316 surfaces were removed into the depth by the sand-blasting. This is because the unprinted surfaces were selectively nitrogen-embedded up to a nitrogen content of 4 to 5 mass% and solid-solution hardened not to be mechanically blasted.

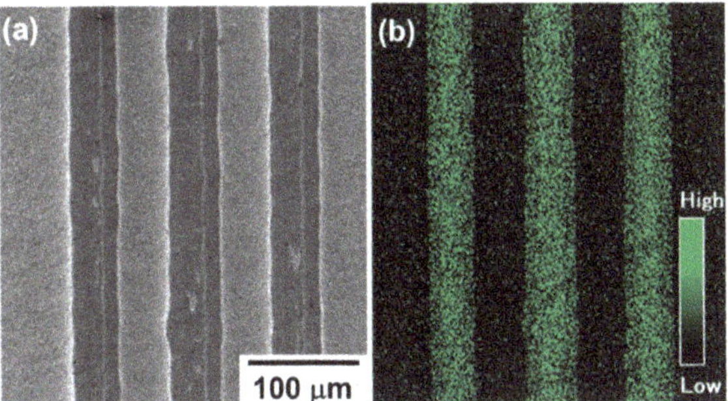

Figure 11. SEM image and nitrogen mapping on the micro-punch heads with three continuous closed loops. (**a**) SEM image and (**b**) nitrogen mapping.

3.3. Micro-Embossing into Copper Substrates

The micro-punch in Figure 9 was utilized for micro-embossing process by a Uniaxial compression testing machine. Figure 12 shows the micro-embossed oxygen-free copper substrate for thermal spreading in the package of hollowed GaN chips. Three continuous closed grooves were formed in the copper substrate to surround its center part for packaging the hollowed GaN chips. Three continuous loop punch heads corresponded to the three closed loop grooves.

Figure 13a shows an SEM image of the three loop grooves at the corner of the copper substrate. Figure 13b depicts the surface profile of the cross-section along the *C–C'* in Figure 13a. The average width of the micro grooves was 58 µm at a measurement depth of 20 µm and a depth of the micro grooves of 35 µm. Three continuous closed loop heads with a line width of 75 µm and height of 117 µm formed the three continuous micro-grooves, with a width of 58 µm and depth of 35 µm through this micro-embossing process at a 10-kN compression load. The difference the between line widths of the three continuous closed loop heads and the embossed three continuous micro-grooves was not considered due to elastic deformation. The surface roughness in line *D–D' with the length of* 100 μm, reached Ra = 0.126 mm by using the cutoff value of 20 μm after JIS B 0601 reached *Ra* 0.126 μm.

Figure 12. Oxygen-free copper substrate for thermal spreading in the package of hollowed GaN chips after micro-embossing.

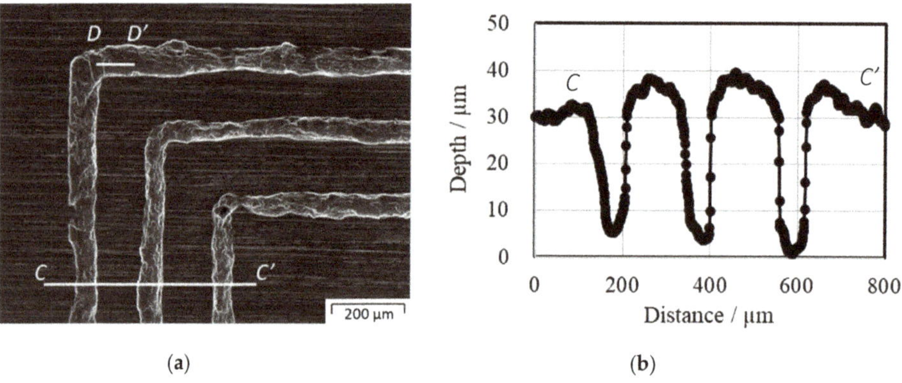

(**a**)   (**b**)

Figure 13. Enlarged SEM image and depth distribution of the micro-embossed oxygen-free copper substrate. (**a**) Enlarged SEM image of micro-embossed oxygen-free copper substrate, (**b**) depth distribution of the micro-embossed oxygen-free copper substrate measured along the line C–C'. The average micro-emboss width at a depth of 20 μm is 58 μm and the average micro-embossing depth is 35 μm.

4. Discussion

A micro-milling process was employed to compare the processing time for the fabrication of the same three closed loop AISI316 punch array as that made by the present plasma printing process. A milling tool with a diameter of 10 μm was prepared to achieve fine corner curvatures of the micro-cavities in Figures 3 and 10b. The average machining speed, cutting depth, as well as cutting distance of a single cutting layer were assumed to be 5 mm/s, 5 μm, and 8 μm, respectively, without fracture of the thin milling tools. The milling time to remove the three closed loop areas of 18 mm × 6.5 mm × 0.1 mm and 19.5 mm × 6.5 mm × 0.1 mm was 7 h. In addition, a conventional punch needs heat treatment and required 8 h after the milling process. Including the takt time taken to prepare the computer aided machining (CAM) data for micro-milling, the practical takt time was nearly 2 h. The present plasma printing required only 10 min at most for screen printing, 5 h for plasma nitriding, including heating and cooling, and 5 min for set-up and blasting. No CAM data were necessary since the CAD data were reflected on the screen. This comparison proves the superiority of plasma printing for fabrication of micro-punch arrays for precision mechanical milling processes.

Picosecond laser machining, as well as fiber–laser machining, have been utilized for the formation of micro-groove textures in each copper plate [12]. On the other hand, a micro-grooved copper substrate was fabricated by micro-embossing the plasma-printed punch. The same multi-arrayed punch, was repeatedly utilized for a series of stamping operations to yield the demanded number of copper substrates for packaging. This takt time was reduced to 10 s, including the setting and stamping durations.

The takt time for the production of the micro-textured copper with comparison to the picosecond laser machining to make a micro-grooved copper plate was as follows. In laser micromachining with a beam spot diameter of 50 μm and repetitive frequency higher than 10 MHz [13], the takt time per path for machining a single line down to a depth of 5 μm was only 1 s, including the on/off operation and beam positioning control. Assuming that no adjustment is needed to form the sharp corners of two crossing micro-grooves, the takt time is estimated to be: (1 s) × (12 lines for three micro-grooves) × (7 paths by 35 μm/5 μm) × (double paths for formation of groove side surfaces) = 168 s. Total takt time to complete three closed micro-grooves into the copper substrate was reduced by 1/17 by using the present stamping approach.

Microgroove textures for joining must be tailored to have geometric compatibility to the substrate size and chip allocation on the substrate. When using laser machining, more takt time is necessary to prepare for CAM data and for actual machining operations. The present plasma printing process has the intrinsic flexibility to transcribe a tailored micropattern to the multi-punch array on the die unit for micro-embossing the microgroove textures onto the copper substrate without increasing the takt time of production. Furthermore, this nitrided multi-punch array has sufficient hardness to prolong the die life in practical micro-embossing operations.

5. Conclusions

Plasma printing was successfully applied to fabricate a micro-punch array with three continuous closed loop heads. Three loops with a line width of 75 μm and a height of 117 μm were accurately micro-embossed into an oxygen-free copper substrate to form three continuous closed micro-grooves. These microgrooves work as a key-wedge for micro-joining between the copper substrate and plastic molds in packaging. This plasma printing can be utilized to fabricate tailored three-dimensional micro-textures and to improve interfacial integrity in packaging.

This plasma printing has sufficient flexibility to form any complex-shaped micro-groove network into copper-alloy base substrates for plastic packaging. Different from precision mechanical milling, there is no increase in takt time since the plasma printing procedure is indifferent to the complexity of a micro-grooving network. No CAM data are necessary to transform the geometric model in CAD to a three-dimensional punch array. The dimensional accuracy, as well as geometrical topology, are preserved by screen printing and the selective nitrogen supersaturation. Micro-embossing to form the micro-groove network in the substrate, with use of CNC (Computer Numerically Controlled) stamping, requires much less takt time than needed for short-pulse laser machining to each substrate. In particular, the present process is favored for mass production of substrates for plastic mold packaging.

Author Contributions: T.S. and T.A. made the conceptual design, and planned and executed the series of experiments, with Y.S. and K.D., and wrote this paper.

Funding: This study was financially supported in part by the METI-Program for Supporting Industries in 2018.

Acknowledgments: The authors would like to express their gratitude to T. Yamaguchi (Sanko-Light Industries, Co., Ltd.), S. Kurozumi (Nano-Coat Film, llc.) and H. Hasegawa (Chuo-Denshi Kogyo, Co., Ltd.) for their help in experiments.

Conflicts of Interest: The authors declare no conflict of interest.

References

1. Prejs, A.; Wood, S.; Pengelly, R.; Pribble, W. Thermal analysis and its application to high power GaN HEMT amplifiers. In Proceedings of the IMS-2009 IEEE, Boston, MA, USA, 7–12 June 2009; IEEE Xplore Digital Library: San Diego, CA, USA, 2009; pp. 917–922.
2. Doughetty, D.; Mahalingam, M.; Viswanathan, V.; Zimmerman, M. Multi-lead Organic Air-Cavity Package for High Power High Frequency RFICs. In Proceedings of the IMS-2009 IEEE, Boston, MA, USA, 7–12 June 2009; IEEE Xplore Digital Library: San Diego, CA, USA, 2009; pp. 473–476.
3. Martin, Q.D. High power plastic packaging with GaN. In Proceedings of the CS MANTECH 2015, Scottsdale, AZ, USA, 18–21 May 2015; CS MANTECH: Beaverton, OH, USA, 2015; pp. 123–126.
4. Longford, A.; Matlis, J.; Lynch, J. Advanced design of using LCP-based pre-molded Lead frame packages for RF and MEMS applications. *Adv. Microelectron.* **2012**, *39*, 8–12.
5. Aizawa, T. Low temperature plasma nitriding of austenitic stainless steels. In *Title of the Stainless Steels and Alloys*; Duriagina, Z., Ed.; IntechOpen: London, UK, 2019; Chapter 3; pp. 31–50.
6. Aizawa, T. Characterization on the properties of nitrided layer in dies and molds (low temperature plasma nitrides stainless steels). *Bulletin JSTP* **2019**, *2*, 411–415.
7. Aizawa, T.; Takashima, T.; Shiratori, T. Plasma printing to fabricate the micro-piercing dies for miniature metal products. In Proceedings of the 8th AWMFT, Suwa, Japan, 15 October 2015; CD-ROM: Suwa, Japan, 2015; pp. J1/1–J1/6. [CrossRef]
8. Shiratori, T.; Aizawa, T.; Saito, Y.; Wasa, K. Plasma printing of an AISI316 micro-meshing punch array for micro-embossing onto copper plates. *Metals* **2019**, *9*, 396. [CrossRef]
9. Aizawa, T.; Yoshihara, S. Microtexturing into AISI420 dies for fine piercing of micropatterns into metallic sheets. *J. JSTP* **2019**, *60*, 53–57. [CrossRef]
10. Aizawa, T.; Yoshino, T. Plastic straining for microstructural refinement in stainless steels by low temperature plasma nitriding. In Proceedings of the 12nd SEATUC Conference, YagYakarta, Indnesia, 12–13 March 2018; UG-Press: Shah Alam, Malaysia, 2018; pp. 121–126.
11. Aizawa, T.; Saito, Y.; Yoshihara, S.-I. Micro-embossing into copper plates for mold-packaging into power semiconducting devices. In Proceedings of the 44th Autumnal Meeting of JSTP, Dalian, China, 27 August–1 September 2017; Nissei-Epro, Co., Ltd.: Minato, Tokyo, Japan, 2018; pp. 3–4.
12. Saito, Y.; Aizawa, T.; Wasa, K.; Nogami, Y. Leak-proof packaging for GaN chip with controlled thermal spreading and transients. In Proceedings of the 2018 IEEE BiCMOS and Compound Semiconductor Integrated Circuits and Technology Symposium, San Diego, CA, USA, 15–17 October 2018; IEEE Xplore Digital Library: San Diego, CA, USA, 2018. [CrossRef]
13. Aizawa, T.; Inohara, T. Pico- and Femtosecond Laser Micromachining for Surface Texturing. In *Title of Micromachining*; Duriagina, Z., Ed.; IntechOpen: London, UK, 2019. [CrossRef]

Article

Impact of the Deionized Water on Making High Aspect Ratio Holes in the Inconel 718 Alloy with the Use of Electrical Discharge Drilling

Magdalena Machno [1],*, Rafał Bogucki [2], Maciej Szkoda [1] and Wojciech Bizoń [3]

[1] Institute of Rail Vehicles, Faculty of Mechanical, Cracow University of Technology, 31-155 Cracow, Poland; maciej.szkoda@pk.edu.pl

[2] Institute of Materials Engineering, Faculty of Materials Engineering and Physics, Cracow University of Technology, 31-155 Cracow, Poland; rbogucki@mech.pk.edu.pl

[3] Institute of Production Engineering, Faculty of Mechanical, Cracow University of Technology, 31-155 Cracow, Poland; wojciech.bizon@pk.edu.pl

* Correspondence: magdalena.machno@pk.edu.pl.; Tel.: +48-12-374-36-56

Received: 5 February 2020; Accepted: 20 March 2020; Published: 24 March 2020

Abstract: Nickel-based superalloys are being increasingly applied to manufacture components in the aviation industry. The materials are classified as difficult-to-machine using conventional methods. Nowadays, manufacturing techniques are needed to drill high aspect ratio holes of above 20:1 (depth-to-diameter ratio) in these materials. One of the most effective methods of making high-aspect-ratio holes is electrical discharge drilling (EDD). While drilling high aspect ratio holes, a crucial issue is the flushing of the gap area and the evacuation of the erosion products. The use of deionized water as the dielectric fluid in the EDD offers a considerable potential. This paper includes an analysis of the influence of the machining parameters (pulse time, current amplitude and discharge voltage) on the process performance (drilling speed, linear tool wear, taper angle, hole's aspect ratio, side gap thickness), during the EDD with the use of deionized water in the Inconel 718 alloy. The obtained through holes were subjected to the extended analysis. The impact of the initial working fluid temperature and pressure on the conditions of the flow through the electrode channel was also subjected to the analysis. The deionized water properties were changed by applying an initial temperature. Based on the results of an analysis of the previous research, the EDD of the through holes was performed for a pre-set initial temperature (~313.15 °K) and initial pressure of the working fluid (8 MPa) and selected process parameters. An analysis of the results indicates increasing of hole's aspect ratio by about 15% (above 30), decreasing the side gap thickness by about 40% and enhanced surface integrity.

Keywords: difficult-to-cut material; Inconel 718 alloy; micro-drilling; EDM; aspect ratio hole; deionized water

---

1. Introduction

Nickel-based superalloys, such as Inconel 718, play an increasingly important role in the development and manufacture of aircraft engines components (such as turbine blades, guide vanes). The industrial use of Inconel 718 alloy started in 1965; hence, it is a relatively recent alloy. The Inconel 718 alloy is characterized by excellent mechanical properties, excellent resistance to creep at temperatures up to 973 °K, and corrosion and oxidation resistance in aggressive environments [1].

The major alloyed composition of the material includes Ni and Cr. Additionally, in the chemical composition, elements such as Al, Ti, Nb, Co, Cu and W occur. Fe can also be added in amounts ranging from 1% to 20% [1]. The alloy elements such as Ni and Cr provide the corrosion resistance of the Inconel 718 and crystallize as a γ phase (the precipitation fine hard and dispersed precipitates, i.e., γ' and γ''). The added Ni element forms hardening precipitates γ'' (Ni3Nb, a body centered tetragonal metastable phase). On the other hand, Ti and Al participate in order to form of intermetallic γ' (Ni3 (Ti, Al), simple cubic crystal). The added C element participates in the forming of MC carbides (M = Ti or Nb), but the C content must be low enough to enable Nb and Ti precipitation in the form of γ' and γ'' particles. In addition, Mo is often the content element of the material increasing the mechanical resistance by solid solution hardening [1,2]. In order to obtain high mechanical properties, the alloy is subjected to heat treatment. It involves annealing in the temperature range of 1273–1473 °K for 1 hour with cooling in water with subsequent aging in the temperature range from 923 to 1173 °K in order to separate the coherent phases γ' and γ'', which are responsible for high hardness and alloy strength [3]. Too low annealing temperature may lead to the formation of undesirable NbC, δ-Ni3Nb and Laves phases, resulting in reduced plasticity, fatigue and creep properties [4]. The high hardness of the alloy (average hardness 414 Hv) combined with the low thermal conductivity (11.4 W/(m·°K)) is the cause of problems during machining [5].

Due to extremely tough nature (such as lower thermal conductivity, high work hardening, presence of abrasive carbide particles, high hardness, affinity to react with tool material, high toughness) of the Inconel 718 superalloy, the making of high aspect ratio holes in its structure involves significant difficulties in the case of conventional machining [5,6]. In order to overcome the limitations, the aerospace industry currently applies non-conventional methods such as electrochemical machining (ECM), laser beam machining (LBM) and electrical discharge machining (EDM) to produce micro-scale cooling holes in superalloy materials [7].

In modern gas turbine engines, in order to improve the turbine's operating efficiency, a high temperature of the gas before the turbine (in the range of 823.15–1373.15 °K) is applied, which decreases the components' resistance. To enhance the additional resistance of these materials of components into high temperature, a considerable number of holes (20,000–40,000) with a diameter in the range of 0.3–5 mm and an aspect ratio in the range of (40–600):1 (depth-to-diameter ratio) are made in their structure [5,8–10]. The main task of the holes (named "cooling holes") is to reduce the temperature of the component material by flow of the cooling factor (gas or liquid) through the holes [7]. The efficiency of the cooling process depends on the dimensional shape accuracy and the quality of the holes' inner surface. One of the most effective methods of drilling high aspect ratio holes in the nickel-based superalloys is electrical discharge drilling (EDD) [11,12].

1.1. Electrical Discharge Drilling Process (EDD)

In the EDD process, the allowance is removed through electrical discharges that occur between two electrodes (one of the electrodes being the workpiece and the other one the tool) in a narrow gap ($\sim\mu$m) filled with the working fluid, and the forces occurring between the tool and the workpiece surface are negligible or do not occur. The transformation of electrical energy into thermal energy leads to the vaporization and melting of the material of the workpiece and of the tool electrode [13–15]. Due to the presence of high temperature (about 10,000 °K) in the machining zone, the quality of the machined surface is unsatisfactory because of the occurrence of erosion micro-craters, re-solidified material, heat affected zone and white layer (recast layer) involving micro-cracks and residual tensile stresses [16–18]. A disadvantage of the process is also a low material removal rate and high tool wear reaching above 50% [19]. However, the process of electrical discharge drilling offers the possibility of making burr-free holes with high-precision (accuracy of <5 μm) in a range of materials regardless of their hardness as long as the material is electrically conductive. Typical drilled hole diameters are in the range of 0.008–0.5 mm and are characterized by a depth-to-diameter ratio of 20:1 or higher [19–21].

In the electrical discharge process, the materials usually selected for making tool electrodes are copper, bronze, zinc, tungsten, graphite. The chosen tool electrode should provide criteria such as being good conductor, high electrode resistance to wear, machinability, enough surface roughness. To provide these criteria, the commonly materials to make tool electrode are copper and graphite [22,23]. In the case of electrical discharge drilling of deep holes in Inconel 718, a copper tube-electrode is more appropriate providing high material removal rate and low surface roughness [24]. The tool electrode made of copper is characterized by a high electrical conductivity (up to 60.9 MS/m) and a high thermal conductivity (388 W/(m·°K)), which are the properties needed in materials used to make tool electrodes in the EDD process. However, when deionized water (consisting of hydrogen and oxygen) is used as working fluid, the conditions of high temperature at the hole bottom can contribute to form a passive layer on the surface of the cooper tool electrode and machined surface of the Inconel 718 alloy. In addition, hydrogen penetrating the structure of copper material reduces the amount of oxides. On the other hand, water vapor forming into the copper structure influences high pressure, which can cause cracks in the material structure. In the case of the workpiece, the content of Cr in the Inconel 718 alloy structure should provide the oxidation resistance at high temperature conditions [25].

In the EDD process, the making of high aspect ratio holes is a challenge due to the difficulties involved in removing the process products such as debris and gas bubbles from the gap area (especially if the drilling hole is deep) [26–28]. Consequently, the debris accumulates at the hole bottom leading to the occurrence of abnormal/secondary discharges (such as arc and/or short circuits) between the debris and the sidewall of the hole. In the result, a poor surface quality (increased roughness parameters *Ra* and *Rz*), decreased hole accuracy and machining speed, excessive tool wear, decreased gap area occur [14,26,27,29,30]. Abnormal discharges can also occur in the corner of the hole bottom. In [31], the analysis of experimental research shows that the crack density is higher at the edge of the hole bottom, which may attest to the occurrence of secondary discharges in the zone. Secondary discharges also cause an excessive wear of the electrode's tip edge [21,32–34].

In addition, due to secondary discharges, unremoved and re-melted debris starts to attach and accumulate on the electrode's surface. Then, electrical discharges can occur between the re-solidified debris on the electrode's surface and the hole wall. In [21], the measurements of the tool electrode after drilling show an increase in the electrode's diameter by about 38 μm. This can affect the decrease of the side gap thickness and contribute to difficulties of the working fluid flowing outside of the hole thus decreasing the process stability. In [35], the authors also analyze the attachment of the debris to the electrode. The attachment of the debris takes place in the central region of the electrode's tip, with a smaller amount at the tip's edge. It suggests that the mechanism of attachment involves re-melting of the debris and does not occur randomly from the dielectric.

Previous papers concerning the experimental research into the EDD process emphasize the significance of efficient flushing of the gap zone [20,32,36]. Generally, a crucial role in the EDD process is played by the working fluid as a flushing agent, which cools the material of both electrodes (also remelted debris) and removes the machining particles from the discharge gap. However, efficient flushing often requires additional fixtures or adjustment of the machine. It is worth underlining that the properties of the working fluid such as electric conductivity and viscosity can substantially affect the working fluid flow, machining efficiency, electrode wear and recast layer thickness [18,37].

1.2. The Working Fluid Flow through the Electrode Channel and the Interelectrode Gap

The process efficiency depends strongly on the removal of the eroded particles, which is highly influenced by the flushing and thus the rate of the flow of the working fluid out of the hole. Flushing affects the quality of the drilled holes (dimensional and shape accuracy, low roughness parameters of the inside of the hole's surface), process performance, electrode tool wear. In [38,39], the authors highlighted the fundamental role of the flushing system during the drilling of holes. Efficient gap zone flushing constitutes the main factor ensuring stability of the EDD process, especially in the case of drilling high aspect ratio holes.

The flushing efficiency can depend on the length and the inside shape (single-channel, multi-channel) of the electrode tool [40] and the working fluid's initial pressure [38]. The dielectric fluid's pressure affects the increase in the metal removal rate (MRR) and the reduction of the surface roughness. In [14], the analysis of the research results shows that the optimum volumetric flow rate can ensure a constant interelectrode gap, which can decrease the number of secondary discharges.

The application of a high-volumetric flow rate (25 l/h) decreases the electrode tool wear due to appropriate cooling of the electrode material and sufficient removal of the debris from the gap. The electrode's diameter can also affect the kind of flow conditions in the electrode channel and the gap area. For the outer diameter of 0.5 mm of the tube-electrode, the laminar flow conditions in circular tube are considered. With the diameter of 1.0 mm, turbulent flow of the working fluid can take place. In addition, an analysis of the results proves that the simple cylindrical flushing channel inside of the electrode provides the best flow performance. The single-channel electrode provides comparatively better removal rates of the erosion products and a lower electrode wear ratio than multi-channel electrodes.

To analyze the flushing, the simulations were also performed. In [41], a mathematical model was developed that considers tool movement in a solid–liquid two-phase gap flow field and a 3D model to simulate the tool movement and debris generation. The analysis of the results of the model developed shows that with an increase in the flushing velocity, the fluid at the bottom collects more debris. In addition, the fluid flow is limited. The inner diameters of tube (single- or multi-channel) electrodes with a diameter of under 1.0 mm are too small to ensure effective gap flushing.

There are several flushing methods such as internal, external, suction-assisted flushing, flushing with different electrode movements or vibration- supported flushing [36]. However, external flushing can contribute to the vibration of a thin and long electrode [21].

1.3. Significance of the Working Fluid in the EDM Process

In electrical discharge machining, the application of deionized water as dielectric fluid indicates considerable potential since it can be treated as slightly conductive electrolyte, and electrodischarge erosion is accompanied by electrochemical dissolution [16,42]. The reaction of electrochemical dissolutions is possible due to resistivity of deionized water ranging widely between 0.1 and 10 MΩ cm based on its purity [43].

The EDM with the use of deionized water is often termed as a hybrid process of simultaneous Electrical Discharge and Electrochemical Machining (SEDCM) or Electrochemical Discharge Machining (ECDM). A major advantage of the process is the removal of the material by simultaneous interaction of electrochemical dissolution and electrical discharges in a single impulse [44]. The electrochemical dissolution improves the process performance (the maximum material removal rate in deionized water can be about 2.5 to 3 times faster) and reduces the tool electrode wear (the tool wear can be reduced up to 96%), in comparison to the application of oil-based dielectric [45]. The electrochemical reaction also enhances the quality of the machined surface by removing erosion micro-craters and re-solidified material on the rim [16,42]. After the EDM with the use of deionized water, micro-parts with a surface roughness parameter of Ra = 22 nm [16] and Ra = 12–43 nm were obtained [43]. Additionally, an analysis of the results of the EDM with the use of deionized water in the Inconel 718 alloy shows that the recast layer thickness can be reduced to the average thickness of 3–8 μm [18]. Overall, for machining efficiency and machined surface quality, deionized water is a better dielectric for electrical discharge machining of nickel-based superalloys.

In the case of reinforced electrochemical dissolution, the machined surface can be damaged, and the dimensional and shape accuracy of the hole can decrease [16,42–46]. In [45], the authors analyze the influence of deionized water on the dimensional accuracy of the machined parts. When applying low resistivity (of about 0.1 MΩ cm), the machined micro-column is tapered due to excessive electrochemical dissolution. The increase in resistivity of up to 12 MΩ cm reduces the tapered shaped of the column. This is because the high resistivity of water reduces the discharging distance and suppressed electrochemical dissolution. In addition, as the resistivity of deionized water decreases, the amount of oxidized materials increases. The oxidized material is experimentally proven to be nonconductive, and hence, there is no effect on the machining process. The similar results were determined in [47]. On the other hand, in [42], the high-frequency bipolar pulse generator that is applied offers the possibility of drilling micro-holes in deionized water without electrolytic corrosion of the surface near the hole top. In addition, the hole's inner surface is improved (surface roughness $Ra = 0.105$ μm) due to a small amount of electrochemical dissolution.

The material chemical composition of the surface after the EDM process using deionized water was analyzed [43]. The analysis of the chemical composition material of the machined surface proves the existence of two defined zones: crater zone and crater-free zone. For the crater zone, the occurrence is noted of 4.12% of the oxygen element and 12.62% of the carbon element on the surface. The carbon element results from deposition of debris particles on the machined surface. The oxygen element stems from the rapid oxidation of the material with the coexistence of high temperature in the plasma channel and oxygen gas disassociated from the deionized water. For the crater-free zone, the oxygen and carbon elements are reduced considerably (from 4.12% to 2.31% and from 12.62% to 9.99%, respectively). These changes probably result from the dissolution of the material from the machined surface through electrochemical reaction.

In the EDM process, it is worth underlining the significance of the temperature conditions occurring in the gap area. When using deionized water as working fluid, the bubbles are much smaller, while a single bubble is formed in the oil. Small bubbles move faster and do not significantly interfere with the erosion process [28,40]. In addition, the area proportion of the bubbles in the discharge gap is smaller in deionized water than in oil [28]. An analysis of the bubbles' movement shows that an increase in the bubbles' diameter is observed with an increase in the depth of the hole. This may lead to an increase in the bubble occupancy in the gap area and the weakening of the insulation strength of the dielectric liquid because, in general, the insulation strength of gas is much smaller than that of liquid [48]. The bubbles' movements are also investigated in [29,49]. This analysis shows that at the beginning of consecutive pulse discharges, the bubbles rapidly remove the debris from the gap bottom. As the discharging continues, the bubbles' ability to remove the debris weakens resulting in debris aggregation at the gap bottom and thus unstable machining [49]. In [29], the authors consider that the presence of bubbles increases the evacuation of eroded particles. If the kinematic viscosity of the dielectric fluid increases, the jump efficiency (defined as the ratio of the eroded particle number out the gaps at the end of the jump to the jump time) decreases, and a smaller number of particles is removed. However, the change of deionized water properties is not related to the value of dielectric fluid temperature, but in [21,29], the authors consider that the gaseous bubbles generated by the secondary discharges push the debris further along the axis of the electrode's feed. Most of the debris is eventually driven out of the hole due to secondary discharges. In addition, the accumulated gas bubbles at the hole bottom can prevent electrochemical reaction [43].

The kind of working fluid also influences the process of forming the bubble around the plasma channel. As a result of a single impulse discharge, the molecules, atoms, ions and electrons formed as a result of evaporation, dissociation and ionization of the working liquid and the electrode material are compressed in a small bubble around the plasma channel [28]. The bubble diameter increases radially and peaks when the pressure inside the bubble reaches its minimum after the finished electrical discharge. Then, the bubble diameter is compressed to the initial diameter. In reality, the dielectric fluid viscosity results in the damping of the bubble diameter increase. It is noted that the bubble diameters in deionized water and oil are almost the same at the beginning of the diameter increase. After the damping, the bubble volume is significantly smaller in water than in oil. The reason factor is gas components of the bubble generated in oil, that is, gaseous hydrogen and hydrocarbon gases (such as methane, ethane, and acetylene), which are dissociated gases of hydrocarbon oil and cannot be recombined. The bubbles generated in deionized water are mainly composed of hydrogen and oxygen, which can be reversibly recombined into water. In addition, the damping coefficient in oil is higher than that in water, which is affected by the higher fluid viscosity in oil (2.4×10^{-3} Pa·s) than in water (1.0×10^{-3} Pa·s). In [50], authors also consider that the viscosity of the working fluid in the EDD process is relevant. The obtained results show that a low viscosity enables more effective flushing of the debris.

The above analysis of the papers shows a significant potential of the EDM process with the use of deionized water as working fluid, particularly for drilling high aspect ratio holes in difficult-to-cut materials. Using deionized water in the process improves the process performance (such as material removal rate, tool wear, surface roughness) and hole accuracy. In the EDD process, the influence of the deionized water properties on the process should be subjected to a more extensive analysis. Properties such as density, viscosity and electrical conductivity can have a major effect on the working fluid flow through the gap area and removal of erosion products. The change of the deionized water properties, especially the electrical conductivity, can contribute to remove an allowance in the similar range of ECM and EDM during the single pulse duration. The increase of the electrical conductivity can be influenced by the applied initial temperature of deionized water. The complexity of the phenomena occurring in the gap area during the process indicates the need for further experimental research.

This paper presents an analysis of the results of experimental research involving electrical discharge drilling (EDD) of aspect ratio holes in the Inconel 718 alloy with the use of deionized water as the dielectric fluid. The aim of the research was to check the impact of the properties of deionized water, such as electrical conductivity, and the working-fluid pressure on the process performance, dimensional and shape accuracy and inner surface integrity of the holes. The properties of deionized water were changed by applying an initial temperature of the deionized water.

The first part of the experimental research comprised an analysis of the impact of the process parameters (pulse time, current amplitude and discharge voltage) on process performance. The process performance was investigated in terms of linear tool wear, drilling speed and the accuracy of the holes (such as taper angle, aspect ratio of hole, side gap thickness). In order to examine the relationship between the process parameters and performance criteria, Analysis of Variance (ANOVA) techniques were applied. The process of drilling the holes was done on a sample consisting of two parts. The drilling was carried out at the junction of the sample parts. Once the parts of the sample were separated, an analysis of the dimensional and shape accuracy and quality of the inner hole surface was carried out. During the next part of the experiments, the impact of the initial temperature and the initial pressure of the deionized water onto the gap area on the conditions of fluid flow via the electrode channel on the volumetric flow rate and the Reynolds number was checked. The obtained through holes were subjected to the extensive analysis. Based on the previous series of experiments, the EDD of through holes for selected initial working fluid temperatures and pressure and selected optimum machining parameters was performed.

2. Materials and Methods

2.1. Materials

The Inconel 718 alloy was used as the workpiece material and a tube-electrode was the tool (single-channel, made of copper) (Figure 1a). A special sample consisting of two parts was designed and produced as the workpiece for the purpose of the experimental test. The holes were drilled at the junction of the sample parts (Figure 1b). The chemical composition and the main physical-mechanical properties of the electrode material are presented in Tables 1 and 2, respectively.

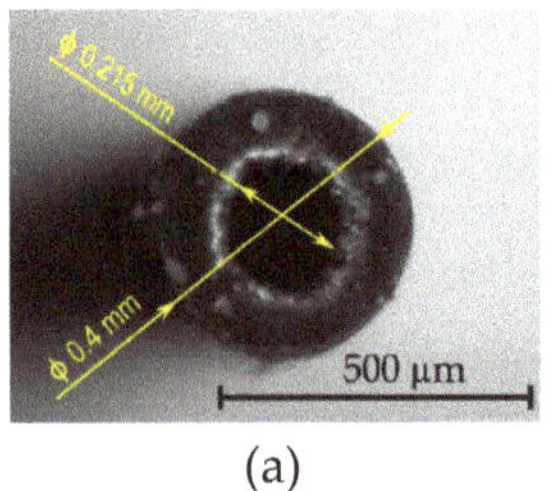

(a)

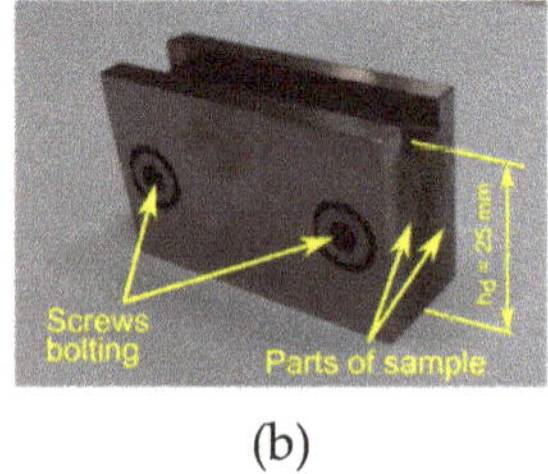

(b)

Figure 1. (**a**) Tool electrode tip and (**b**) photograph of the sample; h_d—the maximum drilling depth.

Table 1. Chemical composition of Inconel 718 (wt. %).

Ni	Cr	Fe	Nb	Mo	Ti	Al	Co	Mn	C	Si	P
50.0–55.0	17.0–21.0	Balance	4.75–5.5	2.8–3.3	0.65–1.15	0.2–0.8	<1.0	<0.35	<0.08	<0.35	<0.015

Table 2. Physical and mechanical properties of the workpiece and of the electrode material, for $T = 298.15\,°\text{K}$ and $T = 291.15\,°\text{K}$, respectively [51–54].

Property	Workpiece	Tool
Density, (kg/m^3)	8190	8960
Heat capacity, (J/(kg °K))	435	385
Thermal conductivity, (W/(m °K))	8.9	388
Melting temperature range, (°K)	1533–1609	1338–1356

2.2. Experimental Procedure

First, electrical discharge drilling of deep holes was carried out on the experimental test stand shown in Figure 2a. In order to avoid problems with the drilling of through holes, an additional technological pad was applied on the underside of the sample (the thickness of the pad was 0.5 mm). In addition, to minimize the impact of electrode vibrations and the clamping eccentricity on the drilling process, an electrode guide system was applied (Figure 2b). The aim of the experimental research was to examine the influence of selected machining parameters on the dimensional accuracy of the hole, the machining efficiency and the tool electrode wear. Table 3 presents the data on the drilling process, and Table 4 shows the adopted ranges of values. The experiments were performed according to the theory of carrying out an experiment with the use of a three-level rotatable research plan that included 20 experimental tests, with six repetitions in the research plan center. The results of the experiments are shown in Table 5. The statistical techniques such as Analysis of Variance (ANOVA) were applied to investigate the relationship between the input and output parameters.

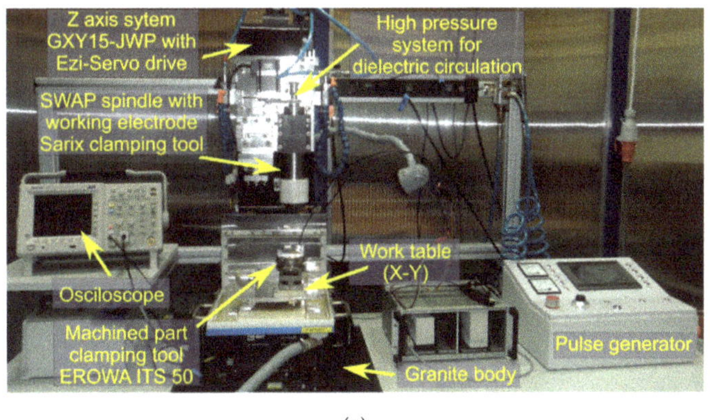

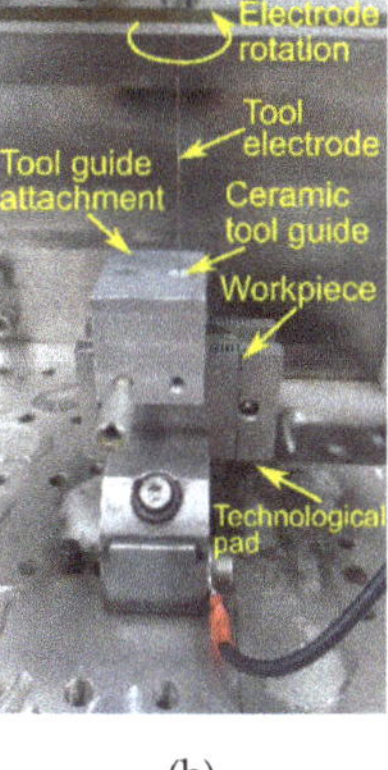

(a) (b)

Figure 2. (**a**) Photograph of the test stand and (**b**) the experimental setup of the electrode guiding system.

Table 3. Machining parameters.

Input Parameters	Output Parameters
Pulse time, t_i (µs)	Drilling speed, v (µm/s)
Current amplitude, I (A)	Linear tool wear, TW (%)
Discharge voltage amplitude, U (V)	Taper angle, tap_α
	Aspect ratio hole, AR
	Side gap thickness, S_b (µm)

Table 4. Process parameters and their levels.

Coded Parameter	Real Parameter	Level				
		1		**2**		**3**
X_1	U (V)	80		100		120
		1	2	3	4	5
X_2	t_i (µs)	100	282	550	818	999
X_3	I (A)	3	3.33	3.83	4.32	4.65

Table 5. Research plan and the results of the experiments.

Experiment No.	X_1	X_2	X_3	U (V)	t_i (µs)	I (A)	v (µm/s)	TW (%)	tap_α	AR	S_b (µm)
1	1	2	2	80	282	3.33	6.69	39.27	−0.00313	25	102
2	1	2	4	80	282	4.32	8.89	51.58	−0.00446	28	128
3	1	4	2	80	818	3.33	5.92	54.95	−0.00247	22	122
4	1	4	4	80	818	4.32	7.90	81.54	−0.00593	26	121
5	3	2	2	120	282	3.33	8.79	45.68	−0.00767	29	120
6	3	2	4	120	282	4.32	12.56	52.70	−0.00903	27	130
7	3	4	2	120	818	3.33	6.62	22.86	−0.0026	24	127
8	3	4	4	120	818	4.32	9.33	73.55	−0.00933	26	129
9	1	3	3	80	550	3.83	7.81	47.30	−0.0036	27	103
10	3	3	3	120	550	3.83	9.06	49.61	−0.00409	31	120
11	2	1	3	100	100	3.83	8.65	39.46	−0.00872	29	89
12	2	5	3	100	999	3.83	7.55	72.08	−0.00424	27	119
13	2	3	1	100	550	3.00	7.37	33.99	−0.00123	31	111
14	2	3	5	100	550	4.65	12.27	51.44	−0.01208	25	139
15	2	3	3	100	550	3.83	7.87	48.90	−0.00456	29	94
16	2	3	3	100	550	3.83	7.94	50.72	−0.00436	29	96
17	2	3	3	100	550	3.83	8.06	55.09	−0.00227	30	107
18	2	3	3	100	550	3.83	8.76	55.10	−0.00188	30	111
19	2	3	3	100	550	3.83	8.43	56.07	−0.00521	29	115
20	2	3	3	100	550	3.83	8.16	53.70	−0.00215	29	112

The following constant parameters were assumed: initial interelectrode gap thickness (S_0 = 50 µm), inlet dielectric fluid pressure (p = 8 MPa), rotational speed of the clamp and the electrode (n = 400 rpm), drilling time of each hole ($t_{drilling}$ = 45 min), pulse off time (t_{off} = t_i) and deionized water as the dielectric fluid. Before each experiment was started, the temperature T (T = 297.15–316.15 °K) and the electrical conductivity κ of the deionized water were measured (κ = 3.8–6.8 µS/cm). The dielectric fluid was flushed down to the gap zone through the interior hole of the tube (Figure 3).

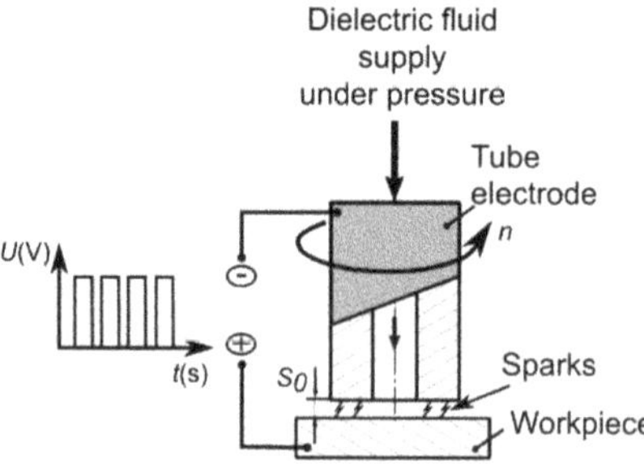

Figure 3. The scheme of the electrical discharge drilling (EDD) process.

Measurements of the diameters along the hole's depth were performed (five for each diameter) with the use of the K-401 stereo microscope with a Common Main Objective (CMO) Infinity optical system

(Motic, Richmond, Canada) and equipped with a Moticam 2300 digital camera with MoticImages Plus system (Motic, Richmond, Canada).

The difference between the relevant diameters of the two parts of sample was in the range 20–40 µm, and the calculated radius deviation of the tool electrode was 0.01 mm (($D_{tool\ guide} - D_{tool}$)/2, $D_{tool\ guide}$—inner diameter of the ceramic tool guide and D_{tool}—outer diameter of the tool electrode). The above factors allowed the assumption of symmetrically drilled holes.

The drilling speed v is calculated from the following equation:

$$v = h/t_{drilling}, \tag{1}$$

where h is the hole depth, and $t_{drilling}$ is the drilling time.

The linear tool wear TW is calculated according to the formula:

$$TW = (h_{tool}/h)\cdot100\%, \tag{2}$$

where h_{tool} is the shortening of the electrode.

The aspect ratio hole AR is given by the following equation:

$$AR = h/D_{average}, \tag{3}$$

where $D_{average} = (D_{top} + D_{bottom})/2$ is the average of the hole diameters, D_{top} is the average top diameter, and D_{bottom} is the average bottom diameter (values of D_{top} and D_{bottom} are the mean values from five measurements).

The taper angle tap_α is calculated from the following equation:

$$tap_\alpha = (D_{top} - D_{bottom})/2h. \tag{4}$$

The side gap thickness S_b is calculated according to the formula:

$$S_b = (D_{top} - D_{tool})/2. \tag{5}$$

2.3. Experimental Procedure for the Flow of the Deionized Water through the Electrode Channel

The second part of experimental research included an analysis of the working fluid's flow through the electrode channel. Deionized water was applied as the working fluid. The aim of the experiment was to examine the impact of the initial working-fluid temperature T the and initial working-fluid pressure p on the volumetric flow rate Q and the Reynolds number Re. Table 6 presents the thermal properties of the working fluid according to its temperature. In Table 3, the boiling point of the working fluid (T = 373.15 °K) is also specified. It is related to reaching the critical conditions in the gap zone (then the deionized water temperature reaches the boiling point), which ensures the occurrence of electrical discharges in a single impulse time. The tool electrode, which was applied, had the same shape and the same dimensional parameters as in the EDD experiment (cross-sectional area of the electrode's channel S = 0.0363 mm², length 150 mm). The gap thickness S_0 was 100 µm. During the experiment, the working fluid was flushed down the interior hole of the tube with the pre-set initial pressure and pre-set initial temperature (designation 1, Figure 4). Next, the working fluid flowed onto the flat surface of a metallic plate (designation 2, Figure 4).

The research was performed according to the experiment theory, with the use of a two-level rotatable research plan that included 11 experimental tests with three repetitions in the research plan's center. Table 7 presents the adopted parameter ranges. The results of the experiment are shown in Table 8.

Table 6. Selected physical properties of the deionized water during saturation pressure.

Physical Properties	Working Fluid Temperature, $\sim T$ (°K)			
	298.15	303.15	313.15	373.15
Density, (kg/m^3)	997.1	995.7	992.2	958.4
Specific heat, (kJ/(kg°K))	4.178	4.176	4.175	4.211
Thermal conductivity, (W/(m°K))	0.606	0.615	0.633	0.682
Dynamic viscosity, (kg/(s·m))·10^6	880.637	792.377	658.026	277.528
Kinematic viscosity, (m^2/s)·10^6	0.896	0.804	0.661	0.296
Electrical conductivity, (µS/cm)	0.05501	0.07101	0.11351	0.79303
Resistivity, (MΩcm)	18.180	14.082	8.810	1.261

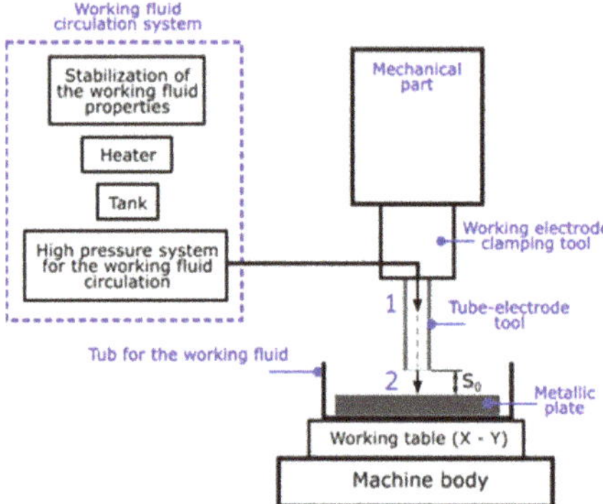

Figure 4. Test stand diagram.

Table 7. Process parameters and their levels.

Coded Factor	Real Parameter	Level		
		1	2	3
X_{11}	Initial working-fluid temperature, $\sim T$ (°K)	298.15	303.15	313.15
X_{22}	Initial working-fluid pressure, p (MPa)	5	7	8

Table 8. Research plan and results of the experiments; * $Re > 2300$ turbulent flow is considered.

Experiment No.	Coded Factor		Real Parameter		Response Variables		
	X_{11}	X_{22}	$\sim T$ (°K)	p (MPa)	Q (m^3/s)·10^7	v_f (m/s)	Re
1	1	1	298.15	5	4.778	13.22	1582
2	3	1	313.15	5	4.333	11.99	1945
3	1	3	298.15	8	6.667	18.44	2208
4	3	3	313.15	8	5.833	16.14	2619*
5	1	2	298.15	7	5.444	15.06	1803
6	3	2	313.15	7	5.417	14.98	2431*
7	2	1	303.15	5	4.722	13.06	1743
8	2	3	303.15	8	6.278	17.37	2317*
9	2	2	303.15	7	5.667	15.67	2091
10	2	2	303.15	7	5.708	15.79	2107
11	2	2	303.15	7	5.817	16.09	2147

In a cylindrical tube, the laminar flow is considered with the Reynolds number of below 2300. For the values of the Reynolds number above 2300, the turbulent flow is estimated.

The volumetric flow rate Q was determined experimentally. In order to calculate the flow conditions, the Reynolds number Re was used, which is defined in Equation (6)

$$Re = (v_f \cdot r)/\gamma, \tag{6}$$

where r was the radius of the channel diameter in the tool electrode, γ was the kinematic viscosity of the working fluid (value applied for the relevant working fluid temperature according to Table 6), and the mean flow rate velocity of the fluid v_f was calculated from the following formula:

$$v_f = Q/S, \tag{7}$$

where S was cross-sectional area of the electrode's channel.

In order to estimate the impact of the initial working-fluid temperature T and the initial working-fluid pressure p on the volumetric flow rate Q and the Reynolds number Re, the *Matlab* software was employed. A second-degree polynomial (with constant, linear, interaction and square terms) was used as the function. The regression equations obtained for the relationships $Q\,(T, p)$ and $Re\,(T, p)$, are described as the following equations:

$$Q\,(T, p) = -3.272{\cdot}10^{-5} + 2.175{\cdot}10^{-7}{\cdot}T + 1.349{\cdot}10^{-7}{\cdot}p - 5.867{\cdot}10^{-10}{\cdot}T{\cdot}p - 3.561{\cdot}10^{-10}{\cdot}T^2 + 7.574{\cdot}10^{-9}{\cdot}p^2, \tag{8}$$

$$Re\,(T, p) = -55753.573 + 359.136{\cdot}T - 921.200{\cdot}p + 2.681{\cdot}T{\cdot}p - 0.560{\cdot}T^2 + 24.147{\cdot}p^2. \tag{9}$$

The values of the coefficient R^2 (R—square Statistic) and R^2 adjusted (Adjusted R—square Statistic) are high (Table 9), which can confirm that the fitted quadratic models are statistically significant for the analyzed relationships of $Q\,(T, p)$ and $Re\,(T, p)$. The values of "p-Value" are less than 0.05 (i.e., 95% of the confidence level), which confirms that the models obtained are statistically significant.

Table 9. Diagnostic statistics for the regression.

	$Q\,(T, p)$	$Re\,(T, p)$
R^2	0.9593	0.9761
R^2 adjusted	0.9187	0.9521
p-Value	0.0017	0.00047

3. Results

3.1. Results of the EDD Analysis

In order to determine the impact of the process parameters on the process performance, the ANOVA analysis was employed. The regression equations $v\,(U, t_i, I)$, $TW\,(U, t_i, I)$, $tap_\alpha\,(U, t_i, I)$, $AR\,(U, t_i, I)$ and $S_b\,(U, t_i, I)$ were determined by the relevant Equations (10)–(14):

$$v\,(U, t_i, I) = 19.588 + 0.1033{\cdot}U + 0.0073{\cdot}t_i - 13.5625{\cdot}I + 2.1517{\cdot}I^2 - 0.0001{\cdot}U{\cdot}t_i, \tag{10}$$

$$TW\,(U, t_i, I) = 26.0443 + 0.4937{\cdot}U - 0.0606{\cdot}t_i - 10.0885{\cdot}I - 0.0012{\cdot}U{\cdot}t_i + 0.0529{\cdot}t_i{\cdot}I, \tag{11}$$

$$tap_\alpha\,(U, t_i, I) = -0.015524 - 0.00003{\cdot}U + 0.00005{\cdot}t_i - 1{\cdot}10^{-7}{\cdot}t_i^2 + 0.00155{\cdot}I - 8{\cdot}10^{-6}{\cdot}t_i{\cdot}I, \tag{12}$$

$$AR\,(U, t_i, I) = -39.3292 + 1.1504{\cdot}U - 0.0055{\cdot}U^2 + 0.0152{\cdot}t_i - 0.000001{\cdot}t_i^2 + 1.561{\cdot}I, \tag{13}$$

$$S_b\,(U, t_i, I) = 375.2799 - 8.2966{\cdot}U + 0.0422{\cdot}U^2 + 0.2552{\cdot}t_i + 34.0705{\cdot}I - 0.0599{\cdot}t_i{\cdot}I. \tag{14}$$

The Equations (10)–(14) involve the drilling parameters with the regression coefficients at the significant level of "p-Value" < 0.05.

The shape of the holes obtained is characterized by reverse conicity. The value of the top diameter is lower than that of the bottom diameter. This may result from process instability due to the accumulated debris at the hole bottom and an insufficient dielectric flow through the machining gap area.

The analysis of the results shows that the high linear tool wear ($TW > 60\%$) takes place where a higher current amplitude ($I > 4.32$ A) and a higher discharge voltage ($U > 100$ V) are applied. As a result, the amount of the removed material depended mainly on the energy in a single discharge. Higher U and I cause a higher single discharge energy, and the removal of a higher amount of material from both electrodes. The insufficient working fluid flow can then cause excessive accumulation of debris at the hole bottom. A lower interelectrode gap can contribute to secondary discharges between the accumulated debris and the electrode's front, causing excessive tool wear. Additionally, a higher TW (above 80%) is observed for longer pulse time $t_i = 818–999$ µs and a higher current amplitude $I = 4.32 – 4.65$ A (Figure 5a). A longer pulse time can result in a longer time of occurrence of electrical discharges in a single pulse. The occurrence of secondary discharges (between debris and hole sidewall) can also be longer, which results in additional wear of the tool electrode. An increase in the pulse time has an irrelevant effect on the drilling speed v (Figure 6a). However, an increase in the discharge voltage U and a decrease in the pulse time t_i cause a decrease in the linear tool wear (Figure 5b) and an increase in the drilling speed (Figure 6b). In the case of application of $U = 100–120$ V and $t_i = 100–282$ µs, the electrochemical dissolution in a single impulse can be enhanced, which improves the drilling speed. A lower value of the pulse time can reduce the time of occurrence of electrical and abnormal discharges, thus decreasing the linear tool wear. Additionally, based on the ANOVA analysis for TW, the selected relevant parameter is the discharge voltage U ("*p*-Value" = 0.171). The results of the ANOVA analysis for drilling speed v and linear tool wear TW are given in Supplementary section S1.

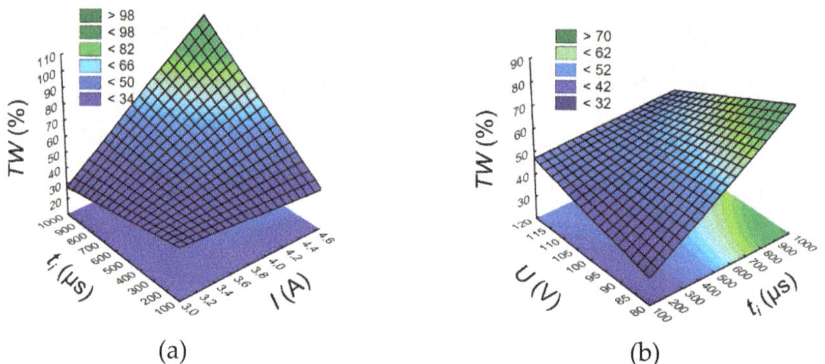

Figure 5. (**a**) Relationship between the linear tool wear TW and the pulse time t_i and current amplitude I, $U = 100$ V and (**b**) discharge voltage amplitude U, $I = 3.83$ A.

The "-" sign for taper angle tap_α indicates a reverse conicity of the hole. For a longer pulse time $t_i = 818–999$ µs and a lower current amplitude $I = 3–3.33$ A, a drop in tap_α was noted (Figure 7a). It is related to the increase in the side gap thickness S_b (Figure 7b). A longer pulse time can cause a longer time of occurrence of electrical and abnormal discharges in a single pulse, which extend the bottom diameter of the hole. An increase in the side gap causes a smaller difference between the top and bottom diameters which decreases tap_α. An increase in S_b, on the other hand, is observed for a lower pulse time $t_i = 100–282$ µs, which was applied and an increase in the current amplitude I (Figure 7b). This may be due to excessive electrochemical dissolution in a single pulse. Electrochemical dissolution also takes place during the flow of the working fluid outside of the hole. An increase in the side gap thickness is also noticed for the higher discharge voltage $U = 100–120$ V, which was applied. The higher discharge voltage and the higher electrical conductivity of deionized water ($\kappa = 3.8–6.8$ µS/cm) could significantly increase the electrochemical dissolution in a single impulse. Based on the ANOVA analysis for the

tap_α, the discharge voltage is regarded as significant parameter ("*p*-Value" = 0.129), but for the S_b, the discharge voltage ("*p*-Value" = 0.233) and the current amplitude ("*p*-Value" = 0.717). The results of the ANOVA analysis for the taper angle tap_α and the side gap thickness S_b are given in Supplementary section S1.

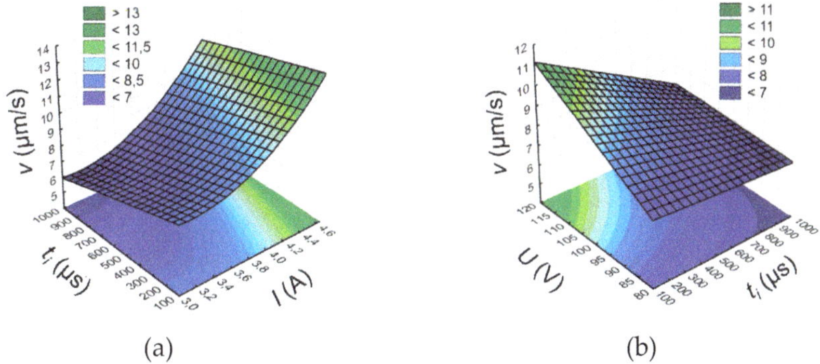

(a)   (b)

Figure 6. Relationship between the drilling speed v and the pulse time t_i and (**a**) current amplitude I, $U = 100$ V and (**b**) discharge voltage amplitude U, $I = 3.83$ A.

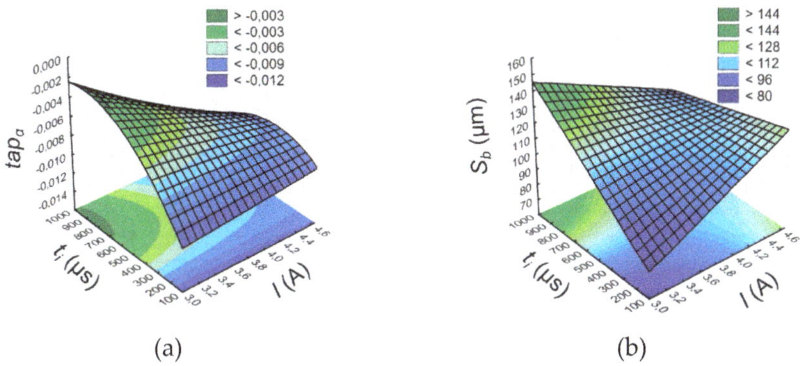

(a)   (b)

Figure 7. (**a**) Relationship between the taper angle tap_α, current amplitude I and pulse time t_i; (**b**) relationship between the side gap thickness S_b, current amplitude I and pulse time t_i; $U = 100$ V.

The aspect ratio hole AR is related to the drilling speed v, which results from the application of the same drilling time for each test. An analysis of the results shows that the extending of the pulse time (Figure 8a,b) causes a decrease in AR. For a longer pulse time in a single impulse, a longer time of electrical and secondary discharges may also occur. As a result, a considerable amount of debris is accumulated at the hole bottom. Where the working fluid flow is insufficient, the process is unstable which reduces the material removal rate. The optimum pulse time affecting the AR increase is within the range of 500–818 μs ($AR > 26$). The ANOVA analysis shows that the process parameters such as the discharge voltage U ("*p*-Value" = 0.116), the pulse time t_i ("*p*-Value" = 0.09) and the current amplitude I ("*p*-Value" = 0.11) should be also regarded as a significant parameter. The results of the ANOVA analysis for the aspect ratio hole AR is given in Supplementary section S1.

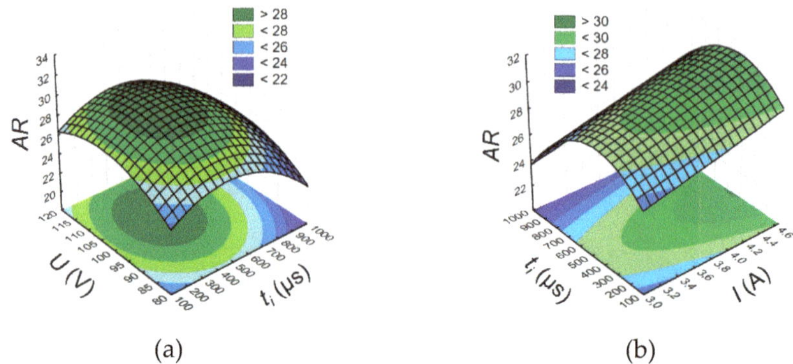

(a)   (b)

Figure 8. (**a**) Relationship between the aspect ratio AR and the pulse time t_i and (**a**) discharge voltage amplitude U, $I = 3.83$ A and (**b**) current amplitude I, $U = 100$ V.

3.2. Results Analysis of the Working Fluid Flow in the Electrode Channel

When deionized water is used as dielectric to obtain electrical discharges in a single pulse, the critical conditions must be achieved within the gap area involving a significant number of gas bubbles. For gas bubbles to appear within the gap area, the deionized water must reach the boiling point (373.15 °K). An increase in the temperature of the deionized water also changes its physical properties such as electrical conductivity, density or viscosity. Density and viscosity values decrease as the temperature increases, while electrical conductivity increases as the temperature increases (Table 6). Lower density and viscosity should improve the working fluid flow through the gap area. However, the higher electrical conductivity of deionized water should reinforce the electrochemical reactions in a single pulse. For this reason, the initial working-fluid temperature is an important parameter influencing the flow conditions and the thermal conditions within the gap area. The other important parameter influencing the fluid flow is the initial working-fluid pressure.

The analysis of the flow results for an electrode with an outer diameter of 0.4 mm shows the Reynolds number below 2300 for $p = 5$ MPa and $p = 7$ MPa, which were applied (Figure 9a). Then, the fluid flow assumes the laminar flow conditions. In the case where the initial working-fluid pressure applied is $p = 8$ MPa and the initial working-fluid temperature is above 300 °K, the Reynolds number is above 2300 and a turbulent flow should be considered. The analysis of the results indicates that the volumetric flow rate is about 1.4 times higher in the case where $p = 8$ MPa is applied than for $p = 5$ MPa (Figure 9b). This may be related to the viscosity and density of deionized water, whose values decrease with temperature increases. Lower viscosity and lower density of deionized water ensure a faster fluid flow through electrode channel.

The consideration of turbulent flow conditions for $p = 8$ MPa and $T > 300$ °K is related to a higher flow rate velocity, which enhances the evacuation of the eroded material from the gap area. It is worth underlining that a turbulent flow may be accompanied by whirls, which may prevent correct working fluid flow with eroded particles and reduce process stability.

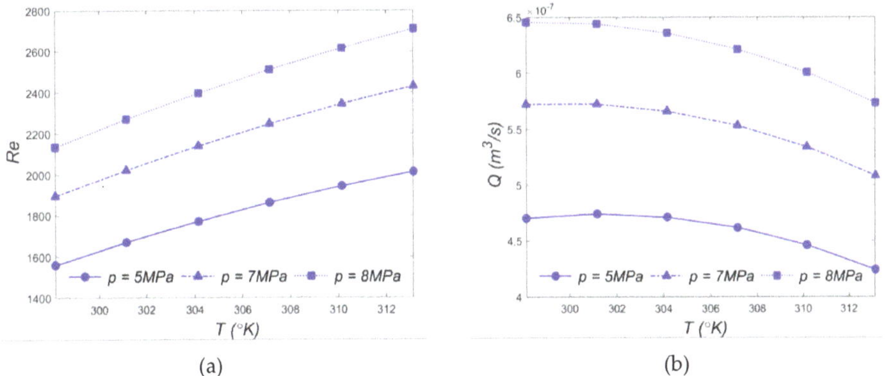

(a)   (b)

Figure 9. Impact of the initial temperature of fluid T and initial working fluid inlet pressure p on (**a**) the Reynolds number Re and (**b**) the volumetric flow rate Q.

3.3. Results of the Analysis of the Through Holes Obtained for the Experiment 1

The experimental research enabled the obtaining of through holes. The sixth test (machining parameters such as $t_i = 282$ µs, $U = 120$ V, $I = 4.32$ A) and the eighth test (machining parameters such as $t_i = 818$ µs, $U = 120$ V, $I = 4.32$ A) provided the drilling of the through holes. In the case of the eighth test, the linear tool wear was significant ($TW > 70\%$), which is due to the application of a longer pulse time. The time of electrical (and secondary) discharges in a single impulse could be too long which resulted in the increased electrode tool wear. Therefore, the machining parameters of the sixth test were chosen to drill additional through holes (two additional through holes were made) (Figure 10). It is worth to underline that the electrical conductivity and the temperature of deionized water while drilling the three through holes were in the range $\kappa = 5.7$–6.3 µS/cm and $T = 302.15$–308.15 °K, respectively. These values of T and κ were higher than for the remained tests.

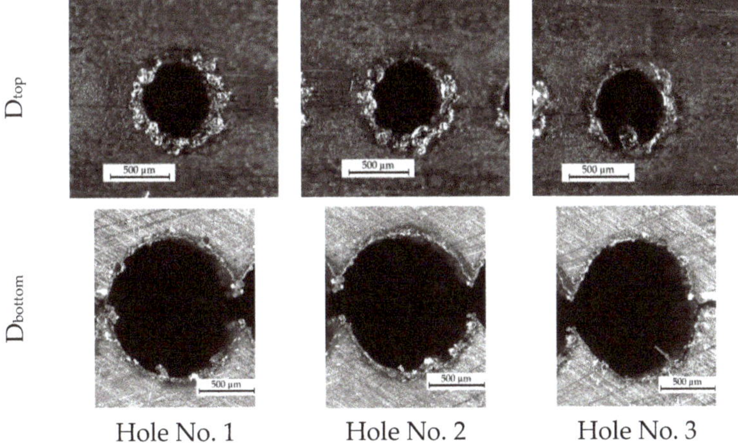

Hole No. 1   Hole No. 2   Hole No. 3

Figure 10. Through holes drilled with the application of the same machining parameters: $t_i = 282$ µs, $U = 120$ V, $I = 4.32$ A.

The dimensional accuracy and the inner surface homogeneity of the drilled holes are insufficient (Figures 10 and 11). The bottom diameter is greater on average by 43% than the top diameter. The profile of the hole based on the average measurements of the diameters along the hole depth for three through holes is presented in Figure 12. The diameter increases along the hole length (from the hole's top to its bottom). Each diameter is larger than the previous one by about 3% on average. This results from an insufficient working fluid flow through the gap area and insufficient debris removal. Secondary discharges can occur between the accumulated debris and the hole sidewall. When the hole depth is significant (above 10,000 μm), the phenomenon of secondary discharges may be excessive. When the hole depth is near 25,000 μm, the bottom diameter is almost twice as large as the top diameter. This confirms that the eroded material is not fully removed and accumulates at the hole bottom.

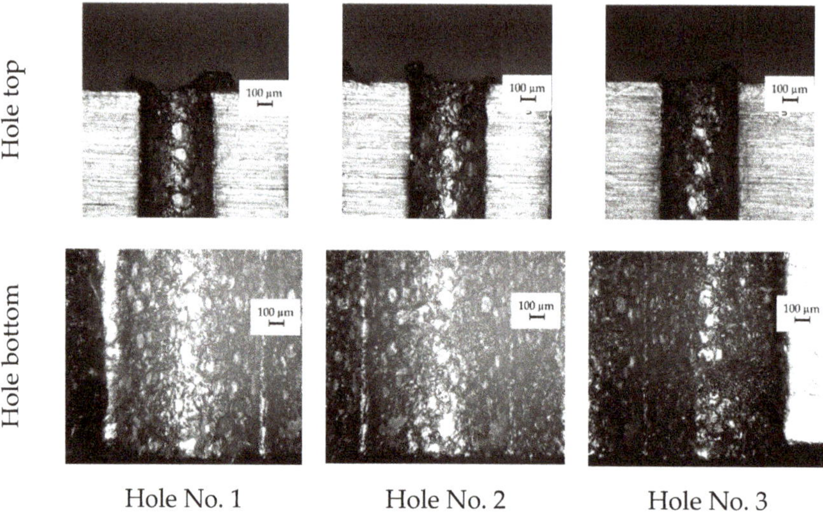

Figure 11. Images of the longitudinal section of three through holes at the hole top and bottom.

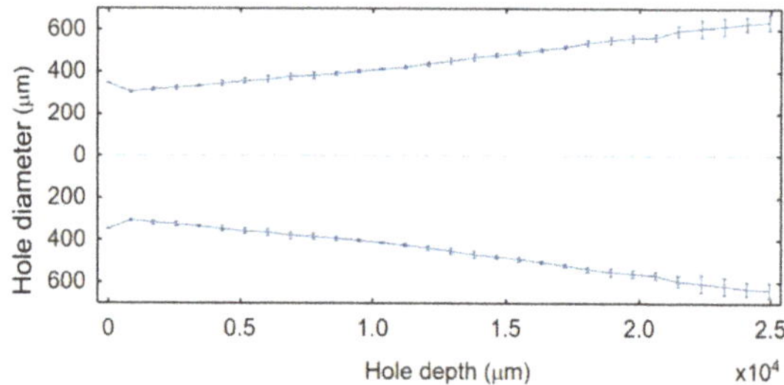

Figure 12. Profile of the through hole along its depth.

On the edge of the bottom diameter, burrs occur. But on the surface of the top diameter, there is a significant amount of re-solidified material and the heat affected zone (Figure 10). After separating the parts of the sample, two top diameters of the hole D_{top1} and D_{top2} (where $D_{top2} > D_{top1}$) can be observed (Figure 13). The determination of the D_{top2} diameter follows from the unremoved and re-solidified material deposited on the hole edge. The average diameter of re-solidified material is larger by about 180 μm than D_{top1}. The top diameter, without the re-solidified material (D_{top1}), should constitute the correct top diameter. The occurrence of the re-solidified material may result from an appropriate working fluid flow at the beginning of the process. The fresh dielectric with a lower temperature could affect the eroded material solidifying it too fast on the surface before it was removed.

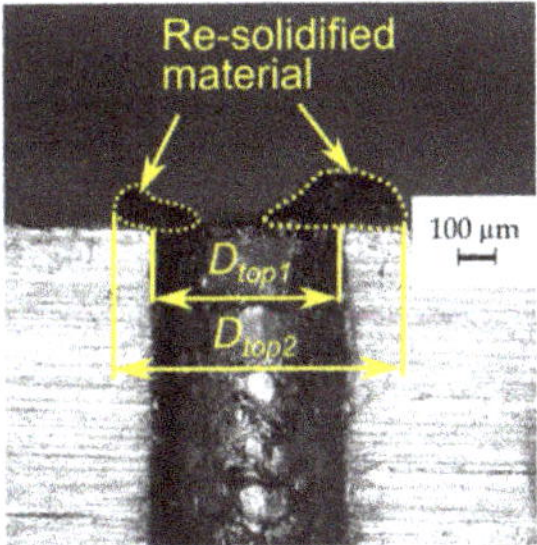

Figure 13. Image of the through hole top of one part of the sample after separation.

At the beginning of drilling the hole, the working fluid flow should be correct. However, on the inner surface of the hole top, a significant amount of melted and re-solidified material is observed (Figure 14), which indicates an insufficient working fluid flow in the gap area. In order to understand this phenomenon, it may be helpful to focus on the conditions contributing to the occurrence of electrical discharges in a single impulse when deionized water is used as a dielectric. The factors which influence these conditions are mainly the discharge voltage and the impulse duration.

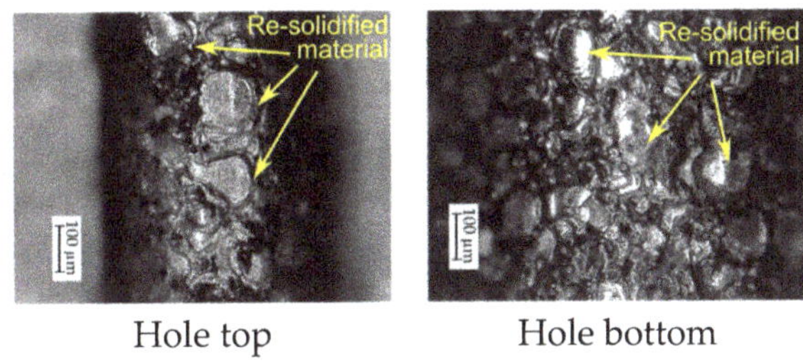

Figure 14. Images of the internal surface of the through hole.

At the beginning of the pulse time (Stage I, Figure 15a), the voltage between the electrodes and the current amplitude increases gradually. The allowance is then removed by electrochemical dissolution. The application of a higher voltage and a working fluid with higher electrical conductivity can contribute to reinforcing the electrochemical dissolution. When the voltage increases (up to φ_2), hydrogen is generated on the tool electrode, which leads to the formation of a gas film around the tool. The bubbles increase their diameters over time until the critical size is achieved. When the number and size of the hydrogen bubbles are sufficient, the resistance on the tool electrode–workpiece interface increases substantially due to the constriction effect. This contributes to increased ohmic heating of the working fluid within the area, causing the bubbles to evaporate. At the critical condition, a large number of gas bubbles cover the maximum possible active surface area of the tool electrode, leading to its blanketing (i.e., isolation between the electrode surface and the working fluid). Consequently, the current decreases in a very short time [55]. At the beginning of Stage II, a sudden decrease in voltage occurs (up to φ) and, at the same time, the current amplitude increases, which may indicate the occurrence of discharges and the allowance is removed in a manner typical of the EDM process. At the end of the pulse time, due to the electric capacity effect, the voltage between the electrodes gradually decreases to zero (Stage III, Figure 15a) [56,57]. The voltage applied should be high enough, and the pulse time should be long enough for the discharges proper to occur in a single impulse. The recorded voltage of the discharge and current amplitude in a single impulse characterizes the typical waveform for the ECDM process (Figure 15b). There is no current increase at the beginning of the pulse time due to the application of the pulse generator setting up the EDM process.

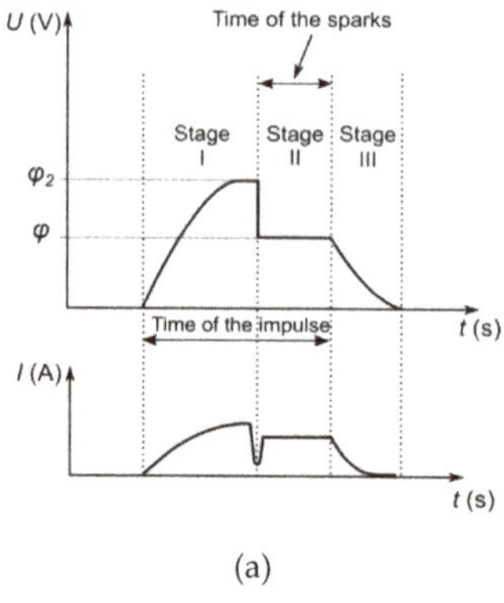

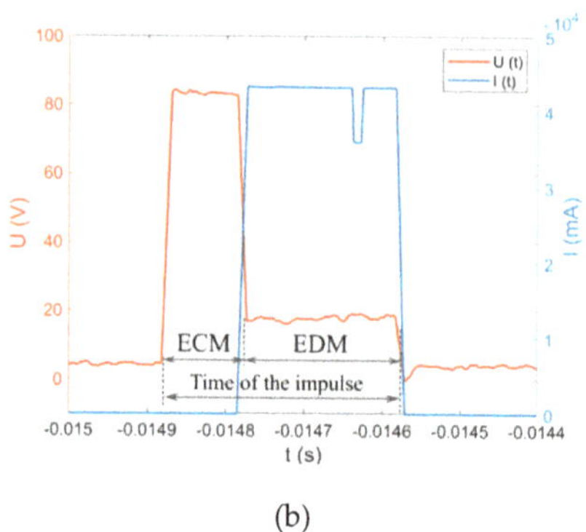

Figure 15. (**a**) Typical voltage and current waveform in a single impulse during the Electrochemical Discharge Machining (ECDM) process; (**b**) the recorded voltage and current waveform for the through hole parameters.

The observation of the surface on the hole top and near the hole bottom can confirm the occurrence of a considerable number of gas bubbles within the gap area, which were not fully removed. However, many more small bubbles might have occurred at the bottom of the hole because of incorrect fluid flow. For this reason, oval traces of re-solidified material on the surface of the hole bottom are smaller than on the surface of the hole top. The oval shape of the traces may be due to the places where the electrical discharges occurred. According to [58], electrical discharges can occur on small areas between bubbles (bubble bridges) and the workpiece surface. This may also explain the occurrence of dark oval traces on the surface of the electrode tool (Figure 16). The gas bubbles creating the gas film on the electrode come and go all the time. For this reason, only darker traces of electrical discharges remained on the electrode surface, and the electrode material did not erode. The lower thermal conductivity of Inconel 718 could also cause a significant amount of heat to remain in the processing area in case of abnormal flow of the liquid. As a result, a considerable number of small bubbles could be present all the time within the gap area creating favorable conditions for the occurrence of electrical and secondary discharges.

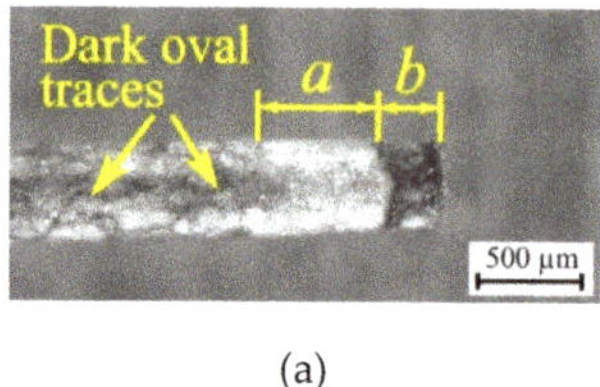

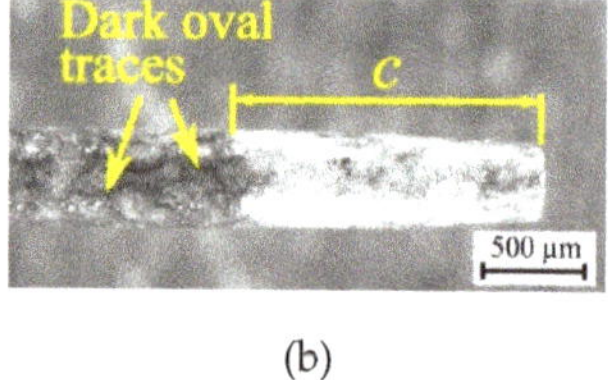

Figure 16. Photographs of tool electrode wear after drilling the holes with the machining parameters applied: (**a**) t_i = 282 μs, U = 120 V, I = 4.32 A; (**b**) t_i = 550 μs, U = 100 V, I = 3.83 A.

The shape of the tool electrode after the drilling did not change significantly. The tool electrode tip wore to a lesser degree after the drilling of the through hole (machining parameters: t_i = 282 μs, U = 120 V, I = 4.32 A, electrode tip diameter smaller by about 10%) than after the drilling with the machining parameters in the research plan center (t_i = 550 μs, U = 100 V, I = 3.83 A, electrode tip diameter smaller by about 20%). The length of the electrode tip surface, which was worn, was also larger after the drilling with the parameters in the research plan center (c = 1229 μm, $t_{drilling}$ = 2700 s, h = 22,044 μm) than after the drilling of the through hole ($a + b$ = 916 μm, $t_{drilling}$ = 2316 s, h = 25,000 μm) (Figure 16a,b). The dark fragment on the electrode tip (b = 269 μm, Figure 16a) may be the result of a significant amount of the eroded particles at the hole bottom and the occurrence of the critical conditions including the generation of gas bubbles and the secondary discharges within the gap area. In addition, the presence of high temperature at the gap area and application of deionized water effected the formation of copper oxide, which provoked a grey–blue color on the electrode surface after the drilling process.

3.4. Results of the Analysis of the Through Holes Obtained for the Experiment 2

Based on an analysis of the results of electrical discharge drilling and the working fluid flow through the electrode channel, the experimental research into the EDD of through holes in the Inconel 718 alloy was done. In contrast to the previous experiment, the following was applied: longer pulse time (t_i = 500 µs), the given initial working-fluid temperature (T = ~313.15 °K) and higher electrical conductivity of deionized water (κ = 7.2–12.8 µS/cm). Three tests were performed with the same machining parameters (Figure 17).

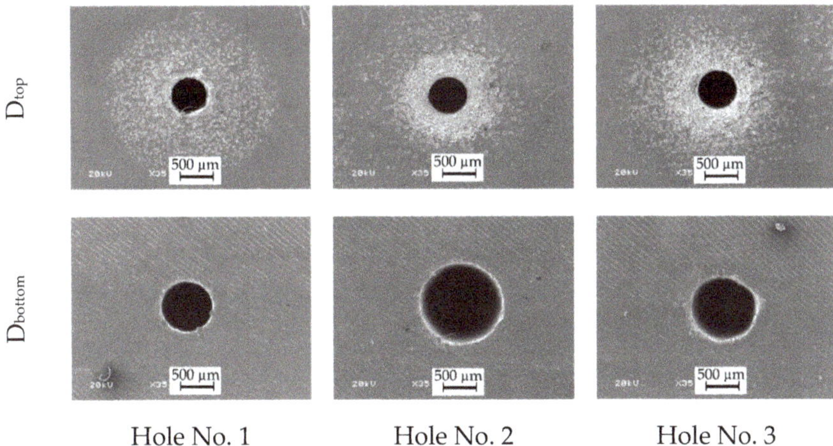

Figure 17. Through holes drilled in the Inconel 718 alloy.

The significant electrical conductivity of deionized water caused the corrosion of the workpiece surface. Serious damage is observed on the surface around the top diameter (Figure 17). The thickness of the damaged surface is about 550 µm.

The damaged surface around the hole was subjected to a qualitative analysis with the use of a scanning electron microscope (SEM) with an energy dispersive spectroscopy system (EDS), manufactured by JEOL Ltd. (Tokyo, Japan). The analysis was made in three marked zones: zone 1 (Z_1), zone 2 (Z_2) and zone 3 (Z_3) (Figure 18). A surface analysis (for deep areas, designation 1 in Figure 18, enlarged at point Z_1) and a spot analysis (for the flat areas, designation 2 and 3 in Figure 18, enlarged at point Z_1) were performed.

On the damaged surface, zones cavities (Figure 18, enlarged at points Z_2 and Z_3, designation I), darker zones near the cavities (Figure 18, enlarged at points Z_2 and Z_3, designation II) and brighter zones between the cavities (Figure 18, enlarged at points Z_2 and Z_3, designation III) can be observed. The analysis carried out for the three areas (marked as Z_1, Z_2, Z_3) indicates the occurrence of the average percentage volume of the main components of the Inconel 718 alloy: Ni-67.47 wt.%, Cr-17.66 wt.%, Fe-9.45 wt.% (Figure 19a–c). Additionally, in the cavities (Figure 19a) and darker zones (Figure 19b) the presence of oxygen was detected. This results from the use of deionized water. In high temperature conditions, a passive layer might have formed on the workpiece surface due to oxidation reaction. The passive layer might have prevented excessive electrochemical dissolution when deionized water was flowing out of the hole.

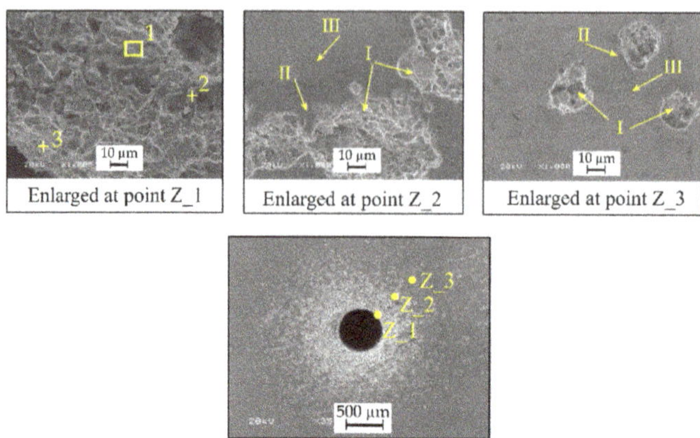

Figure 18. SEM images surface around the D_{top} of hole No. 3.

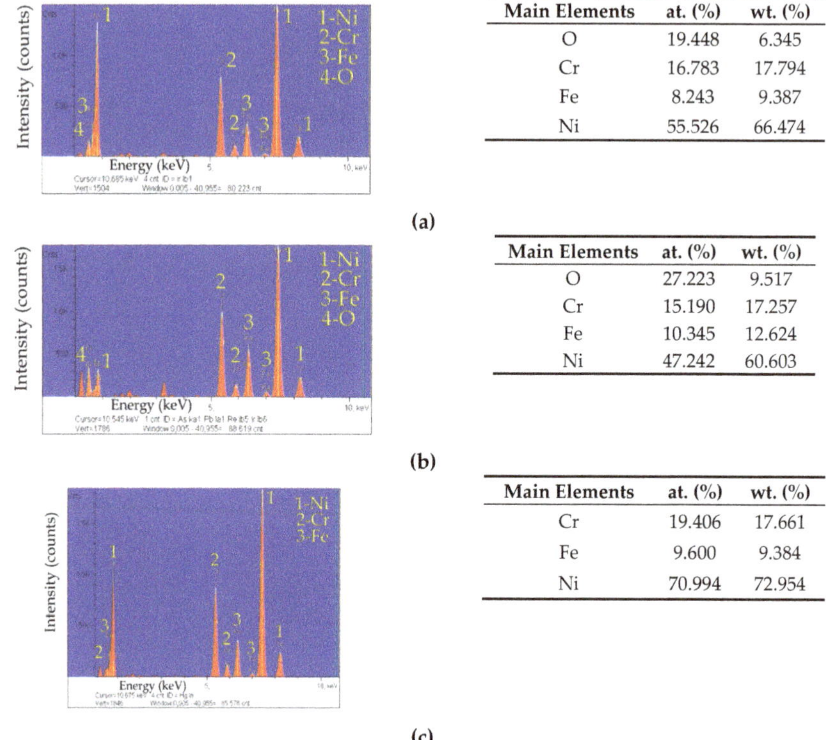

Main Elements	at. (%)	wt. (%)
O	19.448	6.345
Cr	16.783	17.794
Fe	8.243	9.387
Ni	55.526	66.474

(a)

Main Elements	at. (%)	wt. (%)
O	27.223	9.517
Cr	15.190	17.257
Fe	10.345	12.624
Ni	47.242	60.603

(b)

Main Elements	at. (%)	wt. (%)
Cr	19.406	17.661
Fe	9.600	9.384
Ni	70.994	72.954

(c)

Figure 19. Chemical composition of the surface around the top diameter as determined by EDS: (**a**) zone of cavities, (**b**) darker zones near the cavities, (**c**) brighter zones between the cavities.

Observation of the damage that surfaced indicates that electrochemical reactions were the most excessive while the fluid was flowing from the electrode channel (the initial water temperature was about 313.15 °K). The splashing working fluid further on the workpiece surface lost its heat, and its electrical conductivity decreased. This can be inferred from the size of the cavities near and farther away from the hole edge. The cavities located farther are less numerous and smaller (Figure 18, enlarged at points Z_2 and Z_3).

The EDS analysis of the surface near the hole edge indicates the absence of oxygen (Figure 18, enlarged at point Z_1). On this area, the material was removed by simultaneous interaction of electrochemical dissolution and electrical discharges in a single pulse.

Based on Equations (1)–(5), the process performance comprising the drilling speed v, the linear tool wear TW, the aspect ratio hole AR, the taper angle tap_α and the side gap thickness S_b were determined (in Figure 20a–e as Experiment 2). The analysis of the results showed that the drilling speed, the linear tool wear and the taper angle are at a similar level in comparison to the previous experiment (Figure 20a,b,d, respectively). The aspect ratio hole and the side gap thickness improved by about 15% and 40%, respectively (Figure 20c,e).

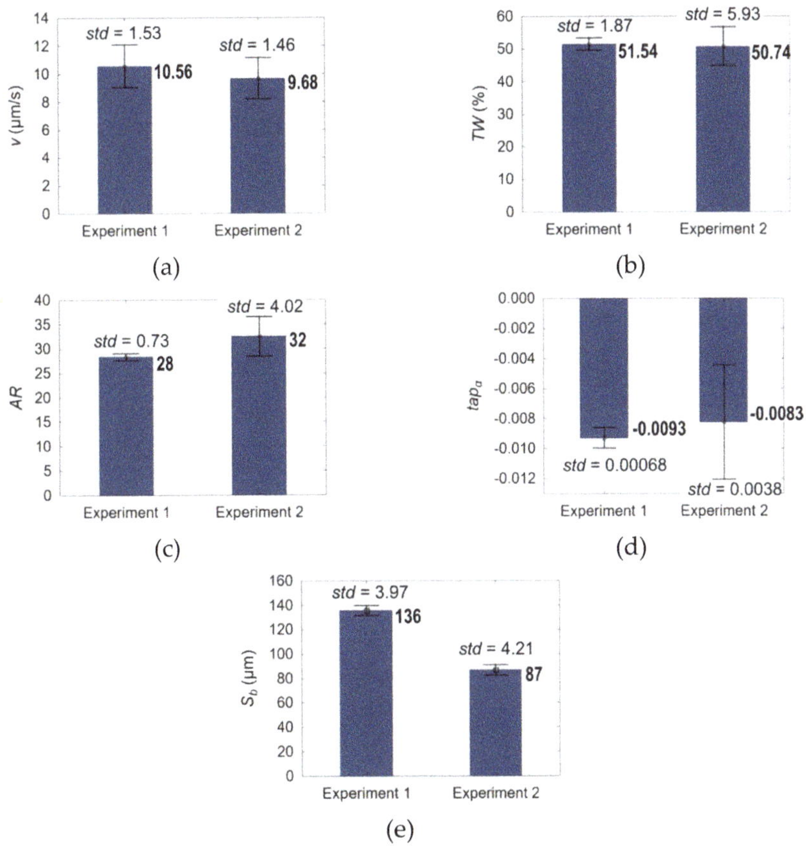

Figure 20. Comparison of the process performance in Experiments 1 and 2 for: (**a**) the drilling speed v, (**b**) the linear tool wear TW, (**c**) the aspect ratio hole AR, (**d**) the taper angle tap_α, and (**e**) the side gap thickness S_b; std – the value of the standard deviation.

The increase in AR results from obtaining a lower average top diameter value. The edge of the top diameter is free of re-solidified material, a heat-affected-zone and burrs, which improves its accuracy (Figures 17 and 21). This may result from excessive electrochemical dissolution in a single pulse due to a higher electrical conductivity of deionized water resulting from the application of a higher initial working-fluid temperature (about 313.15 °K). In addition, the higher discharge voltage ($U = 120$ V) additionally enhances the electrochemical reactions in a single pulse. The longer pulse off time (about twice as long as during Experiment 1), on the other hand, results in more efficient removal of the eroded material. The conditions in the working gap might also have improved due to a higher fluid flow because of lower viscosity and lower density of deionized water. A smaller top diameter influences the improvement of AR and S_b.

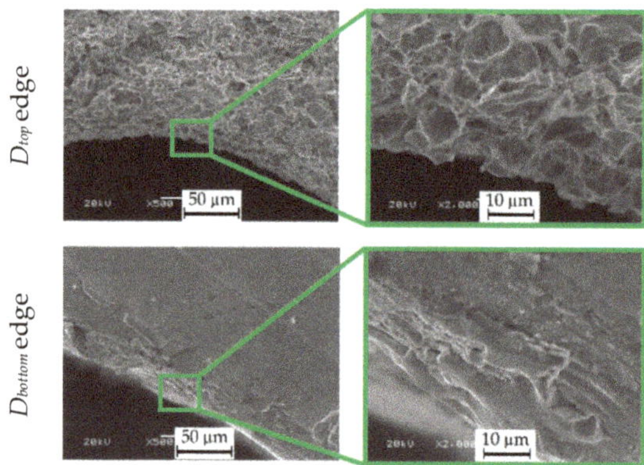

Figure 21. SEM images of the surface of D_{top} and D_{bottom} edge, hole No. 3.

The surface around the top diameter is smoother and without microcracks. The bottom diameter is characterized by a significantly smaller number of burrs than after Experiment 1 (Figures 17 and 21). The white layer around the bottom diameter is present, but the diameter quality is enhanced. The average value of the bottom diameter is still large (about 1000 µm) in comparison with the average value of the top hole diameter (about 600 µm), what causes the hole conicity (Figure 20d). The increased bottom diameter result from the accumulation of debris at the hole bottom and the occurrence of abnormal discharges. Additionally, the application of a higher initial working-fluid temperature (about 313.15 °K) and a higher initial working-fluid pressure ($p = 8$ MPa) could cause a turbulent flow and the occurrence of whirls at the hole bottom. The presence of whirls could cause difficulties removing debris and bubbles from the bottom of the hole.

In [43], it was observed that gas bubbles, which were occurring, can prevent electrochemical dissolution. The higher electrical conductivity of deionized water and higher discharge voltage have contributed to achieving critical conditions faster in the gap area and faster occurrence of the bubbles. The higher electrical conductivity also has caused the increase in the fluid temperature by generating Joule heat. The considerable number of gas bubbles with the gap area may explain the absence of drilling speed improvement. However, in [21,29], it is considered that the bubbles push the debris and help remove them from the gap. However, the extending diameter along the hole depth attests to the occurrence of a significant amount of debris within the side gap area near the hole bottom. This may confirm the occurrence of whirls resulting from a turbulent flow. The whirls could hinder proper working fluid flow and, additionally, push the unremoved debris towards the corner of the hole bottom.

The inner hole surface is significantly less rough and more homogenous (Figure 22a), in comparison with the surface after the electrochemical machining process only (Figure 22b) and after the previous experiment (Figure 10). There are no micro-craters, re-solidified material or micro-cracks on the surface. The surface structure is similar to the structure that was obtained around the hole top. This is due to parallel removal of the material as a result of electrochemical reactions and electrical discharges. When the working fluid was flowing out of the hole, electrochemical dissolution might have occurred as well, which additionally improved the homogeneity of the inner surface of the hole.

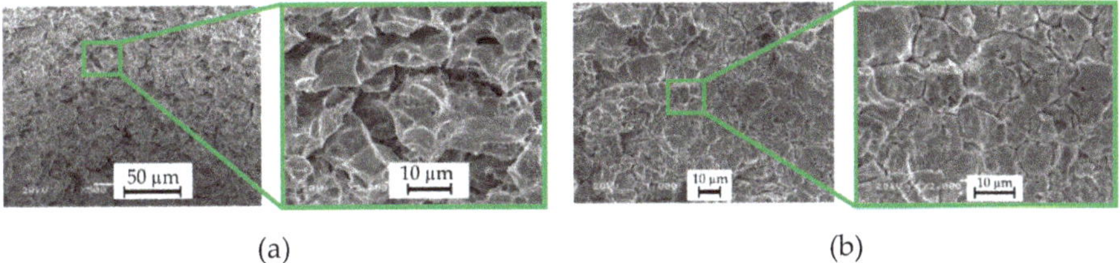

Figure 22. SEM images of (**a**) the inner surface of the hole (hole No. 3) and (**b**) the surface after electrochemical dissolution with the use of deionized water (machining parameters: U = 25 V, t_i = 300 µs).

4. Discussion

The results analysis of the experimental research indicates that the applied higher temperature of deionized water (T = ~313.15 °K), used as working fluid, effected the improvement of shape dimensional drilling the through holes. The inner surface homogeneity of holes also significantly improved. On the inner of the through holes observed a lack presence of re-solidified material, of which a significant amount was present on the surface after the first part of the research (Figures 10 and 13). This is result using higher temperature of deionized water had an influence in increasing electrical conductivity during the experiment 2. The electrical conductivity increase caused the enhanced electrochemical dissolution at the beginning of the single pulse time (removing an allowance in the similar ranges by ECM and EDM during the single pulse duration). The reinforced electrochemical dissolution was enough to remove the erosion craters and to avoid extending the top diameter of holes simultaneously. In addition, using of the longer pulse off time (t_{off} = 500 µs, above twofold longer in comparison to the first experiment), could also cause to avoid re-solidified material on the hole edge. During the experiment 1 the pulse off time (t_{off} = t_i) equaled 282 µs, which was too short time a fresh working fluid to flow onto the gap area. The insufficient decrease in the temperature of the working fluid in the gap area could cause a too fast reoccurrence of conditions enough to spark (particularly, in case of using the higher initial working–fluid temperature T = 313.15 °K). As the result, the amount of re-solidified material could be occurred on the machined inner surface of hole.

Indicating potential turbulent flow conditions for applying T = ~313.15 °K and p = 8 MPa is possible during the drilling process, but whirls accompanying the turbulent flow can hinder sufficient working fluid flow via the gap area and the side gap near the hole bottom. During the experiment 2 the properties of deionized water such as density and viscosity were reduced by using the higher deionized water temperature, but during the conditions did not influence the improvement removing of debris from the gap area. Hence, the presence of whirls could be possibility.

The results analysis shows that the properties of deionized water influence significantly the thermal phenomena present in the interelectrode area. The obtained higher electrical conductivity of deionized water by use of the higher initial temperature (T = ~313.15 °K), enabled removing the material in similar ranges of both processes (ECM and EDM) during single pulse time. Further studies should focus on more detailed analysis impact of the deionized water properties on the EDD process and also the influence of the physical-mechanical properties of both electrode materials.

5. Conclusions

The use of electrical discharge machining with the application of deionized water to drill high aspect ratio holes in the Inconel 718 alloy demonstrates a significant potential. One of the advantages of the application of this kind of working fluid is the possibility of removing material by simultaneous electrochemical dissolution and electrical discharges in a single pulse. Where deep holes are made using the EDD process, an important factor is effective removal of erosion products from the gap area (such as eroded particles and gas bubbles). Insufficient flushing of the gap area contributes to decreasing the dimensional and shape accuracy (aspect ratio, conicity), the inner surface homogeneity of the holes, and the process performance (drilling speed, linear tool wear). In the experimental research conducted, the change in the properties of the working fluid under the influence of temperature and the working fluid pressure affect the nature of the fluid flow through the machining area. The analysis of the results enables the formulation of the following conclusions:

1. The EDD experimental research with the use of deionized water (with given the initial temperature ~313.15 °K and the initial pressure 8 MPa) in the Inconel 718 alloy confirmed the possibility of making high aspect ratio holes ($AR > 30$) and decreasing the side gap thickness by about 40%.

2. For drilling high aspect ratio through holes, the optimum machining parameters were pulse time $t_i = 282$ µs, discharge voltage amplitude $U = 120$ V, current amplitude $I = 4.32$ A and pulse of time $t_{off} = 500$ µs.

3. The setting of a higher initial working-fluid temperature (~313.15 °K) increased the electrical conductivity of deionized water, which reinforced the electrochemical reactions in a single pulse. As a result, in the diameter of the hole top, there were no burrs or heat affected zones. The homogeneity of the inner surface of the hole was also improved (without micro-cracks and erosion micro-craters).

4. On the surface around the hole top, the material is seriously damaged as a result of excessive electrochemical dissolution. The presence of oxygen was detected in cavities and darker zones between the cavities. This indicates oxidation of the machined surface and the creation of a passive layer.

5. The bottom diameter is too large and larger than the top diameter by about 40%. This is due to the accumulation of debris at the hole bottom and the occurrence of secondary discharges. Removal of eroded material from the hole bottom might have been made difficult by the whirls resulting from a turbulent fluid flow.

6. The dark oval traces on the electrode surface and the oval shape of the cavities on the inner surface of the hole can indicate the occurrence of electrical discharges on the small areas between the gas bubbles.

7. The complex nature of the phenomena within the gap area during the EDD process with the use of deionized water requires further analysis and better understanding. Further experimental research should focus on the impact of the physical and mechanical properties of the Inconel 718 alloy and tool electrode material on the process.

Supplementary Materials: The following are available online http://www.mdpi.com/1996-1944/13/6/1476/s1. Supplementary section S1 presenting the ANOVA analysis for: drilling speed v, linear tool wear TW, the taper angle tap_α, the side gap thickness S_b and the aspect ratio hole AR.

Author Contributions: Conceptualization, M.M.; methodology, M.M.; validation, M.M., R.B.; formal analysis, R.B.; investigation, W.B. and M.M.; writing—original draft preparation, M.M. and W.B.; writing—review and editing, M.Sz.; visualization, M.M.; supervision, M.S. and M.M. All authors have read and agreed to the published version of the manuscript.

Funding: The APC was funded by the Faculty of Mechanical, Cracow University of Technology.

Acknowledgments: We would like to thank Institute of Production Engineering (Faculty of Mechanical, Cracow University of Technology) for carrying out the experiments and the technical support.

Conflicts of Interest: The authors declare no conflict of interest.

References

1. Thomasa, A.; El-Wahabi, M.; Cabrera, J.M.; Prado, J.M. High temperature deformation of Inconel 718. *J. Mater. Process. Technol.* **2006**, *177*, 469–472. [CrossRef]

2. Maja, P.; Błyskuna, P.; Kutb, S.; Romelczyk-Baishyaa, B.; Mrugałac, T.; Adamczyka, J.; Mizeraa, B. Flow forming and heat-treatment of Inconel 718 cylinders. *J. Mater. Process. Technol.* **2018**, *253*, 64–71. [CrossRef]

3. Sundararaman, M.; Mukhopadhyay, P.; Banerjee, S. Some aspects of the precipitation of metastable intermetallic phases in INCONEL. *Metall. Trans. A* **1992**, *23*, 2015–2028. [CrossRef]

4. Janaki Ram, G.D.; Venugopal Reddy, A.; Prasad Rao, K.; Reddy, G.M. Microstructure and mechanical properties of Inconel 718 electron beam welds. *Mater. Sci. Technol.* **2005**, *21*, 1132–1138.

5. Rajesha, S.; Sharma, A.K.; Kumar, P. On Electro Discharge Machining of Inconel 718 with Hollow Tool. *J. Mater. Eng. Perform.* **2012**, *21*, 882–891. [CrossRef]

6. Thakur, D.G.; Ramamoorthy, B.; Vijayaraghavan, L. Study on the machinability characteristics of superalloy Inconel 718 during high speed turning. *Mater. Design* **2009**, *30*, 1718–1725. [CrossRef]

7. Kliuev, M.; Boccadoro, M.; Perez, R.; Dal Bó, W.; Stirnimann, J.; Kuster, F.; Wegener, K. EDM drilling and shaping of cooling holes in Inconel 718 turbine blades. *Procedia CIRP* **2016**, *42*, 322–327. [CrossRef]

8. Antar, M.; Chantzis, D.; Marimuthu, S.; Hayward, P. High speed EDM and laser drilling of aerospace alloys. *Procedia CIRP* **2016**, *42*, 526–531. [CrossRef]

9. Klocke, F.; Klink, A.; Veselovac, D.; Aspinwall, D.K.; Soo, S.L.; Schmidt, M.; Schilp, J.; Levy, G.; Kruth, J.P. Turbomachinery component manufacture by application of electrochemical, electro–physical and photonicprocesses. *CIRP Ann. Manuf. Technol.* **2014**, *63*, 703–726. [CrossRef]

10. Lipiec, P.; Machno, M.; Skoczypiec, S. The Experimental Research on Electrodischarge Drilling of High Aspect Ratio Holes in Inconel 718. In Proceedings of the AIP Conference Proceedings, Palermo, Italy, 23–25 April 2018; Volume 1960, pp. 1–6.

11. Goiogana, M.; Sarasua, J.A.; Ramos, J.M. Ultrasonic assisted electrical discharge machining for high aspect ratio blind holes. *Procedia CIRP* **2018**, *68*, 81–85. [CrossRef]

12. Unune, D.R.; Nirala, C.K.; Mali, H.S. Accuracy and quality of micro-holes in vibration assisted micro-electro-discharge drilling of Inconel 718. *Measurement* **2019**, *135*, 424–437. [CrossRef]

13. Ekmekci, B.; Sayar, A. Debris and consequences in micro electricdischarge machining of micro–holes. *Int. J. Mach. Tools Manuf.* **2013**, *65*, 58–67. [CrossRef]

14. Munz, M.; Risto, M.; Haas, R. Specifics of flushing in electrical discharge drilling. *Procedia CIRP* **2013**, *6*, 83–88. [CrossRef]

15. D'Urso, G.; Merla, C. Workpiece and electrode influence on micro-EDM drilling performance. *Precis. Eng.* **2014**, *38*, 903–914. [CrossRef]

16. Nguyen, M.D.; Rahman, M.; Wong, Y.S. Enhanced surface integrity and dimensional accuracy by simultaneous micro-ED/EC milling. *CIRP Ann. Manuf. Technol.* **2012**, *61*, 191–194. [CrossRef]

17. Xu, Z.; Zhang, C. A tube electrode high-speed electrochemical discharge drilling method without recast layer. *Procedia CIRP* **2018**, *68*, 778–782. [CrossRef]

18. Li, Ch.; Xu, X.; Li, Y.; Tong, H.; Ding, S.; Kong, Q.; Zhao, L.; Ding, J. Effects of dielectric fluids on surface integrity for the recast layer in high speed EDM drilling of nickel alloy. *J. Alloys Comp.* **2019**, *783*, 95–102. [CrossRef]

19. Skoczypiec, S.; Ruszaj, A. A sequential electrochemical—Electrodischarge process for micropart Manufacturing. *Precis. Eng.* **2014**, *38*, 680–690. [CrossRef]

20. Bamberg, E.; Heamawatanachai, S. Orbital electrode actuation to improve efficiency of drillingmicro-holes by micro-EDM. *J. Mater. Process. Technol.* **2009**, *209*, 1826–1834. [CrossRef]

21. Yu, Z.Y.; Rajurkar, K.P.; Shen, H. High aspect ratio and complex shaped blind micro holes by micro EDM. *CIRP Ann. Manuf. Technol.* **2002**, *51*, 359–362. [CrossRef]

22. Amorim, F.L.; Weingaertner, W.L. The Behavior of Graphite and Copper Electrodes on the Finish Die-Sinking Electrical Discharge Machining (EDM) of AISI P20 Tool Steel. *J. Braz. Soc. Mech. Sci. Eng.* **2007**, *XXIX*, 1–6. [CrossRef]

23. Jaharah, A.G.; Liang, C.G.; Wahid, S.Z.; Ab Rahman, M.N.; Che Hassan, C.H. Performance of copper electrode in electrical discharge machining (EDM) of AISI H13 harden steel. *Int. J. Mech. Mater. Eng.* **2008**, *3*, 25–29.

24. Kuppan, P.; Narayanan, S.; Oyyaravelu, R.; Balan, A.S.S. Performance evaluation of electrode materials in electric discharge deep hole drilling of Inconel 718 superalloy. *Procedia Eng.* **2017**, *174*, 53–59. [CrossRef]

25. Dobrzański, L.A. *Metals and Their Alloys*; ASKLEPIOS: Gliwice, Poland, 2017.

26. Rajurkar, K.P.; Levy, G.; Malshe, A.; Sundaram, M.M.; McGeough, J.; Hu, X.; Resnick, R.; DeSilva, A. Micro and nano machining by electro-physical and chemical processes. *Ann. CIRP* **2006**, *55*, 643–666. [CrossRef]

27. Yahagi, Y.; Koyano, T.; Kunieda, M.; Yang, X. Micro Drilling EDM with high rotation speed of tool electrode using the electrostatic induction feeding method. *Procedia CIRP* **2012**, *1*, 162–165. [CrossRef]

28. Kunieda, M.; Kitamura, T. Observation of difference of EDM gap phenomena in water and oil using transparent electrode. *Procedia CIRP* **2018**, *68*, 342–346. [CrossRef]
29. Pontelandolfo, P.; Haas, P.; Perez, R. Particle hydrodynamics of the electrical discharge machining process. Part 2: Die sinking process. *Procedia CIRP* **2013**, *6*, 47–52. [CrossRef]
30. Liu, Y.; Chang, H.; Zhang, W.; Ma, F.; Sha, Z.; Zhang, S. A simulation study of debris removal process in ultrasonic vibration assisted electrical discharge machining (EDM) of deep holes. *Micromachines* **2018**, *9*, 378. [CrossRef]
31. Wong, Y.S.; Lim, L.C.; Lee, L.C. Effects of flushing on electro-discharge machined surfaces. *J. Mater. Process. Technol.* **1995**, *48*, 299–305. [CrossRef]
32. Ferraris, E.; Castiglioni, V.; Ceyssens, F.; Annoni, M.; Lauwers, B.; Reynaerts, D. EDM drilling of ultra-high aspect ratio micro holes with insulated tools. *CIRP Ann. Manuf. Technol.* **2013**, *62*, 191–194. [CrossRef]
33. Zhang, W.; Liu, Y.; Zhang, S.; Ma, F.; Wang, P.; Yan, C. Research on the gap flow simulation of debris removal process for small hole EDM machining with Ti alloy. In Proceedings of the 4th International Conference on Mechatronics, Materials, Chemistry and Computer Engineering (ICMMCCE), Xi'an, China, 12–13 December 2015; Atlantis Press: Paris, France, 2015.
34. Haas, P.; Pontelandolfo, P.; Perez, R. Particle hydrodynamics of the electrical discharge machining process. Part 1: Physical considerations and wire EDM process improvement. *Procedia CIRP* **2013**, *6*, 41–46. [CrossRef]
35. Murray, J.; Zdebski, D.; Clare, A.T. Workpiece debris deposition on tool electrodes and secondary discharge phenomena in micro-EDM. *J. Mater. Process. Technol.* **2012**, *212*, 1537–1547. [CrossRef]
36. Tanjilul, M.; Ahmed, A.; Senthil Kumar, A.; Rahman, M. A study on EDM debris particle size and flushing mechanism for efficient debris removal in EDM-drilling of Inconel 718. *J. Mater. Process. Technol.* **2018**, *255*, 263–274. [CrossRef]
37. Ayesta, I.; Flaño, O.; Izquierdo, B.; Sanchez, J.A.; Plaza, S. Experimental study on debris evacuation during slot EDMing. *Procedia CIRP* **2016**, *42*, 6–11. [CrossRef]
38. Arantes, L.J.; da Silva, E.R.; dos Santos, R.F.; Sales, W.F.; Raslan, A.A. The electrical discharge machining process aided by abrasive jet. *Int. J. Adv. Manuf. Technol.* **2016**, *87*, 411–420. [CrossRef]
39. Machno, M. Impact of process parameters on the quality of deep holes drilled in Inconel 718 using EDD. *Materials* **2019**, *12*, 2298. [CrossRef] [PubMed]
40. Kliuev, M.; Baumgart, C.; Wegener, K. Fluid Dynamics in electrode flushing channel and electrode-workpiece gap during EDM drilling. *Procedia CIRP* **2018**, *68*, 254–259. [CrossRef]
41. Zhang, S.; Zhang, W.; Liu, Y.; Ma, F.; Su, C.H.; Sha, Z. Study on the gap flow simulation in EDM small hole machining with Ti Alloy. *Adv. Mater. Sci. Eng.* **2017**, *2017*, 8408793.
42. Chung, D.K.; Shin, H.S.; Park, M.S.; Chu, C.N. Machining characteristics of micro EDM in water using high frequency bipolar pulse. *Int. J. Precis. Eng. Manuf.* **2011**, *12*, 195–201. [CrossRef]
43. Nguyen, M.D.; Rahman, M.; Wong, Y.S. Simultaneous micro-EDM and micro-ECM in low-resistivity deionized water. *Int. J. Mach. Tools Manuf.* **2012**, *54*, 55–65. [CrossRef]
44. Dong, S.; Wang, Z.; Wang, Y.; Liu, H. An experimental investigation of enhancement surface quality of microholes for Be-Cu alloys using micro-EDM with multi-diameter electrode and different dielectrics. *Procedia CIRP* **2016**, *42*, 257–262. [CrossRef]
45. Kim, B.H.; Chu, C.N. Micro electrical discharge milling using deionized water as a dielectric fluid. *J. Micromech. Microeng.* **2007**, *17*, 867–874.
46. Modica, F.; Marrocco, V.; Valori, M.; Viganò, F.; Annoni, M.; Fassi, I. Study about the Influence of powder mixed water based fluid on micro-EDM process. *Procedia CIRP* **2018**, *68*, 789–795. [CrossRef]
47. Shin, H.S.; Kim, B.H.; Park, M.S.; Chu, C.N. Surface finishing of micro-EDM holes using deionized water. *J. Micromech. Microeng.* **2009**, *19*, 045025.
48. Li, G.; Natsu, W.; Yu, Z. Study on quantitative estimation of bubble behavior in micro hole drilling with EDM. *Int. J. Mach. Tools Manuf.* **2019**, *146*, 103437. [CrossRef]
49. Wang, J.; Han, F.; Cheng, G.; Zhao, F. Debris and bubble movements during electrical discharge machining. *Int. J. Mach. Tools Manuf.* **2012**, *58*, 11–18. [CrossRef]
50. Li, G.; Natsu, W. Realization of micro EDM drilling with high machining speed and accuracy by using mist deionized water jet. *Precis. Eng.* **2020**, *61*, 136–146. [CrossRef]

51. Hernándes, M.; Amriz, R.R.; Cortès, R.; Gómora, C.M.; Plascencia, G.; Jaramillo, D. Assessment of gas tungsten arc welding thermal cycles on Inconel 718 alloy. *Trans. Nonferrous Met. Soc. China* **2019**, *29*, 579–587. [CrossRef]

52. Pan, Z.; Feng, Y.; Hung, T.P.; Jiang, Y.C.; Hsu, F.C.; Wu, L.T.; Lin, C.F.; Lu, Y.C.; Liang, S.Y. Heat affected zone in the laser-assisted milling of Inconel 718. *J. Manuf. Processes* **2017**, *30*, 141–147. [CrossRef]

53. Khan, A.A. Electrode wear and material removal rate during EDM of aluminum and mild steel using copper and brass electrodes. *Int. J. Adv. Manuf. Technol.* **2008**, *39*, 482–487. [CrossRef]

54. Gholipoor, A.; Baseri, H.; Shabgard, M.R. Investigation of near dry EDM compared with wet and dry EDM processes. *J. Mech. Sci. Technol.* **2018**, *29*, 2213–2218. [CrossRef]

55. Basak, I.; Ghosh, A. Mechanism of spark generation during electrochemical discharge machining: A theoretical model and experimental verification. *J. Mater. Process. Technol.* **1996**, *62*, 46–53. [CrossRef]

56. Liu, J.W.; Yue, T.M.; Guo, Z.N. An analysis of the discharge mechanism in electrochemical discharge machining of particulate reinforced metal matrix composites. *Int. J. Mach. Tools Manuf.* **2010**, *50*, 86–96. [CrossRef]

57. Crichton, I.M.; McGeough, J.A. Studies of the discharge mechanism in electrochemical arc machining. *J. Appl. Electrochem.* **1985**, *15*, 113–119. [CrossRef]

58. Basak, I.; Ghosh, A. Mechanism of material removal in electrochemical discharge machining: A theoretical model and experimental verification. *J. Mater. Process. Technol.* **1997**, *71*, 3350–3359. [CrossRef]

MDPI

St. Alban-Anlage 66

4052 Basel

Switzerland

Tel. +41 61 683 77 34

Fax +41 61 302 89 18

www.mdpi.com

Materials Editorial Office

E-mail: materials@mdpi.com

www.mdpi.com/journal/materials